Stream and
Watershed
Restoration

Companion website

This book has a companion website:

www.wiley.com/go/roni/streamrestoration

with Figures and Tables from the book

Stream and Watershed Restoration

A Guide to Restoring Riverine Processes and Habitats

Philip Roni and Tim Beechie

Fish Ecology Division
Northwest Fisheries Science Center
National Marine Fisheries Service
National Oceanic and Atmospheric Administration
2725 Montlake Boulevard East
Seattle, Washington 98112,
USA

ADVANCING RIVER RESTORATION AND MANAGEMENT

WILEY-BLACKWELL

A John Wiley & Sons, Ltd., Publication

Library of Congress Cataloging-in-Publication Data
Roni, Philip.
  Stream and watershed restoration : a guide to restoring riverine processes and habitats / Philip Roni and Tim Beechie.
      p. cm.
  Includes bibliographical references and index.
  ISBN 978-1-4051-9955-1 (cloth) – ISBN 978-1-4051-9956-8 (pbk.)   1. Watershed restoration. 2. Stream restoration.   3. Aquatic ecology.   4. Restoration ecology.   I. Beechie, T. J. (Tim J.) II. Title.
  QH541.15.R45R66 2013
  627'.5–dc23
                2012016383

Contents

Companion website

This book has a companion website:

www.wiley.com/go/roni/streamrestoration

with Figures and Tables from the book

List of Contributors

Tim Beechie   Fish Ecology Division, Northwest Fisheries Science Center, National Marine Fisheries Service, National Oceanic and Atmospheric Administration, 2725 Montlake Boulevard East, Seattle, Washington 98112, USA
tim.beechie@noaa.gov

Lucy Butler   Eden Rivers Trust, Dunmail Building, Newton Rigg College, Penrith, Cumbria CA11 0AH, UK
lucy@edenriverstrust.org.uk

Janine Castro   US Fish and Wildlife Service, 2600 SE 98th Avenue, Suite 100, Portland, Oregon 97266, USA
janine_m_castro@fws.gov

Rickie Chen   University of Washington, School of Forest Resources, Box 352100, Seattle, Washington 98195, USA
rickietw@gmail.com

Steve Clayton   CH2M HILL, 322 East Front Street, Suite 200, Boise, Idaho 83702, USA
Steve.Clayton@CH2M.com

Brian Cluer   Southwest Region, National Marine Fisheries Service, National Oceanic and Atmospheric Administration, 777 Sonoma Ave, Rm. 325, Santa Rosa, California 95404, USA
brian.cluer@noaa.gov

Peter Downs   School of Geography, Plymouth University, Room 001, 7 Kirkby Place, Drake Circus, Plymouth, Devon PL4 8AA, UK
peter.downs@plymouth.ac.uk

Angela M. Gurnell School of Geography, Queen Mary, University of London, Mile End Road, London E1 4NS, UK
a.m.gurnell@qmul.ac.uk

Karrie Hanson   Fish Ecology Division, Northwest Fisheries Science Center, National Marine Fisheries Service, National Oceanic and Atmospheric Administration, 2725 Montlake Boulevard East, Seattle, Washington 98112, USA
karrie.hanson@noaa.gov

Martin Liermann   Fish Ecology Division, Northwest Fisheries Science Center, National Marine Fisheries Service, National Oceanic and Atmospheric Administration, 2725 Montlake Boulevard East, Seattle, Washington 98112, USA
martin.liermann@noaa.gov

Alistair Maltby   The Rivers Trust, Rain-Charm House, Kyl Cober Parc, Stoke Climsland, Callington, Cornwall PL17 8PH, UK
alistair@theriverstrust.org

Sarah Morley   Fish Ecology Division, Northwest Fisheries Science Center, National Marine Fisheries Service, National Oceanic and Atmospheric Administration, 2725 Montlake Boulevard East, Seattle, Washington 98112, USA
sarah.morley@noaa.gov

Susanne Muhar   Department of Water, Atmosphere and Environment, Institute of Hydrobiology and Aquatic Ecosystem Management, University of Natural Resources and Life Sciences, Max-Emanuelstraβe 17, 1180 Vienna, Austria
susanne.muhar@boku.ac.at

Clint Muhlfeld   US Geological Survey, Northern Rocky Mountain Science Center, Glacier Field Station, Glacier National Park, West Glacier, Montana, USA
cmuhlfeld@usgs.gov

Junjiro Negishi   Hokkaido University, N10W5, Sapporo, Hokkaido 060-0810, Japan
negishi@ees.hokudai.ac.jp

Michael Pearsons   Pearson Ecological, 2840 Lougheed Highway, Agassiz, British Columbia, V0M 1A1, Canada
mike@pearsonecological.com

George Pess   Fish Ecology Division, Northwest Fisheries Science Center, National Marine Fisheries Service, National Oceanic and Atmospheric Administration, 2725 Montlake Boulevard East, Seattle, Washington 98112, USA
george.pess@noaa.gov

John S. Richardson   Department of Forest Sciences, 3041- 2424 Main Mall, University of British Columbia, Vancouver, British Columbia, V6T 1Z4, Canada
John.Richardson@ubc.ca

Philip Roni   Fish Ecology Division, Northwest Fisheries Science Center, National Marine Fisheries Service, National Oceanic and Atmospheric Administration, 2725 Montlake Boulevard East, Seattle, Washington 98112, USA
phil.roni@noaa.gov

Stefan Schmutz   Department of Water, Atmosphere and Environment, Institute of Hydrobiology and Aquatic Ecosystem Management, University of Natural Resources and Life Sciences, Max-Emanuelstraβe 17, 1180 Vienna, Austria
stefan.schmutz@boku.ac.at

Conor Shea   US Fish and Wildlife Service, 1655 Heindon Road, Arcata, California 95521, USA
conor_shea@fws.gov

Peter Skidmore   Skidmore Restoration Consulting, LLC, 323 North Plum Avenue, Bozeman, Montana 59715, USA
restoringrivers@yahoo.com

Jon Souder   Coos Watershed Association, PO Box 5860, Charleston, Oregon 97420, USA
jsouder@cooswatershed.org

Colin Thorne   Chair of Physical Geography, University of Nottingham, Room B26f Sir Clive Granger Building, University Park, Nottingham NG7 2RD, UK
colin.thorne@nottingham.ac.uk

Foreword

Our technical and engineering capacity increased tremendously in the last century, allowing us to manipulate rivers to meet demands for electricity, construction, navigation, and human safety. In this period of intense development, we strongly regulated rivers and damaged them. We were able to address a large number of human needs, thus improving human well-being. Other needs were compromised however, and river channel reactions to our actions have had more negative consequences than were ever anticipated.

We entered into a new period of 'sustainable' development in the mid-1990s, recognizing the value of riverscapes and the ecosystem services provided by rivers with little or no regulation. After decades of diking, some countries promote 'more space' or 'room' for rivers and diverse schemes of dike setback have been undertaken, both approaches aiming to protect populations. In urban and suburban areas, riverfronts and corridors are increasingly valued for their potential contribution to the quality of city life. In this context, new questions emerge around environmental ethics and justice, and public actions to mitigate, improve, enhance, restore, and repair the rivers are underway.

Repairing rivers is becoming a real challenge in different parts of the world, and varying strategies are proposed by decision makers to promote this new objective. In Europe, we must reach a good ecological status for our rivers by 2015. This effort is often associated with actions on the physical conditions: so-called hydromorphological measures.

River restoration is a newly emerging practical science based on the principles of engineering ecology. It is still a work in progress. We need to learn a lot, partly from our previous errors. We have passed the 'we did it' step, where the fact of doing was already an achievement in itself. Now we are climbing the 'we did this and it was successful' step. We are realizing more humbly that restoration is not so easy. We need to experiment, try, and innovate in a domain where uncertainty is high and patience is required; nature does not react obligingly to our requirements. After almost two decades of river res-

toration some scientists are now deeply involved in this new domain, becoming restoration specialists and providing feedback to the scientific community and practitioners. Good practice guidelines are therefore needed to allow us to move forward and improve decision-making procedures, techniques, and *savoir faire*.

What is the problem? Why is it degraded? These may seem basic questions, but their answers allow us to know whether we act on the disease or its symptoms. Rivers are not only water canals but also complex corridors with water and other features. Riparian vegetation is completely integrated within this environment, contributing to its good health. Flooding, erosion, and sediment deposition are the engines of this natural infrastructure. A good repair is based on a good diagnosis.

Where should we repair? When thinking about restoration, it is important to look at the big picture. The regional level is strategic here because it allows consideration of the different geographical contexts that control river functioning. Restoration measures can be valuable in a given regional context but not in another. Considering this level is therefore critical for improving the success of restoration. The regional level is also appropriate for policy making to plan and target actions. At this level we can prioritize where restoration would be most beneficial. Where is the most damage? Where would restoration most clearly satisfy needs?

What should we do when we repair? How can we design a restoration project? We need to consider geographical complexity; rivers have different sensitivities to change and can react differently to our actions. This can be helpful at times, but it may also have counter effects if we do not properly appreciate these properties. We also need to consider different timescales; sustainable solutions for very active and reactive rivers will be different from other cases, and self-restoration may take time to propagate its effect downstream. Process-based thinking provides a framework for preventing unexpected river responses. When we play with nature, there are rules we must know. Monitoring is also needed because we do not know enough: we must better define and characterize what we call a restoration success. How should we do this? What can we learn from previous experiences? How can we capture this valuable feedback? All these are questions to be answered in the near future.

Restoration is a management action and a socioeconomic challenge. It aims to provide benefits for society. Such action is therefore conducted in a collective framework and there is a need to reconsider why we are acting. Working not only with nature but also with society has become necessary. Public participation is important because these approaches are new, not always intuitive and sometimes contradictory to historical practices. Discussing conflicting aspects in order to understand different viewpoints opens ways to co-construct a river's future. Environmental education is equally a means for preparing society, especially younger generations, to develop with nature. Cost-effective pragmatic measures should be a goal shared by all participants.

All these questions are addressed in this very valuable contribution by a team of authors known internationally for their competence in the field of river restoration. Their motivation is to transfer their knowledge and help society answer this new challenge. Opportunistic restoration must give way to a more strategic framework for prioritizing, designing and implementing actions in an iterative way. The questions are complex. The authors' response here is interdisciplinary, crossing examples and experiences from North America and Europe where pioneer experiences provide discussion material. This book is a very well-illustrated and exemplified update step. It comprehensively summarizes previous results, then moves on to promote principles, strategies, methods, and techniques to improve our practices and answer social demands for a better environment.

Hervé Piégay,
Research Director at the Centre National de la
Recherche Scientifique, Lyon, France

Series Foreword

Advancing River Restoration and Management

The field of river restoration and management has evolved enormously in recent decades, driven largely by increased recognition of the ecological values, river functions, and ecosystem services. Many conventional river management techniques, emphasizing hard structural controls, have proven difficult to maintain over time, resulting in sometimes spectacular failures, and often degraded river environment. More sustainable results are likely from a holistic framework, which requires viewing the 'problem' at a larger catchment scale and involves the application of tools from diverse fields. Success often hinges on understanding the sometimes complex interactions among physical, ecological and social processes.

Thus, effective river restoration and management requires nurturing the interdisciplinary conversation, testing and refining our scientific theories, reducing uncertainties, designing future scenarios for evaluating the best options, and better understanding the divide between nature and culture that conditions human actions. It also implies that scientists better communicate with managers and practitioners, so that new insights from research can guide management, and so that results from implemented projects can in turn, inform research directions.

The series provides a forum for 'integrative sciences' to improve rivers. It highlights innovative approaches, from the underlying science, concepts, methodologies, new technologies, and new practices, to help managers and scientists alike improve our understanding of river processes, and to inform our efforts to better steward and restore our fluvial resources for more harmonious coexistence of humans with their fluvial environment.

G. Mathias Kondolf,
University of California, Berkeley
Hervé Piégay
University of Lyon, CNRS

Preface

This book was borne out of the clear need for a comprehensive resource for developing stream and watershed restoration programs at regional (provincial), watershed, reach, and project scales. Many restoration efforts have failed to meet their objectives because they have not adequately addressed the root cause of habitat degradation, or because they do not recognize the role that watershed and riverine processes play in determining the outcome of restoration actions. Over our many years of experience in watershed research, we have repeatedly seen the need for a systematic process-based approach to planning, prioritizing, designing, and evaluating habitat restoration programs and projects. In the chapters that follow, we strive to meet this need. This book is a synthesis of our previous efforts on restoration that have been published as manuscripts, books, and technical reports, as well as our experience teaching practitioners and students in workshops and university courses. We focus primarily on restoration of physical processes and habitat and draw heavily from our experiences in North America and Europe, the continents where considerable habitat restoration and research has occurred. However, the principles and methods covered in this volume are applicable to stream, river, and watershed restoration anywhere in the world and useful for programs that focus on improving degraded water quality and reducing contaminants.

This book is intended as a guide for practitioners, an instructional manual for educators and students, and a general reference for those interested or active in the field of aquatic and restoration ecology. It is organized in a stepwise fashion covering the key aspects of aquatic restoration including: assessing watershed and riverine processes and conditions; identifying restoration opportunities; choosing appropriate restoration techniques; prioritizing restoration actions; and monitoring and implementation. For the educator and student, it is set up so that each chapter can be covered as a section of a course on stream and watershed restoration. Ideally, an instructor will use our text along with data from a local watershed to create realistic and relevant assignments and exercises for the students. For those new to restoration, we recommend reading through most of the chapters in the order presented before embarking on a project. For the experienced restoration practitioner or those with a general interest in the topic, we recommend reading the introduction and watershed processes chapters (1 and 2) and then the remaining chapters in the order that most suits your needs and interests.

This book would not have been possible without the assistance of numerous individuals and organizations. First, our employer the Northwest Fisheries Science Center and current and former supervisors John Ferguson, Tracy Collier, and Doug Dey deserve special thanks for allowing us to pursue and work on this project. We would also like to thank all those who assisted us by reviewing chapters including: Peter Kiffney, John Klochak, Keith Hendry, Mason Bryant, Tom O'Brien, Matt Hudson, Jennifer Steger, Lauren Senkry, Erik Michelson, Martin O'Grady, Sarah Miller, Chris James, Robin Jenkinson, Pauliina Louhi, Ray White, Martin Janes, Jenny Mant, Mary Raines, and Dana Warren. Karrie Hanson, Ed Quimby, Bert Tarrant and JoAnne Butzerin provided much-needed assistance with technical editing. We thank Su Kim, Clemens Trautwein and Hiroo Imaki for assistance in developing figures and numerous individuals for providing photos of restoration used in one or more chapters. Finally, we would like to thank all those who have and continue to dedicate their lives to environmental restoration. We hope that this book will serve you well in your challenge to protect and restore streams, rivers, and watersheds.

Philip Roni and Tim Beechie
June 2012

1

Introduction to Restoration: Key Steps for Designing Effective Programs and Projects

Philip Roni & Tim Beechie

Northwest Fisheries Science Center, National Oceanic and Atmospheric Administration, USA

1.1 Introduction

The restoration of streams, rivers, and watersheds has become a growth industry in North America and Europe in the 21st century, with an estimated $1 billion spent annually in the United States alone (Bernhardt *et al.* 2005). This comes with a growing appreciation from the general public of the importance of water, watersheds, and natural places not only for their wildlife and fisheries, but also for social, cultural, economic, and spiritual reasons. With this increased emphasis on restoration has come the need for new techniques and guidance for assessing stream and watershed conditions, identifying factors degrading aquatic habitats, selecting appropriate restoration actions, and monitoring and evaluating restoration actions at appropriate scales. All these require detailed consideration of not only the latest scientific information but also regulations and socioeconomic constraints at local, regional, and national levels. Thus the challenges facing watershed restoration in the 21st century are multifaceted, including both technical and non-technical issues.

As interest in aquatic restoration has increased, several texts have been produced over the last few decades to assist with various aspects of river restoration. Most have focused on habitat improvement techniques specific to trout and salmon (e.g. Hunter 1991; Mills 1991; Hunt 1993; O'Grady 2006) or design considerations for specific techniques (e.g. Brookes & Shields 1996; Slaney & Zoldakis 1997; RRC 2002). A few have provided more comprehensive regional overviews of riverine restoration planning and techniques (Ward *et al.* 1994 in UK; Cowx & Welcomme 1998 in Europe; FISRWG 1998 in USA; CIRF 2006 in Italy). Still others have published overviews of key concepts and principles (e.g. Brierley & Fryirs 2008; Clewell & Aronson 2008). Collectively these publications cover many of the tools, techniques, and concepts needed for restoration planning, but no single book covers the full restoration process from initial assessment to monitoring of results and adaptive management. In this book, we strive to meet the need for a comprehensive guide and educational tool that covers the key steps in this process and provide a text that links watershed assessment and problem identification to identification of appropriate restoration measures, project selection, prioritization, project implementation, and effectiveness monitoring (Figure 1.1). Each of these steps is discussed in detail in subsequent chapters. In addition, we discuss the human dimension and how one can best work with citizens, government bodies, and private companies to develop restoration projects and goals. In this introductory

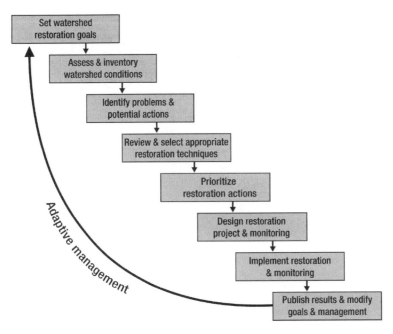

Figure 1.1 Major steps in the restoration process required to develop a comprehensive restoration program and well-designed restoration projects.

chapter we provide important background on the need for restoration, its relatively short history, and the major steps and considerations for planning and implementing restoration actions.

1.2 What is restoration?

Restoration ecology is a relatively young field with considerable confusion over its terminology (Buijse *et al.* 2002; Omerod 2004; Young *et al.* 2005). The terms restoration, rehabilitation, enhancement, improvement, mitigation, reclamation, full and partial restoration, passive and active restoration, and others have been used to describe various activities meant to restore ecological processes or improve aquatic habitats (Table 1.1). These represent a gradient of activities from creating new habitats, to mitigating for lost habitat, to full restoration of ecosystem processes and functions and even protection. In practice, the term restoration is used to refer to any of the above activities. To avoid further confusion over terminology, we therefore use the term in this sense throughout this text. Where appropriate, we distinguish between full restoration, partial restoration and habitat improvement or creation (Table 1.1).

We focus most of our discussion on 'active restoration,' which are restoration efforts that take on the ground action to restore or improve conditions. However, regulations, laws, land-use practices, and other forms of 'passive restoration' that eliminate or prevent human disturbance or impacts to allow recovery of the environment are equally important. For example, most of the improvements in water quality and habitat condition in the USA, Europe, and elsewhere would not have occurred without legislation and regulation. Similarly, habitat protection, while not typically included in definitions of restoration, is a critical watershed conservation and restoration strategy that should not be overlooked. Given the continued pressure on aquatic ecosystems, including a growing human population and climate change, habitat loss will continue and even outpace restoration efforts unless protection of high-quality functioning habitats is a high-priority component of restoration plans. In fact, habitat protection in many cases is a type of passive restoration that allows ecosystems to recover following disturbance. Ultimately, it is much more cost-effective to protect functioning habitats from degradation than it is to try to restore them once they have been damaged.

Table 1.1 Commonly used restoration terminology and general definitions. In this book and in practice, the term restoration is used to encompass all these activities with the exception of protection and mitigation. Where appropriate, we distinguish between restoration in its strictest sense (full restoration), rehabilitation (partial restoration), and habitat improvement or creation. Modified from Roni (2005), Roni *et al.* (2005), and Beechie *et al.* (2010).

Term	Definition
Protection	Creating laws or other mechanisms to safeguard and protect areas of intact habitat from degradation.
Restoration	Returning an aquatic system or habitat to its original, undisturbed state. This is sometimes called 'full restoration,' and can be further divided into passive (removal of human disturbance to allow recovery) and active restoration (active manipulations to restore processes or conditions).
Rehabilitation	Restoring or improving some aspects or an ecosystem but not fully restoring all components. It is also called 'partial restoration' and may also be used as a general term for a variety of restoration and improvement activities.
Improvement	Improving the quality of a habitat through direct manipulation (e.g. placement of instream structures, addition of nutrients). Sometimes referred to as habitat enhancement and sometimes also considered as 'partial restoration' or rehabilitation.
Reclamation	Returning an area to its previous habitat type but not necessarily fully restoring all functions (e.g. removal of fill to expose historic estuary, removal of a levee to allow river to periodically inundate a historic wetland). Sometimes referred to as compensation.
Creation	Constructing a new habitat or ecosystem where it did not previously exist (e.g. creating new estuarine habitat, or excavating an off-channel pond). This is often part of mitigation activities.
Mitigation	Taking action to alleviate or compensate for potentially adverse effects on aquatic habitat that have been modified or lost through human activity (e.g. creating of new habitats to replace those lost by a land development).

1.3 Why is restoration needed?

It may seem obvious to people living in densely popu-lated and developed areas why one might seek to restore streams or watersheds, but the level of human impact and the reasons for restoration vary widely among stream reaches, watersheds, regions, and countries. Human impacts to watersheds began well before recorded history. Archeological evidence indicates that localized deforesta-tion and subsequent impacts to watersheds occurred in populated areas throughout the world even prior to 1000 BC (Williams 2001). For example, forest removal or con-version to agricultural lands occurred in the Mesolithic and Neolithic periods (c. 9000–3000 BC) in parts of Greece and Britain (van Andel *et al.* 1990; Brown 2002). Deforestation expanded during both the Bronze and Iron Age (c. 3000 BC to 500 AD) when metal tools replaced stone tools and made clearing of forests and plowing of lands easier. Extensive hillslope erosion and subsequent sedimentation and aggradation of river valleys in Greece and other areas in the eastern Mediterranean is attributed to deforestation and intensive agriculture during the

Bronze Age (van Andel *et al.* 1990; Montgomery 2007). This was followed by diversion of rivers, draining wetlands, and harnessing waterpower in some areas of Europe and the Mediterranean with the rise of the Roman Empire (Cowx & Welcomme 1998). Deforestation, which often leads to increased silt loads, expanded rapidly during the Middle Ages not only in Europe but also in China and elsewhere, resulting in filling of coastal and low-lying areas and presumably other impacts to streams. During medieval times and through the Renaissance (c. 1000 to 1700 AD), extensive deforestation and conversion of lands to agriculture in Europe and the Mediterranean were common (Cowx & Welcomme 1998; Williams 2001). This occurred somewhat later in the New World and elsewhere following European colonization. More dramatic changes to rivers and watersheds occurred during the Industrial Revolution, as construction of dams and weirs to power industry and rapid industrialization caused the pollution of many waters. In parts of Europe, the mass production of drainage tiles and other technologies led to the drainage and conversion of vast wetlands to agricultural land (Vought & Lacoursière 2006). Increasing urban and agricultural activities resulted in some local channelization of rivers and streams. The combination of migration barriers (dams) and pollution due to industry and the rapidly growing human population led to the decline of several migratory fishes in Europe and eastern North America.

The most severe impacts to aquatic systems in North America, Europe and elsewhere arguably occurred in the late 19th and during the 20th century. Increasingly mechanized societies channelized and dredged rivers, drained wetlands, cut down entire forests, intensified agriculture, and built dams for power, irrigation, and flood control. In the UK, Ireland, Europe, the USA, and elsewhere, large river channelization and wetland drainage programs occurred from the early part of the 20th century up until the 1970s (Cowx & Welcomme 1998; O'Grady 2006). This history of land and water uses along with other human activities produced the degraded conditions we see on the landscape today. For example, it is estimated that worldwide over 50% of wetlands may have been lost (Goudie 2006). Coastal wetland loss in some US states and Europe countries exceed 80% (Dahl & Allord 1999; Airoldi & Beck 2007). Estimates suggest that globally more than 75% of riverine habitats are degraded (Benke 1990; Dynesius & Nelsson 1994; Muhar *et al.* 2000; Vörösmarty *et al.* 2010).

The above factors, coupled with an increasing human population, have led to increased air pollution, highly modified and polluted rivers, and a rapid increase in number of threatened, endangered, or extinct species (Figure 1.2; Goudie 2006). The World Water Council esti-

mates that more than half the world's rivers are polluted or at risk of running dry, and less than 20% of the world's freshwaters are considered pristine (World Water Council 2000; UN Water 2009). Moreover, 80% of human water supplies are threatened by watershed disturbance, pollution, water resource development or other factors (Vörösmarty *et al.* 2010). As recently as 2004, 44% of the stream miles in the USA were considered too polluted to support fishing or swimming (EPA 2009). Current species extinction rates are estimated to be more than 100–1000 times background (prehistoric) rates (Baillie *et al.* 2004), and some studies suggest that modern rates are more than 25,000 times background rates (Wilson 1992). Extinction rates for freshwater fauna are thought to be 4–5 times that of terrestrial species (Ricciardi & Rasmussen 1999), and habitat loss and degradation are believed to be the primary cause of extinctions (Baillie *et al.* 2004). A suite of human activities has led to degradation of streams and watersheds and impaired their use for biota (including humans), and therefore stream and watershed restoration has become critically important worldwide.

1.4 History of the environmental movement

The rapid modification of our natural environment was recognized centuries ago. Limited protection of forests for hunting and timber production occurred in the ancient times, middle ages (c. 500–1500 AD), and the early modern period (c. 1500–1800 AD). Ancient empires such as Assyria, Babylon, and Persia set aside hunting reserves and the Roman Empire set up a system of protected areas for wildlife (Brockington *et al.* 2008). The Emperor Hadrian set half of Mount Lebanon aside in the 2nd century AD to protect cedar forests (Brockington *et al.* 2008). As early as the 11th century in Scotland and 13th century in England, laws and fishing seasons were set to protect salmon (Montgomery 2003). However, large-scale environmental movements did not start until the late 19th and early 20th century in the UK, Europe, the USA, Australia, New Zealand and elsewhere (Hutton & Connors 1999). The late 1800s saw the establishment of some of the first national parks such as Yellowstone National Park in the USA, Rocky Mountain National Park in Canada, and Royal National Park in Australia. During the same period, the Audubon Society, the Sierra Club, the Wilderness Society in America, and the Royal Society for Protection of Birds in the UK were formed and began pushing for greater protection of wild lands and wildlife.

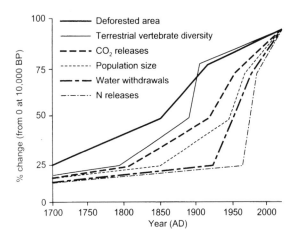

Figure 1.2 Increase in selected human impacts during the last 300 years (percent increased compared to 10,000 BP). From Goudie (2006). Reproduced by permission of John Wiley & Sons.

The modern environmental movement began in the 1960s, initially focusing on water and air quality issues. In the USA, key publications on increasing environmental problems such as Rachel Carson's *Silent Spring* (Carson 1962) and a series of environmental disasters led to a large environmental movement and a series of laws to protect the environment in the 1960s and 1970s. These laws included the Wilderness Act (1964), the National Environmental Policy Act (1969), the Clean Air Act (1970), the Water Pollution Control Act (1972), and the Endangered Species Act (1973). Similar legislation was passed in the 1970s, 1980s, and 1990s in other industrialized countries (e.g. German Federal Nature Conservation Act 1976, Swiss Environmental Protection Law 1983, Canadian Fisheries Act 1985, Canadian Water Act 1985, Japanese Act on Conservation of Endangered Species of Wild Fauna and Flora 1992, Australian Endangered Species Protection Act 1992). In 2000, the European Union (EU) passed the Water Framework Directive (WFD), arguably the most sweeping legislation for the protection and restoration of watersheds and aquatic biota. The WFD combined with other EU Directives for the conservation of nature and biodiversity such as the Birds Directive (79/409/EEC) and the Habitats Directive (92/43/EEC) provide a legal basis to implement comprehensive, interdisciplinary basin-wide restoration programs.

Another key environmental aspect is the importance and economic value of ecosystem goods and services. Until recently the value of ecosystems was only based on the goods they might produce (e.g. harvestable fish, food, timber), but in recent years the services or benefits we derive directly or indirectly from ecosystem functions have also been recognized. These other services include waste processing, carbon sequestering, regulation of atmospheric gases, water regulation, climate regulation, genetic resources, and many others (Costanza *et al.* 1997; Cunningham 2002). In fact, the economic value of ecosystem services globally has been estimated to be 2–3 times that of the total global gross domestic product from world economies (Costanza *et al.* 1997). This realization of the importance of functioning ecosystems for our economic prosperity and our very existence has led to further emphasis on protecting and restoring natural ecosystems globally.

1.5 History of stream and watershed restoration

Similar to the environmental movement, the earliest stream restoration efforts were largely undertaken by hunters and fishermen. While efforts to minimize erosion and protect water supplies and agricultural land date back thousands of years (Riley 1998), the first substantial efforts to restore streams are thought to have been made in the late 1800s by local fishing clubs in the USA and river keepers on British estates interested in improving salmon or trout fishing (Thompson & Stull 2002; White 2002). As early as 1885, Van Cleef called for the restoration and protection of trout streams in the Eastern USA (Van Cleef 1885). There is also evidence of early restoration efforts in Germany and Norway (Walter 1912; Thompson & Stull 2002). These early efforts often included stocking of fish and killing of predatory birds, fish and mammals, actions that today would be frowned upon (White 2002).

More formalized efforts to restore streams were undertaken in the USA in the early part of the 20th century (Thompson & Stull 2002). The Civilian Conservation Corps and some smaller state-sponsored stream and land restoration programs began implementing restoration projects on miles of small streams in the Midwest, Rocky Mountains and elsewhere during the Great Depression, partly to combat soil and bank erosion. These efforts tended to focus on planting trees, fencing out livestock, bank protection and stabilization, installing small log structures or weirs to create pools, and even excavation of pools. The latter three techniques were largely engineering approaches attempting to create pool habitat or a static stream channel, and often treated symptoms (lack of pools) rather than underlying problems (e.g. excess sediment, lack of riparian vegetation and woody debris) (White 1996; Riley 1998). It is however important to remember that, during this period, streams were highly degraded from decades of severe overgrazing and removal of streamside vegetation and it was not yet fully understood how quickly riparian banks and vegetation might recover once they were protected (White 2002). The 1940s and 1950s witnessed an increased emphasis on planting of vegetation to stabilize banks; however, these efforts were often not viewed as favorably as instream structures and hardening of banks, which were seen as quicker and more permanent (White 1996). Both before and after World War II in Europe there were efforts to stabilize banks using plantings and bioengineering approaches, but again these were largely to create static channels and prevent streams from moving.

Expansion of state and federal stream restoration programs in the USA continued from the 1950s through the 1980s. Following years of overgrazing and other human activities, riparian vegetation began to recover along numerous streams in the USA and Canada (White 2002). During this period, there was also an increased focus on placement of log and boulder cover structures, based largely on promising results from trout stream restoration in Wisconsin and Michigan. However, these structural techniques were largely pioneered in low-energy Eastern and Midwestern streams and met with mixed results when applied elsewhere, particularly in higher-gradient higher-energy streams of the mountainous western North America. Several of these techniques were subsequently applied in European streams in the 1980s and 1990s with varying degrees of success. Despite the emphasis on structural treatments, the key stream restoration manual (White & Brynildson 1967) recommended protecting riparian vegetation before installing instream structures. Unfortunately, this sage advice was largely ignored until recently when the importance of watershed processes became more widely accepted (Chovanec et al. 2000; Hillman & Brierely 2005; Beechie et al. 2010). Fortunately, as early as the 1960s some states were acquiring land along streams to let riparian vegetation and streams recover naturally. There was also an increasing understanding of riverine processes – partly based on Leopold et al. (1964) – which biologists were attempting to incorporate into stream restoration projects.

The late 1980s and early 1990s saw rising awareness in the importance of riparian areas, the physical and ecological importance of large wood, and a better understanding of physical and biological processes and how land use and human activities impact those processes and fish habitat (White 2002). This was initially based on extensive studies on forested streams in the Pacific Northwest of North America, but was later based on studies in a range of land uses and ecoregions. The results of these studies led to recommendations for a watershed or ecosystem approach to management and a growing call for looking beyond an individual stream reach when planning restoration (Beechie & Bolton 1999; Roni et al. 2002; Hillman & Brierely 2005). From the 1990s until today, restoration efforts have slowly been changing from a focus on localized habitat improvement actions at a site or reach scale (which often overlooked the root causes of habitat degradation) to a more holistic watershed or ecosystem approach which tries to treat the underlying problem that has led to the habitat degradation (to be discussed in great detail in the following chapters). This is not to say that certain habitat improvement techniques are not widely used or are ineffective, but rather that greater emphasis has been placed on restoring

whole watersheds through improving land use, reducing sediment sources, protecting riparian areas, and other restoration efforts focused on restoring the processes that create and maintain stream habitats and health.

European river restoration efforts largely began in the 1980s and increased dramatically during the 1990s (Cowx & Welcomme 1998), focusing mostly on rehabilitation of channelized, straightened and engineered channels and floodplains. In fact, the science of floodplain restoration and remeandering of rivers was largely developed in Europe, and much of the literature on this topic comes from European case studies (e.g. Brookes 1992, 1996; Iversen et al. 1993). With the exception of some early erosion reduction efforts to reduce declining production of agricultural lands in the 1970s, restoration efforts in Australia and New Zealand and other developed countries also began in the 1980s and 1990s (Gippel & Collier 1998).

The number and scale of watershed restoration efforts, along with spending on restoration, has increased rapidly in the last few decades in North America, Europe, Australia, and elsewhere. This has been partly driven by increasing environmental awareness, stronger environmental regulations, and declines in species of fish and aquatic organisms that are of high socioeconomic and cultural value. As discussed in the Section 1.4, legal mechanisms have been developed to restore water quality, individual species, and riverine ecosystems in developed countries. Perhaps the most commonly recognized legal mandates are those requiring protection or restoration of specific species under national laws such as the Endangered Species Act in the USA, the Canadian Species at Risk Act, or the European Red List. These legislative actions are generally reactive and drive attempts to restore habitats for listed species. While the legislation behind these lists generally calls for conservation and restoration of the ecosystems upon which these species depend, restoration actions are commonly focused on restoring specific habitats deemed important for one species or another. In the USA and Canada, for example, massive efforts to restore watersheds in the Pacific Northwest of North America are almost exclusively focused on recovering threatened and endangered salmon and trout populations (Katz et al. 2007), although restoration actions such as sediment reduction and riparian restoration also benefit other species. Beyond endangered species concerns, many nations have also passed legislation aimed at more holistic attempts to restore riverine ecosystems (e.g. the Clean Water Act in the USA or the Water Framework Directive in the EU) which seek to improve more broadly defined hydromorphological, chemical, and biological conditions of rivers.

In conjunction with changing drivers of restoration and an increasingly holistic approach to restoring watersheds, the expertise needed to plan and implement projects has also evolved. Early restoration efforts were often initiated by outdoorsmen or fisheries biologists and later by engineers, and focused on structural treatments or bank stabilization. The greater emphases on watershed processes in the USA and Europe has also led to improved design of more traditional habitat improvement techniques and greater emphasis on addressing root causes of degradation. Given that streams integrate both terrestrial and aquatic processes at multiple scales, the practice of restoring processes or improving habitats of an aquatic ecosystem requires an interdisciplinary approach to be successful. This often requires the collaboration of those with expertise in fish and aquatic biology, riparian and stream ecology, geology, hydrology and water management, geomorphology, landscape architecture, and even public policy, economics, and other social sciences. That is not to say that all projects will require expertise in all these fields, but most will benefit from an interdisciplinary team; this will certainly be essential for large or comprehensive restoration projects or programs to achieve their goals. Another aim of this book is therefore to provide a common basis and level of knowledge for individuals from various backgrounds to work together on developing and implementing successful restoration programs.

1.6 Key steps for planning and implementing restoration

Despite large financial investments in what has recently been called the 'restoration economy' (Cunningham 2002) and increasing literature on restoration planning, numerous watershed councils, river trusts, agencies, and other restoration practitioners do not follow a systematic approach for planning restoration projects throughout a watershed or basin. As a result, a number of restoration efforts fail or fall short of their objectives. Some of the most common problems or reasons for failure of a restoration program or project include:
• not addressing the root cause of habitat or water quality degradation;
• not recognizing upstream processes or downstream barriers to connectivity;
• inappropriate uses of common techniques (one size fits all);

• an inconsistent (or complete lack of an) approach for sequencing or prioritizing projects;
• poor or improper project design;
• failure to get adequate support from public and private organizations; and
• inadequate monitoring to determine project effectiveness.

These challenges and problems can be overcome by systematically following several logical steps that are critical to developing a successful restoration program or project (Figure 1.1). This book is designed to cover these steps in detail to assist with improving the design and evaluation of stream and watershed restoration plans and projects. We begin with a discussion of watershed processes and process-based restoration (Chapter 2), as these basic concepts underlie the restoration steps in subsequent chapters. The following chapters then explain the key steps, including: assessing watershed conditions and identifying restoration needs (Chapter 3); selecting appropriate restoration actions to address restoration needs (Chapter 5); identifying a prioritization strategy for prioritizing actions (Chapter 6); planning and implementing projects (Chapter 7); and developing a monitoring and evaluation program (Chapter 8). Goals and objectives need to be set at multiple stages of the restoration process, and there are multiple steps within each stage which we will discuss within each chapter. In addition, the human and socioeconomic aspects need to be considered throughout the planning and design process (Chapter 4). We close with a discussion of how to synthesize all these pieces to develop restoration plans and proposals (Chapter 9).

Throughout this book we emphasize the concept of process-based restoration (Chapter 2), which aims to address root causes of habitat and ecosystem degradation (Sear 1994; Roni *et al.* 2002; Beechie *et al.* 2010). Our purpose in doing so is to help guide river and watershed restoration efforts toward actions that will have long-lasting positive effects on riverine ecosystems and to ensure that, when habitat improvement is undertaken, the site potential and watershed processes are considered. We also emphasize the importance of recognizing socioeconomic and political considerations such as involving landowners and other stakeholders, permit and land-use issues, and education and outreach to the general public to build continued support for restoration (Chapter 4). Failure to consider these factors and involve stakeholders early on can prevent even the most worthwhile and feasible projects from being implemented. The following chapters go into detail on each of the steps for planning and implementing successful stream and watershed restoration programs and projects.

1.7 References

Airoldi, L. & Beck, M.W. (2007) Loss, status, and trends for coastal marine habitats of Europe. *Oceanography and Marine Biology* **45**, 345–405.

Baillie, J.E.M., Hilton-Taylor, C. & Stuart, S.N. (editors) (2004) *2004 IUCN Red List of Threatened Species™, A Global Species Assessment.* IUCN, Cambridge, UK. (Also available at http://data.iucn.org/dbtw-wpd/commande/downpdf.aspx?id=10588&url=http://www.iucn.org/dbtw-wpd/edocs/RL-2004-001.pdf.)

Beechie, T. & Bolton, S. (1999) An approach to restoring salmonid habitat-forming processes in Pacific Northwest watersheds. *Fisheries* **24**(4), 6–15.

Beechie, T.J., Sear, D., Olden, J. *et al.* (2010) Process-based principles for restoring river ecosystems. *BioScience* **60**, 209–222.

Benke, A.C. (1990) A perspective on America's vanishing streams. *Journal of the North American Benthological Society* **9**, 77–88.

Bernhardt, E.S., Palmer, M.A. Allan, J.D. *et al.* (2005) Synthesizing U.S. River Restoration Efforts. *Science* **308**, 636–637.

Brierley, G.J. & Fryirs, K.A. (2008) *River Futures: An Integrative Scientific Approach to River Repair.* Island Press, Washington D.C.

Brockington, D., Duffy, R. & Igoe, J. (2008). *Nature Unbound: Conservation, Capitalism and the Future of Protected Areas.* Earthscan, London.

Brookes, A. (1992) Recovery and restoration of some engineered British River Channels. In: Boon, P.J., Calow, J. & Petts, G.E. (eds) *River Conservation and Management.* John Wiley & Sons Ltd., Chichester, England, pp. 337–352.

Brookes, A. (1996) Floodplain restoration and rehabilitation. In: Anderson, M.G., Walling, D.E. & Bates, P.D. (eds) *Floodplain Processes.* John Wiley & Sons Ltd., Chichester, England, pp. 553–576.

Brookes, A. & Shields, D. (1996) *River Channel Restoration: Guiding Principles for Sustainable Projects.* John Wiley & Sons Ltd., Chichester, England.

Brown, T. (2002) Clearances and clearings: deforestation in Mesolithic/Neolithic Britain. *Oxford Journal of Archaeology* **16**, 133–146.

Buijse, A.D., Coops, H., Staras, M. *et al.* (2002) Restoration strategies for river floodplains along large lowland rivers in Europe. *Freshwater Biology* **47**, 889–907.

Carson, R. (1962) *Silent Spring.* Houghton Mifflin, Boston.

Chovanec, A., Jager, P. *et al.* (2000) The Austrian way of assessing the ecological integrity of running waters: a contribution to the EU Water Framework Directive *Hydrobiologia* **422**, 445–452.

CIRF (Centro Italiano per La Riqualificazione Fluviale) (2006) *La Riqualificazione Fluviale in Italia.* Mazzanti Editori, Venice, Italy.

Clewell, A. & Aronson, J. (2008) *Ecological Restoration: Principles, Values, and Structure of an Emerging Profession.* Island Press, Washington.

Costanza, R., Arge de Groot, R. *et al.* (1997) The value of the world's ecosystems and natural capital. *Nature* **387**, 253–260.

Cowx, I.G. & Welcomme, R.L. (1998) *Rehabilitation of Rivers for Fish.* Fishing News Books, Oxford.

Cunningham, S. (2002). *The Restoration Economy.* Berrett-Koehler Publishing, San Francisco.

Dahl, T.E. & Allord, G.J. (1999) *History of Wetlands in the Conterminous United States.* United States Geological Survey Water Supply Paper 2425. US Geological Survey available at http://water.usgs.gov/nwsum/WSP2425/history.html (accessed 12/23/09).

Dynesius, M. & Nilsson, C. (1994) Fragmentation and flow regulation of the river systems in the northern third of the world. *Science,* **266**, 753–762.

EPA (Environmental Protection Agency) (2009) *National Water Quality Inventory: Report to Congress 2004 Reporting Cycle.* EPA 841-R-08-001, U.S. EPA Office of Water, Washington, DC.

FISRWG (Federal Interagency Stream Restoration Working Group) (1998) *Stream Corridor Restoration: Principles, Processes, and Practices.* GPO Item No. 0120-A. USDA, Washington, DC.

Gippel, C.J. & Collier, K.J. (1998). Degradation and rehabilitation of waterways in Australia and New Zealand. In: de Waal, L.C., Large, A.R.G. & Wade, P.M. (eds) *Rehabilitation of Rivers: Principles and implementation,* Wiley, Chichester, England, pp. 269–300.

Goudie, A. (2006) *The Human Impact on the Natural Environment,* sixth edition. Blackwell Publishing Ltd, Oxford, U.K.

Hillman, M. & Brierley, G. (2005) A critical review of catchment-scale stream rehabilitation programmes. *Progress in Physical Geography* **29**(1), 50–76.

Hunt, R.L. (1993) *Trout Stream Therapy.* University of Wisconsin Press, Madison, Wisconsin.

Hunter, C.J. (1991) *Better Trout Habitat: A Guide to Stream Restoration and Management.* Island Press, Washington DC.

Hutton, D. & Connors, L. (1999) *A History of the Australian Environmental Movement*. Cambridge University Press, New York.

Iversen, T.M., Kronvang, B., Madsen, B.L., Markmann, P. & Nielsen, M.B. (1993) Re-establishment of Danish streams: restoration and maintenance measures. *Aquatic Conservation: Marine and Freshwater Ecosystems* 3(2), 73–92.

Katz, S.L., Barnas, K., Hicks, R., Cowen, J. & Jenkinson, R. (2007) Freshwater habitat restoration in the Pacific Northwest: a decade's investment in habitat improvement. *Restoration Ecology* 15, 494–505.

Leopold, L.B., Wolman, M.G. & Miller, J.P. (1964) *Fluvial Processes in Geomorphology*. WH Freeman and Company, San Francisco.

Mills, D. (1991) *Strategies for the Rehabilitation of Salmon Rivers*. The Atlantic Salmon Trust, The Institute of Fisheries Management & The Linnaean Society of London, London.

Montgomery, D.R. (2003) *The King of Fish: The Thousand Year Run of Salmon*. Westview Press, Cambridge, Massachusetts.

Montgomery, D.R. (2007) *Dirt: The Erosion of Civilizations*. University of California Press, Berkely, California.

Muhar, S., Schwarz, M., Schmutz, S. & Jungwirth, M. (2000) Identification of rivers with high and good habitat quality: methodological approach and applications in Austria. *Hydrobiologia* 422(423), 343–358.

O'Grady, M. (2006) *Channels & Challenges: The Enhancement of Salmonid Rivers*. Irish Freshwater Fisheries Ecology and Management Series: Number 4. Central Fisheries Board, Dublin, Ireland.

Omerod, S.J. (2004) The golden age of restoration science. *Aquatic Conservation: Marine and Freshwater Ecosystems* 14, 543–549.

Ricciardi, A. & Rasmussen, J.B. (1999) Extinction rates of North American freshwater fauna. *Conservation Biology* 13(5), 1220–1222

Riley, A.L. (1998) *Restoring Streams in Cities*. Island Press, Washington D.C.

RRC (River Restoration Centre) 2002. *Manual of Techniques*. The River Restoration Centre, Beds, UK.

Roni, P. (2005) *Monitoring Stream and Watershed Restoration*. American Fisheries Society, Bethesda, Maryland.

Roni, P., Beechie, T.J., Bilby, R.E., Leonetti, F.E., Pollock, M.M. & Pess, G.R. (2002) A review of stream restoration techniques and a hierarchical strategy for prioritizing restoration in Pacific Northwest watersheds. *North American Journal of Fisheries Management* 22, 1–20.

Roni, P., Hanson, K., Pess, G., Beechie, T., Pollock, M. & Bartley, D. (2005) *Habitat Rehabilitation for Inland Fisheries: Global Review of Effectiveness and Guidance for Restoration of Freshwater Ecosystems*. Fisheries Technical Paper 484. Food and Agriculture Organization of the United Nations, Rome, Italy, pp. 116.

Sear, D.A. (1994) River restoration and geomorphology. *Aquatic Conservation* 4, 169–177.

Slaney, P.A. & Zaldokas, D. (1997) *Fish Habitat Rehabilitation Procedures*. British Columbia, Ministry of Environment, Lands and Parks and Ministry of Forests, Vancouver, BC, Canada.

Thompson, D.M. & Stull, G.N. (2002) The development and historic use of habitat structures in channel restoration in the United States: the grand experiment in fisheries management. *Geographie physique et Quaternaire* 56, 45–60.

UN Water (2009) *Water in a Changing World – World Water Development Report 3*. World Water Assessment Program, United Nations Education, Scientific and Educational Organization. Available at http://www.unesco.org/water/wwap/wwdr/wwdr3/.

van Andel, J.H., Zangger, E. & Demitrack, A. (1990) Land use and soil erosion in prehistoric and historical Greece. *Journal of Field Archaeology* 17(4), 379–396

Van Cleef, J.S. (1885) How to restore our trout streams. *Transactions of the American Fisheries Society* 14, 51–55.

Vörösmarty, C.J., McIntyre, P.B., Gessner, M.O. *et al.* (2010). Global threats to human water security and river biodiversity. *Nature* 467, 555–561.

Vought, L.B.M. & Lacoursière, J.O. (2006) Restoration of streams in agricultural landscapes. In: Eiseltová, M. (ed.) *Restoration of Lakes, Streams, Floodplains and Bogs in Europe*. Springer, New York, pp. 225–242.

Walter, E. (1912) Die bewirtschaftung des forellenbaches: eine anleitung zur pflege der bachforelle in freien gewässer für berufs- und sportfischer. (*Management of the Trout Stream: an Introduction to the Stewardship of Brown Trout in Open Waters for the Commercial and Sport fisher*) J. Neumann, Neudamm, Germany.

Ward, D., Holmes, N. & José, P. (1994) *The New Rivers & Wildlife Handbook*. The Royal Society for the Protection of Birds, Bedfordshire, UK.

White, R.J. (1996) Growth and development of North American stream habitat management for fish. *Canadian Journal of Fisheries and Aquatic Sciences* 53, 342–363.

White, R.J. (2002) Restoring streams for salmonids: Where have we been? Where are we going? In: M. O'Grady (ed.) *Proceedings of the 13th International*

Salmonid Habitat Enhancement Workshop, Westport, County Mayo, Ireland, September 2002. Central Fisheries Board, Dublin, Ireland, pp 1–31.

White, R.J. & Brynildson, O.M. (1967) *Guidelines for Management of Trout Stream Habitat in Wisconsin.* Wisconsin Department of Natural Resources Technical Bulletin 39, Madison.

Williams, M. (2001) The history of deforestation. *History Today* **51**, 30–37.

Wilson, E.O. (1992) *The Diversity of Life.* The Belknap Press, Cambridge, Massachusetts.

World Water Council (2000) *World Water Vision.* Earthscan Publications, London. (Also available at http://www.worldwatercouncil.org/.)

Young, T.P., Petersen, D.A. & Clary, J.J. (2005) The ecology of restoration: historical links, emerging issues and unexplored realms. *Ecological Letters* **8**, 662–673.

2

Watershed Processes, Human Impacts, and Process-based Restoration

Tim Beechie[1], John S. Richardson[2], Angela M. Gurnell[3] & Junjiro Negishi[4]

[1]Northwest Fisheries Science Center, National Oceanic and Atmospheric Administration, USA
[2]University of British Columbia, Canada
[3]Queen Mary, University of London, UK
[4]Hokkaido University, Japan

2.1 Introduction

Effective planning, design, and implementation of river restoration efforts each require an understanding of how watershed processes drive the structure and functions of riverine ecosystems, as well as how those processes support a wide variety of ecosystem services. In this book, the term 'watershed process' generally refers to movements of landscape or ecosystem components into and through river systems, which are typically measured as rates (Beechie & Bolton 1999). For example, erosion is a process that moves sediment from hillslopes to river channels, while sediment transport processes move sediment through stream and river channels to deltas and estuaries. Erosion is measured in units of mass/area/time, whereas sediment transport is commonly measured in units of mass/time. We do not restrict the term 'processes' to geomorphological or hydrological processes, but instead refer to a wide range of processes including erosion and sediment transport, storage and routing of water, plant growth and successional processes, delivery of nutrients and organic matter, inputs of thermal energy, trophic interactions, species interactions, and population dynamics. Understanding these processes and relationships between them is critical to the success of river restoration programs.

These driving processes influence states and dynamics of biological communities through a sequence of cause–effect linkages that connect watershed processes to habitat conditions, and habitat conditions to biota (Figure 2.1). 'Habitat conditions' here refers to physical, chemical, and thermal features of the river environment and 'biota' refers to ecological systems and functions that respond to habitat features. Humans alter watershed processes in many ways, leading to changes in habitat conditions, food webs, and biological communities (Allan 2004). Process-based restoration focuses on correcting anthropogenic disruptions to driving processes, thereby leading to recovery of habitats and biota (Sear 1994; Beechie & Bolton 1999). Restoration of critical processes also confers added resilience to river–floodplain ecosystems, as functioning processes allow the system to respond to future disturbances through natural physical and biological adjustments (Brierley et al. 2002; Beechie et al. 2010). While restoration of processes and habitats is critical to ecosystem recovery, correcting ecosystem degradation that results from lack of key species, introduction of non-native species, or poor water quality is also critical to ecosystem recovery (Karr 2006). Because these factors

Stream and Watershed Restoration: A Guide to Restoring Riverine Processes and Habitats, First Edition. Philip Roni and Tim Beechie.
© 2013 John Wiley & Sons, Ltd. Published 2013 by John Wiley & Sons, Ltd.

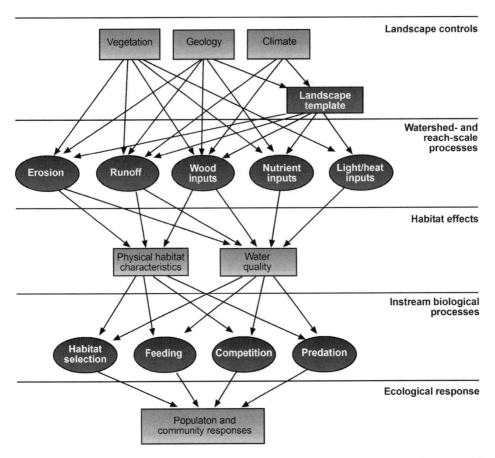

Figure 2.1 Illustration of process linkages between watershed processes, instream processes, and biological responses (adapted from Beechie & Bolton 1999; Beechie *et al.* 2009).

must also often be addressed to achieve restoration goals, we briefly address restoration of other ecosystem alterations and describe a comprehensive set of both processes and ecosystem features that may be important in restoring river ecosystems.

In this chapter we first describe the hierarchical suite of processes that drive riverine ecosystems. We identify and describe the main processes driving riverine habitat dynamics and biota, focusing on processes that are commonly targeted by stream and watershed restoration activities. We also describe how watershed and reach-scale processes drive the expression and dynamics of habitat types in each reach of a river network, and illustrate how habitat conditions control expression and dynamics of biological communities. Throughout, our main purpose is to illustrate how these processes are hierarchically

nested, and to show that higher-level controls set limits on the expression of habitat features or ecosystem attributes as influenced by lower-level controls (e.g. Beechie *et al.* 2010). We then describe the landscape setting, and proceed to watershed-scale processes, reach-scale processes, habitat dynamics, and instream biological processes. We also briefly describe ways in which watershed processes may be altered, thereby affecting the productivity, resilience, and functions of river ecosystems. Alterations to processes are further discussed in the chapter on watershed assessments (Chapter 3), and restoration actions that restore processes are discussed in Chapter 5. Finally, we describe process-based restoration (which is a central theme of river and watershed restoration in this book) and we present four process-based principles to help guide restoration planning and implementation.

2.2 The hierarchical structure of watersheds and riverine ecosystems

Physical and biological features of riverine ecosystems are controlled by a hierarchy of physical, chemical, and biological processes operating across a wide range of space- and timescales (Figure 2.2). These processes control the arrangement of channel and habitat types across the riverine landscape, such as reach-scale channel types, or pool and riffle units at smaller scales (e.g. Frissell et al. 1986; Fausch et al. 2002; Allan 2004). Biological features are also controlled by a hierarchy of processes, including community composition of riparian or aquatic species and locations of suitable habitats for individual species (e.g. Beechie et al. 2008a; Naiman et al. 2010). For example, the typical sequence of channel patterns from headwaters to lower river begins with steep cascades in tributaries and progresses through step-pool, plane-bed, and pool-riffle channels (Montgomery & Buffington 1997) (Figure 2.3). As tributaries coalesce the channels become larger and more complex, and are classified as braided, straight, island-braided, or meandering (Beechie et al. 2006a). This arrangement of reach-level channel types, as well as habitat features within reaches, is determined by six main variables: channel slope, valley confinement, discharge, sediment supply and size, bank strength (including root strength), and wood supply (Figure 2.4).

Two of these variables – channel slope and valley constraint (the 'ultimate controls') – are generally unchanging over human time frames, as the tectonic and erosional processes controlling these variables act over long time frames and across large areas ($>10^2$ years, $>1\,km^2$) (Naiman et al. 1992; Montgomery & Buffington 1997; Figure 2.5). Broad valley forms are defined by characteristics such as valley shape (V-shaped or U-shaped), the presence or absence of terraces, valley slope, and confinement of the stream (floodplain width relative to stream width). Valley forms are also relatively immutable over management time frames (<100 years) (Naiman et al. 1992), and are also controlled by past geological processes such as uplift, glaciation, and river erosion (Bishop et al. 1985; Benda et al. 1992; Montgomery 2002). While reach-scale channel slopes can change locally over shorter time frames, they can only do so within a fairly narrow range set by the valley slope, which is essentially immutable over hundreds of years. Hence, these landscape features (valley slope and constraint) control the range of channel types that can be expressed within valley segments (Naiman et al. 1992).

The third and fourth controlling variables are discharge (or stream flow) and sediment supply. These two variables are controlled at the watershed scale, and the drainage boundary defines the region within which erosion and runoff processes control regimes of sediment supply and stream flow (Figure 2.5). Sediment supply and stream flow largely determine stream size and channel forms (e.g. pools, riffles, gravel or sand bars), within the limits set by valley confinement and channel slope. To some extent, patterns of stream size and sediment supply are also controlled by the structure of the river network (the arrangement of stream channels and confluences) which influence riverine habitats in two ways. First, the general downstream pattern of increasing stream flow and decreasing slope drives a corresponding shift in habitat characteristics (Vannote et al. 1980; Minshall et al. 1985; Naiman et al. 1987). Second, the network structure, arrangement of confluences, nick points, and alternating valley and canyon reaches interrupt gradual downstream trends to create local nodes or reaches with unique geomorphic features (Poole 2002; Benda et al. 2004). For example, sediment supply tends to be higher below large tributaries, creating locally complex habitat features and higher biological diversity. Together, these interacting patterns create a generally predictable arrangement of physical attributes in river systems.

At the reach level, channel forms are mutable and controlled by shorter-term variation in processes such as sediment supply regime, flow regime, and wood and plant propagule supply (Gurnell et al. 2005, 2006a; Figure 2.5). Because these processes naturally vary from year to year as a function of storms and floods, channel locations can change over relatively short time frames (years to decades) while the dominant reach type is usually relatively constant over decadal periods (Beechie et al. 2006a). Moreover, the dominant channel forms tend to exhibit a characteristic arrangement within watersheds, although geological features and the influence of river confluences can interrupt this pattern (e.g. Montgomery 1999; Benda et al. 2004; Beechie et al. 2006a). Within reaches, habitat units (e.g. pools, riffles, ponds) are even more variable in space and time (Frissell et al. 1986), often shifting positions between years as a result of wood movement. However, the relative abundance of habitats is relatively constant in the absence of major changes in driving variables (e.g. Zanoni et al. 2008).

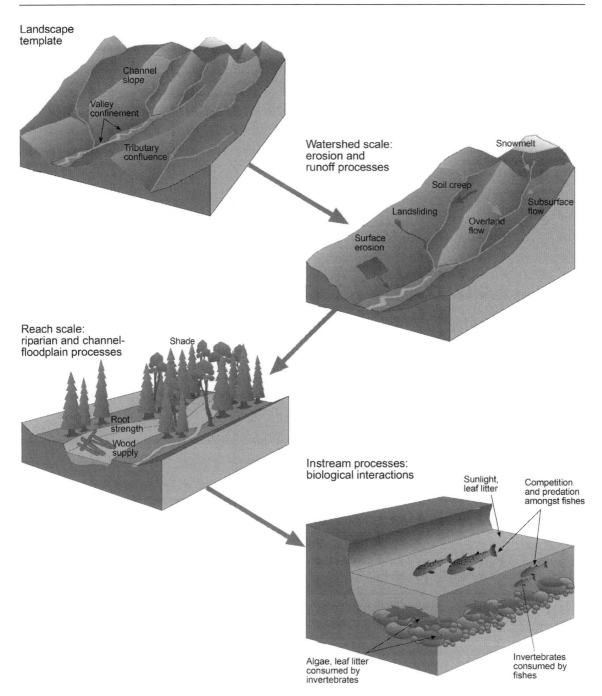

Landscape template

Channel slope

Valley confinement

Tributary confluence

Watershed scale: erosion and runoff processes

Snowmelt

Soil creep

Landsliding

Subsurface flow

Overland flow

Surface erosion

Reach scale: riparian and channel-floodplain processes

Shade

Root strength

Wood supply

Instream processes: biological interactions

Sunlight, leaf litter

Competition and predation amongst fishes

Algae, leaf litter consumed by invertebrates

Invertebrates consumed by fishes

Figure 2.2 Hierarchical nesting of processes controlling population and community responses of riverine biota. Higher-level controls set limits on the types of habitat features or ecosystem attributes that can be expressed at lower levels, and lower-level processes control the expression of attributes within those limits (based on Beechie *et al.* 2010). (See Colour Plate 1)

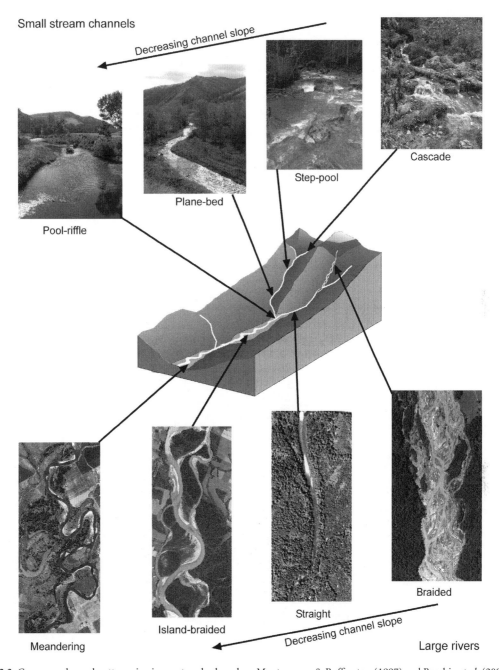

Figure 2.3 Common channel patterns in river networks, based on Montgomery & Buffington (1997) and Beechie *et al.* (2006a).

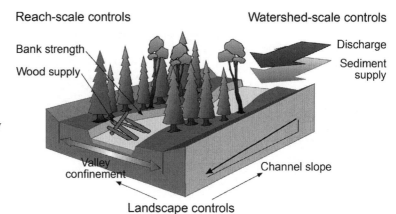

Figure 2.4 Illustration of the six primary controls on channel form, including physical and biological processes that control physical habitat conditions in streams and rivers: channel slope, valley constraint, discharge, sediment supply, bank strength (including root reinforcement), and wood supply.

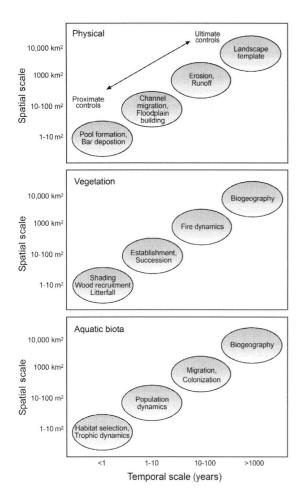

Figure 2.5 Spatial and temporal scales of processes that control physical features, vegetation, and aquatic biology in river ecosystems (adapted from Naiman *et al.* 1992; Beechie *et al.* 2010).

The biological structure of riverine ecosystems is also hierarchically controlled, partly by the physical controls described above and partly by a suite of biological processes. At the landscape scale, biogeography of species controls the pool of species available to form riparian and aquatic communities (Figure 2.5). For example, native riparian species found in a watershed are limited to those that are endemic to the region and its climate, and native fishes are limited to those that can migrate to suitable habitats within their native range. Hence, in the absence of human interventions, local riparian and aquatic communities only comprise species present in the native species pool. At the watershed scale, both riparian and aquatic communities exhibit characteristic shifts in composition in the downstream direction and biological zones are often classified by the dominant species present, or by community composition and species richness (Huet 1959; Sheldon 1968; Ibarra *et al.* 2005). These shifts are partly due to a gradual increase in river size (e.g. the river continuum concept, Vannote *et al.* 1980; Minshall *et al.* 1985; Naiman *et al.* 1987; Figure 2.6). However, differences in channel slope, width, or pattern and dynamics are also influenced by changes in lithology or by network structure, which create localized discontinuities in this downstream trend. At the site scale, biological attributes are largely controlled by local habitat conditions, as well as by species interactions. For example, spawning locations of individual fish species are controlled by locations of suitably sized substrate, water depth, and velocity for the species (Beechie *et al.* 2008a), whereas productivity and species composition of riparian forests are controlled by local soil and moisture conditions (Naiman *et al.* 2010).

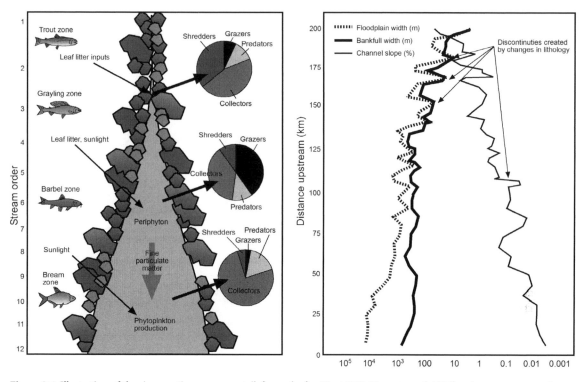

Figure 2.6 Illustration of the river continuum concept (left panel, after Huet 1959; Vannote *et al.* 1980) and measurements of downstream trends in channel slope, bankfull channel width, and floodplain width (right panel, data from Skagit River, USA). Measured trends indicate a systematic downstream trend in river size and slope as indicated by the river continuum concept, as well as local discontinuities created by changes in lithology that create alternating steep canyon reaches and low-gradient valley reaches.

2.3 The landscape template and biogeography

Landscapes and watersheds are physically defined by their topography, geology, and climate, and can be classified by landscape units that stratify broad suites of habitat-forming processes and disturbance regimes (Montgomery 1999; Montgomery & Bolton 2003). This geologic, tectonic, and climatic template controls the arrangement of reach types in the network, locations of tributary junctions, and alternating canyon and floodplain reaches (Montgomery 1999; Benda *et al.* 2004; Brierley & Fryirs 2005; Beechie *et al.* 2006a). That is, the landscape template sets limits on the range of physical and biological attributes that any reach in the network is capable of expressing, and therefore sets limits on what can be achieved through restoration. For example, a steep headwater reach can only develop specific habitat types ranging from boulder cascades to short alluvial reaches

upstream of wood jams, whereas a large low-gradient floodplain reach can develop a suite of habitat types ranging from mainstem pools and riffles, to bank and bar habitats, to a wide range of lentic and lotic floodplain habitats (Brierley *et al.* 2002; Beechie *et al.* 2005a).

Similarly, the landscape template influences hydrologic and sediment supply regimes that control channel forms. For example, regions with steep slopes, shallow soils, and humid climates may be characterized by high landslide rates and coarse sediment loads that tend to create braided or island-braided channels (Sidle *et al.* 1985; Hovius *et al.* 1997; Imaizumi *et al.* 2008). By contrast, a drier region with deep soils and gentle slopes may be characterized by high fine sediment loads and more stable anastomosing or meandering channels (Rust 1981; Beechie *et al.* 2006a). These differences in process regimes create diverse ranges of potential physical conditions that restoration can achieve in streams, but do not describe specific local habitat or biological conditions that are controlled by smaller-scale processes. The timescale of processes that create the landscape template also captures past climatic changes that formed regional geologic features such as glacial terraces (e.g. Benda *et al.* 1992), as well as large but rare disturbances such as volcanic eruptions or mega-floods that alter valley floor morphology (Beechie *et al.* 2001; Brown *et al.* 2001).

Key biological processes that drive the biogeography of species operate at similarly long space- and timescales. That is, processes such as migration, colonization, extinction, and evolution have – over tens of thousands of years – resulted in biotic assemblages that are adapted to the local geographic and climate settings of individual river systems and reaches within systems (Taberlet *et al.* 1998; Waples *et al.* 2004). Examples of these processes are migration and evolution of Pacific salmon (*Oncorhynchus* spp.), and colonization and establishment of plant species after glaciations (Bennett 1986; Waples *et al.* 2008). These processes result in species pools that limit the natural suite of riparian and aquatic communities that can be expressed in each reach. However, short-term climate variations can alter the spatial distribution or relative abundance of species over relatively short time frames (hundreds to thousands of years), yet the broad pool of species available in ecosystems generally changes much more slowly in the absence of human interventions (Pess *et al.* 2003; Bekker 2005).

2.4 Watershed-scale processes

Within limits set by the landscape template, runoff and erosion processes modify channel conditions through time, and energy and nutrient supply processes influence productivity levels. Each of these watershed-scale processes influences the short-term expression of physical and biological conditions within reaches (Table 2.1; Figure 2.2). Understanding the influences of these processes on riverine ecosystems helps explain associations between watershed-scale land use and biological community measures (e.g. Death & Collier 2010), and determines the kinds of assessments needed to identify necessary restoration actions (Chapter 3).

2.4.1 Runoff and stream flow

Stream flow regimes are defined by the magnitude, frequency, duration, timing, and rate of change of flow events (Poff *et al.* 1997). These five components are primarily controlled by the timing and magnitude of precipitation or snowmelt events, but are also moderated by interception, infiltration, and evapotranspiration

processes. There are three main runoff pathways: overland flow (water running over the soil surface); shallow subsurface flow (water flowing through the soil layer); and groundwater flow (water flowing through deep flow paths below the soil layer) (Tague & Grant 2004; Figure 2.7). Interception of rain by vegetation reduces the amount of water reaching the soil, and evapotranspiration removes water from the soil during the growing season. In colder areas, the magnitude of snowmelt runoff can also be reduced by interception, as snow captured in a forest canopy reduces total snow accumulation and snow in the canopy tends to sublimate (a direct transition from ice to water vapor) before melting and reaching the ground to contribute to stream flow. This process is particularly important in watersheds with transitional snowpack (i.e. where snow may melt and accumulate several times a year) because snow interception reduces rapid snowmelt and rain-on-snow flooding in regions with forest cover. In both snow and rainfall zones, the rate of infiltration into the soil determines how much of the runoff is by overland flow or by subsurface flow. Areas with low infiltration rates have more rapid surface runoff, higher peaks, and shorter duration floods, whereas areas with high infiltration rates have slower runoff, lower peak flows, and longer duration floods.

Annual patterns of stream flow (referred to as flow regimes) are controlled by annual patterns of precipitation and temperature, which largely control the timing of runoff (Figure 2.8). Cold regions receive most precipitation as snow, and most runoff occurs during spring or summer snowmelt (Wohl 2000). In the absence of forest cover, snow accumulation is higher and more water is available for runoff during spring melt. In warm regions, precipitation falls as rain and runoff patterns follow the timing of rain storms. In these rainfall-dominated areas, runoff is moderated by evapotranspiration and interception, although intense rainstorms overwhelm these moderating processes and flooding responses strongly reflect the pattern of rainfall. In regions with moderate temperatures, flow regimes may exhibit the transitional runoff pattern which has both rainfall and snowmelt runoff peaks (Mount 1995; Beechie *et al.* 2006b). These double-peaked hydrographs tend to occur at intermediate elevations, or in large rivers receiving flow from both snowmelt and rainfall-dominated tributaries. Finally, where water enters deep aquifers and emerges in streams after long groundwater residence, rivers tend to have muted hydrographs that show little response to either rainfall or snowmelt patterns (Sear *et al.* 1999; Tague *et al.* 2008). Streams and rivers in arid or semi-arid regions may

Table 2.1 Key watershed-scale and reach-scale processes that drive habitat formation and biological responses in river ecosystems.

Process group	Specific processes	Description
Watershed-scale processes		
Runoff and stream flow	Interception	Rainfall captured in tree canopy where it evaporates
	Snow accumulation and melt	Storage of water as snow through winter and release to streams during spring or summer melt
	Surface runoff	Water delivered to streams by overland flow
	Subsurface flow	Water delivered to streams by flow through the soil layer
	Groundwater flow	Water delivered to streams via flow below the soil layer
Erosion and sediment supply	Surface erosion	Erosion of the soil surface by rain splash (dislodging of soil particles by rain) or overland flow
	Mass wasting	Mass movement of soil by landslides, debris flows, and gullying
	Soil creep	Gradual downslope movement of the soil mantle by gravity
Nutrient delivery	Nutrient production and delivery	Nutrient delivery to streams via litter fall, photosynthesis, dissolved nutrients, or anadromous fishes (marine-derived nutrients)
Reach-scale processes		
Riparian processes	Shading	Blockage of solar insolation by vegetation
	Root reinforcement of banks	Additional soil cohesion of river banks provided by roots
	Wood supply	Delivery of dead trees to streams and rivers
	Sediment retention	Trapping of sediment on bars or floodplains by vegetation
	Litter fall	Leaf litter, needles and branches delivered to streams
Stream flow and flood storage	Routing and stream flow	Movement of water through stream and river channels
	Flood storage	Slowing and temporary storage of flood waters on floodplains and in side-channels
Sediment transport and storage	Sediment transport	Movement of sediment by river flow, either in suspension or as bedload
	Sediment storage or retention	Deposition and storage of suspended sediment or bedload sediment in the river channel, sometimes induced by wood jams, aquatic vegetation, or beaver dams
	Floodplain building	Deposition of suspended sediments on floodplain surfaces, sometimes augmented by the influence of vegetation
Channel, floodplain, and habitat dynamics	Pool or bar formation	Formation of pools or bars by hydraulic scour and deposition, often influenced by wood accumulations
	Channel movement	Channel movement by bank erosion (lateral migration) and avulsion
	Pond formation	Construction of beaver dams creates ponds
Organic matter transport and storage	Transport and storage of seeds and plant propagules	Seeds and plant propagules transported by stream flow, and trapped in backwaters and on bars
	Transport and storage of detritus	Organic detritus (e.g. leaves, twigs, needles) transported by stream flow and trapped by bed material, wood jams, and in pools or backwaters
Instream biological processes	Primary production	Algae and aquatic plant production by photosynthesis can drive aquatic food webs
	Secondary production	Production of aquatic invertebrates that consume algae and plants, or leaf litter and other allochthonous organic matter
	Feeding/predation	Consumption of algae, plants, or invertebrates by fishes and other organisms; also predation of fishes by other fishes
	Competition	Competition among taxa (e.g. plants, invertebrates, or fishes) for space or food resources

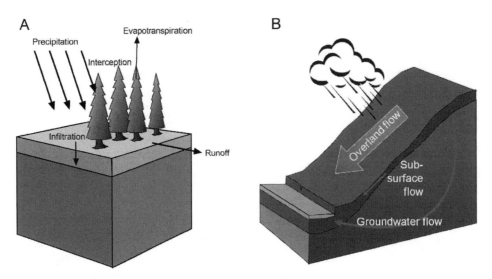

Figure 2.7 (A) Key hydrological processes include precipitation, infiltration, interception, and evapotranspiration, and (B) dominant runoff pathways include surface flow, subsurface flow through the soil layer, and deep groundwater flow below the soil layer.

exhibit any of these patterns, although their hydrographs are most often dominated by long periods of no stream flow with short intense flood flows during and after rain storms.

2.4.2 Erosion and sediment supply

Erosion processes are commonly classified as three types: soil creep (the gradual downslope movement of the soil mantle by gravity); surface erosion (including sheetwash, rilling, and gullying); and mass wasting (e.g. landslides) (Figure 2.9; Dunne & Leopold 1978). We focus on mass wasting and surface erosion because they are influenced by many human activities including logging and road building, tilling and grazing practices, and land clearing and construction activities (Sidle *et al.* 1985; Bradford & Huang 1994; Imaizumi *et al.* 2008). By contrast, human influences on soil creep have not been shown to impact streams and are not a target of restoration activity. Mass-wasting processes such as landsliding are episodic, driven by storm event frequencies and high spatial variation in number and failure potential of landslide sites (Bergstrom 1982; Benda & Dunne 1997a). Failure potentials of landslide sites are primarily a function of slope angle, soil depth and cohesion, root strength, and the amount of water in the soil. Steeper sites with deeper soils and less root strength are most prone to failure, whereas low-slope sites with shallow soils rarely fail. During small storm

events, landslides tend to be very rare and sediment supply low because soil failure thresholds are reached at only a few potential landslide sites (i.e. only the steepest sites with deep soils fail). As storm intensity and duration increase, more potential landslide sites fail and sediment delivery to streams is higher and more widely distributed on the landscape. The combination of variation in storm intensity and site susceptibilities to failure lead to high spatial and temporal variation in sediment supply by mass wasting (Benda & Dunne 1997a).

Surface erosion on bare soils is more predictable than mass wasting because it occurs during virtually all rainstorms and snowmelt, and the severity of erosion varies predictably with rainfall intensity, slope, and soil type (Dunne & Leopold 1978). Surface erosion is relatively rare in naturally forested and grassland environments, but more common in semi-arid to arid zones or in alpine areas that are unvegetated. Wildfire events in relatively dry areas with high soil erodibility can dramatically increase sediment supply because removal of vegetation and decay of roots leads to higher surface erosion rates (Prosser & Williams 1998; Wondzell & King 2003). However, pulses of fine sediment from surface erosion tend to be relatively rare in semi-arid to arid environments as there are often long intervals between intense rainstorms (e.g. Rice 1982). By contrast, fine sediment from alpine areas – especially glaciated areas – is observed consistently during spring snowmelt.

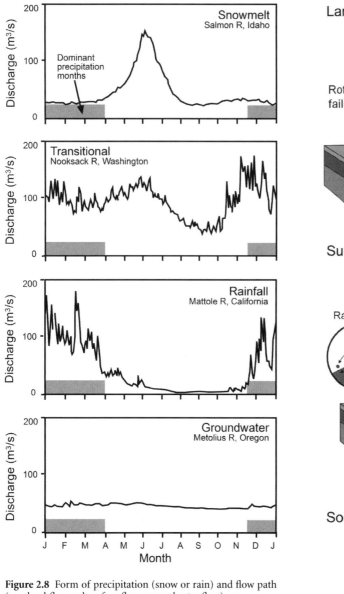

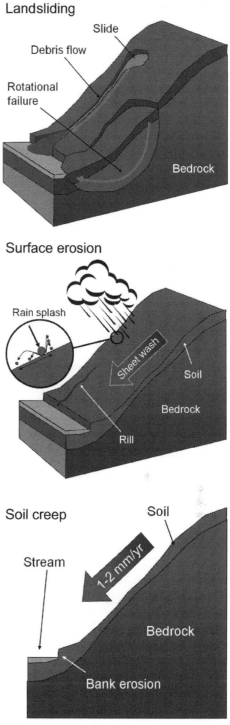

Figure 2.8 Form of precipitation (snow or rain) and flow path (overland flow, subsurface flow, groundwater flow) are dominant controls on the shape of annual hydrographs. Precipitation timing (gray shaded months) is the same in each of the four basins illustrated here, but the proportion of precipitation falling as snow or rain controls the shape of the annual hydrograph when runoff is dominated by subsurface flow or overland flow (upper three panels). When the main flow path is through deep groundwater, rainfall and snowmelt signals are muted and the hydrograph is relatively constant through the year (lower panel). Adapted from Skidmore *et al.* 2011.

Figure 2.9 Erosion processes are classified into three main groups: landsliding, surface erosion, and soil creep. Landsliding mechanisms include both shallow and deep failure processes, as well as debris flows. Surface erosion comprises rainsplash, sheet wash, and rilling. Soil creep is the gradual downslope movement of the soil mantle.

In mountain headwater channels sediment supply is strongly reflected in channel form, as headwater channels are typically supply limited and have a high probability of exhibiting either cascade or bedrock morphologies, but periodic delivery of sediment may create localized alluvial reaches. In mid-network, temporal variability in sediment supply is dampened by bedload transport processes that attenuate and disperse sediment pulses (Benda & Dunne 1997b; Beechie *et al.* 2005b), and small channels adopt step-pool forms where cobble and boulder sorting creates steps or plane-bed forms where substrate is smaller and does not create distinct bedforms (Montgomery & Buffington 1997). Temporal variation in sediment supply is lowest in larger low-gradient streams and rivers as channels become transport-limited and pool-riffle or floodplain channels tend to dominate. Moreover, decreasing flow energy (scaled to channel width) allows aquatic and riparian plants to form increasingly important sediment retention structures (Gurnell *et al.* 2005, 2010). In this zone the supply of sediment interacts with channel slope and size to determine channel form, with high-supply channels tending toward a braided pattern. Island-braided and meandering channels are a reflection of a decreasing channel slope, flow strength, and bedload supply (Schumm 1985; Church 2002; Gurnell *et al.* 2009).

2.4.3 Nutrients

Nutrient dynamics are governed by many watershed-scale features and processes, including parent geology, landforms, precipitation and runoff, and vegetative cover. Parent geology influences land forms, such as depressions or ridges that control nutrient delivery pathways, as well as rates of weathering and concentrations of elements such as calcium and phosphorus. Areas with higher precipitation typically have faster weathering rates and faster rates of delivery of nutrients to the channel, and areas of higher erosion rates also tend to contribute greater nutrient supply to streams. Leaf-fall from riparian trees is a dominant process of nutrient delivery to streams in forested regions, and nitrogen-fixing species such as alder (*Alnus* spp.) can increase nitrogen concentrations in riparian soils and streams (e.g. Compton *et al.* 2003). A variety of factors influence uptake and release of nutrients from the landscape, including land cover type, presence of riparian wetlands, and algal dynamics. Land cover type influences the uptake and release rates of nutrients, with forest lands retaining nutrients, while natural wildfire can reduce the uptake of nutrients and increase rates of nutrient delivery to a stream channel (e.g. Nitschke

2005). Extensive wetlands also influence uptake and production of nutrients in a river system, creating locally higher or lower nutrient availability within reaches. Finally, algal dynamics can reduce nutrient uptake rates and increase concentrations in the water (Hill *et al.* 2001).

2.5 Reach-scale processes

In the previous section we described how watershed-scale processes such as stream flow and erosion shape the physical characteristics of a channel, within the constraints set by the landscape template. That is, the landscape template sets a limited range of conditions that might be expressed in each reach, and inter-annual variations in watershed-scale processes (e.g. erosion and runoff) determine stream flow and sediment supply to the reach. Channel characteristics are further modified by reach-level processes such as sediment routing, flow routing and storage, channel migration, riparian processes, and instream biological processes (Table 2.1). These reach-scale processes then determine specific conditions that may be present at any point in time. We discuss each of these reach-scale processes separately in the following sections.

2.5.1 Riparian processes

Riparian vegetation communities are structured by two main processes: colonization and succession (Hughes 1997). Colonization may result from germination of seedlings or by vegetative reproduction (sprouting from plant fragments), allowing riparian vegetation to become established on bars and developing floodplains. Reproduction by both riparian and aquatic plants is remarkably adapted to the disturbed river corridor environment (Karrenberg *et al.* 2002; Haslam 2006). For example, riparian willows (*Salix* spp.) and poplars (*Populus* spp.) produce enormous quantities of very small short-lived light seeds in spring and early summer. These seeds are dispersed widely by wind and water, ensuring that some fall on the exposed, moist, alluvial sediments required for germination (Karrenberg *et al.* 2002). Once germinated, the seedlings require very particular conditions for survival which depend on the river's flow regime (Rood *et al.* 2005). Specifically, they require a period without flood disturbance and a gradually falling alluvial water table, termed the 'recruitment box' by Mahoney & Rood (1998).

Vegetative fragments from these species essentially require the same conditions as seedlings for sprouting and establishment. However, vegetative fragments can be produced and dispersed by floods at any time of the year and have larger internal resources to support sprouting and establishment. This mode of reproduction is therefore particularly well suited to harsh and frequently disturbed riparian environments. Rapid growth is also crucial for the survival of young riparian plants because plant size is the main factor contributing to their ability to survive fluvial disturbances of differing magnitudes (Corenblit *et al.* 2007; Gurnell *et al.* 2009). Growth rates of riparian trees are heavily dependent upon moisture supply (Gurnell & Petts 2006) and, as a result, there is complex interdependence between physical processes and colonization and succession of riparian vegetation.

Succession is the gradual transition from colonizing species to late successional species (Naiman *et al.* 2005) and, if succession proceeds without a stand-replacing disturbance, then pioneer stands gradually develop into mature vegetation communities (Gurnell *et al.* 2006a; Naiman *et al.* 2010). Succession may be a relatively simple transition from one or a few species to another or several other species, or it may include multiple species transitions. These transitions depend on seed dispersal distances, erosion and depositional processes, and flooding frequency and duration (e.g. Fastie 1995; and various examples in Naiman *et al.* 2005). In forests of the Pacific Northwest USA, for example, the common successional pathway progresses from pioneer willow species to red alder (*Alnus rubra*) or black cottonwood (*Populus trichocarpa*) and then to conifer species such as Sitka spruce (*Picea sitchensis*) and western hemlock (*Tsuga heterophylla*) (Naiman *et al.* 2010). However, some floodplain forest stands are eroded into the river before they reach the mature forest stage, while floodplain formation in other areas leads to colonization and succession of new stands that eventually become late-seral stands or may once again be eroded into the river during stand development (Hughes 1997; Naiman *et al.* 2010). The interplay of physical, hydrological, and successional processes therefore creates a patchwork of forest ages and successional states within the riparian zone (Gregory *et al.* 1991; Corenblit *et al.* 2007; Osterkamp & Hupp 2010).

When viewed at the landscape scale, these interactions between vegetation colonization and succession and fluvial processes lead to distinct, dynamic assemblages of plants and landforms along river systems. For example, colonization and succession processes lead to predominantly mature vegetation along headwater streams, with a relatively narrow fringe of young plants along the channel margins affected by flood disturbances (Agee 1988). Floodplain forests along these small, non-migrating channels are often structured by microtopography and variability in depth to the water table, frequency of flooding, or light availability. On larger rivers that migrate across their floodplains (i.e. braided, island-braided, and meandering channels), floodplain forests predominantly comprise colonizing species on braided channels, late successional species on straight channels, and a high diversity of both species and stand ages on meandering and island-braided channels (Beechie *et al.* 2006a; Naiman *et al.* 2010). These processes of colonization and succession also interact with stream characteristics and flow regimes to create variation in riparian characteristics among physiographic or climatic regions. For example, in regions with predominantly low-gradient streams, riparian areas may be dominated by wetland communities because riparian areas are persistently flooded or wet (Pollock *et al.* 1998) or streams in semi-arid or arid regions may be dominated by small woody species that are adapted to flashy streamflow patterns and episodic riparian disturbance regimes (e.g. Stromberg 2001).

Riparian processes and functions that affect stream ecosystems include root reinforcement of banks, wood supply to streams, sediment retention, leaf litter supply, and shading (Figure 2.10). Delivery and retention of seeds and vegetative propagules also eventually contribute to the physical processes of sediment trapping, root reinforcement and wood delivery. Indeed, two-way interactions between riparian vegetation and physical processes are increasingly being recognized as crucial to fluvial landform development, with vegetation becoming increasingly important along a gradient of decreasing fluvial disturbance (Corenblit *et al.* 2007; Perucca *et al.* 2007). Furthermore, variability in the growth and vigor of riparian trees and shrubs can result in alternation in space and time between predominantly island-braided and bar-braided reaches (Bertoldi *et al.* 2011a, 2011b) or can contribute to transitions from meandering to braiding channel styles triggered by destructive floods (Smith 2004).

At the reach level, riparian processes influence channel form primarily through reinforcement of banks and bars by roots (Bertoldi *et al.* 2009), or – in low-energy channels – by reinforcement of the river bed by aquatic macrophytes, roots, and rhizomes (Liffen *et al.* 2011; Pollen-Bankhead *et al.* 2011). Bank reinforcement by roots influences channel planform primarily in depositional reaches (Figure 2.11), most notably by reducing or preventing lateral migration (Micheli & Kirchener 2002; Beechie *et al.* 2006a). Bank reinforcement often forces

Figure 2.10 (A) Schematic diagram of riparian functions, and photographs of (B) wood-formed pools and bank protection in the Pacific Northwest, USA (photo Tim Beechie), (C) shading of stream by deciduous trees in the western desert ecoregion, USA (photo Tim Beechie), and (D) sediment trapped by riparian tree seedlings in the Tagliamento River, Italy (photo Angela Gurnell).

braided channels into single-thread or island-braided forms (Millar 2000) and, in the lowest energy reaches, reinforces sand and finer sediments to form channel-margin benches that in turn induce erosion of the opposite banks and channel migration and curvature (Gurnell *et al.* 2006b; Bennett *et al.* 2008; Liffen *et al.* 2011). Root reinforcement of banks is not restricted to forested environments, and occurs in most river environments ranging from wet meadows in semi-arid or desert environments to wetland systems to a variety of forested environments.

Sediment trapping by plants on bars and floodplains leads to aggradation of floodplain, bank, and bar surfaces, as does the delivery of wood to the channel (Naiman *et al.* 2010). Sediment trapping by plants and large wood can accelerate floodplain, river margin and in-channel

landform development, as well as channel adjustments to discharge and sediment regimes (Gurnell *et al.* 2009). These processes can then alter the morphology of reaches by changing the spatial extent and biomass of colonizing vegetation, even among reaches where the discharge and sediment regimes are essentially the same (Bertoldi *et al.* 2011b). Indeed some plants, particularly certain riparian tree and emergent aquatic macrophyte species, can perform the role of physical ecosystem engineers by trapping plant propagules and sediment, thereby constructing physical habitats suitable for the germination and growth of other plant species (Kollmann *et al.* 1999; Gurnell *et al.* 2005, 2007; Francis *et al.* 2008).

While interactions between vegetation and channel morphology are generally greater in small streams, delivery of wood to channels can influence channel

Figure 2.11 Examples of root reinforcement of banks by (A) grasses and sedges in a desert ecosystem of Nevada, USA (photo Tim Beechie), (B) deciduous riparian trees along the Tagliamento River, Italy (photo Angela Gurnell), and (C) understory bamboo, Hokkaido, Japan (photo Tim Beechie).

morphology throughout the river network. Wood can create alluvial reaches from bedrock reaches in headwater channels (Montgomery *et al.* 1996a), force formation of pools in mid-network channels (Montgomery *et al.* 1995; Beechie & Sibley 1997; Montgomery & Buffington 1997) and create floodplain hard points that force island formation and channel switching in large floodplain channels (Fetherston *et al.* 1995; Gurnell *et al.* 2001; Abbe & Montgomery 2003; Montgomery & Abbe 2006). Reach-level wood dynamics are perhaps most influential in floodplain reaches where lateral channel migration or avulsion erodes floodplain patches and recruits wood to the channel, and wood accumulations force island formation. Accumulations of wood also contribute to vegetation colonization and growth in floodplain reaches, which ultimately induces or accelerates landform development, helps initiate vegetation colonization, and supports creation of forest patches that are ultimately eroded and recruited as wood to the channel.

Riparian vegetation also shades streams and reduces summer stream temperatures in small streams, especially in densely forested environments. The main mechanism by which riparian forests reduce stream temperature is by blocking solar radiation (Moore *et al.* 2005), but reduced ambient air temperature beneath a forest canopy also has some effect on stream temperatures (Pollock *et al.* 2009). In non-forested areas, brush and grasses may have little effect on stream temperature, but still provide other ecosystem functions such as supply of organic matter, contributions of terrestrial invertebrates, and trapping fine sediments on the floodplain. In low-energy unshaded reaches, abundant aquatic vegetation can moderate summer stream temperatures, induce complex flow patterns, and trap fine sediments into discrete patches (Sand-Jensen 1998; Clarke 2002; Gurnell *et al.* 2006c, 2007; Asaeda *et al.* 2010). Moreover, areas without canopy cover have increased photosynthetically active radiation which leads to increased primary (e.g. algae) and secondary productivity (Kiffney *et al.* 2003).

Finally, fluxes of allochthonous organic matter (e.g. leaf litter) and terrestrial invertebrates to streams are critical drivers of stream food webs. Small streams with closed canopies may have food webs that are driven by leaf litter inputs, with stream invertebrate communities dominated by shredders or filterers that feed on organic particulates. In many cases, organic matter input rates are higher in forested environments than in grasslands (Kawaguchi & Nakano 2001), but riparian functions of grasses and shrubs are nonetheless critical to the health of riverine ecosystems in drier environments.

2.5.2 Fluvial processes: Stream flow and flood storage

Stream flow and hydrologic regime exert strong influences on potential life history strategies and community structure of riparian and aquatic species and ecosystems (Figure 2.12) (Schlosser 1985; Allan 1995; Doyle *et al.* 2005). At the reach scale, the five key attributes of stream flow (magnitude, frequency, duration, timing, and rate of change) influence a variety of physical and ecological functions in rivers and floodplains (Karr 1991; Bertoldi *et al.* 2009). For example, low- and high-flow magnitudes each influence a wide range of ecological processes including riparian vegetation establishment and maintenance, development of floodplain habitats, formation of in-channel habitats, and structure of ecological communities (Poff *et al.* 1997; Richter *et al.* 2003; Beechie *et al.* 2006b). Each of the five key flow attributes is controlled not only by the watershed-scale hydrologic processes described earlier, but also by two key processes affecting routing of flow through channels at the reach scale: flow resistance due to channel roughness and flood storage on floodplains.

Stream flow routing in channels is a function of channel size and channel 'roughness' which represents the degree to which flows are slowed due to grain roughness (bed particle size relative to flow depth), bed form roughness such as bars, pools, or meander bends, and vegetation roughness such as macrophytes, trees, and roots. High-gradient mountain rivers typically have large bed particle sizes (boulders and cobbles) which creates substantial turbulence in the flow and significantly reduces velocity. In these channels vegetation and bed form roughness may be relatively small compared to the grain roughness. By contrast, in small low-gradient streams, bed form and vegetation roughness are often high relative to grain roughness because bed particle sizes are small compared to flow depth. Most of the flow resistance is therefore attributed to bed forms or vegetation. In large rivers, both grain roughness and vegetation roughness are typically low, and flow resistance is largely attributable to bed form roughness, channel meandering, or large accumulations of wood. A primary consequence of these patterns of roughness in river networks is that flow velocities at flood discharges tend to increase in the downstream direction (e.g. Leopold *et al.* 1964). In-channel roughness also contributes to habitat and biological diversity in both small and large channels, as local variation in depth and velocity contributes to nutrient retention, spawning and rearing habitat availability for fishes, and algal and invertebrate colonization and establishment. Sediment particle sorting also creates diverse habitats for benthic organisms (Gooderham *et al.* 2007).

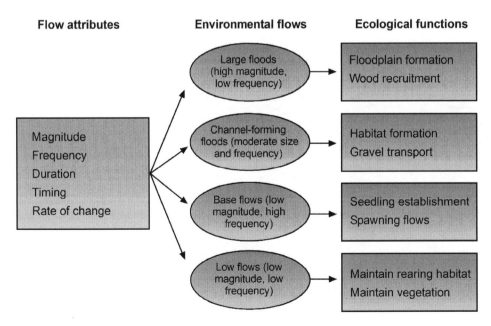

Figure 2.12 Conceptual diagram illustrating flow attributes that characterize key environmental flow components, which in turn influence ecological functions of streams and rivers (based in part on Poff *et al.* 1997).

Routing and storage in floodplain rivers differs from that of non-floodplain channels in that flood storage on the floodplain significantly attenuates many flood peaks. That is, over-bank flooding occupies areas of very high vegetation roughness, which slows flood waters and reduces flood flows in the main channel. The main effects of contiguous, intact floodplains on river flows is to reduce the downstream increase in flood discharge by 'storing' water in floodplain ponds, lakes, and backwaters, as well as in shallow, slow floodplain flows. Ecologically, this floodplain storage creates a variety of lentic habitats, providing spawning and rearing areas, localized nutrient cycling, and opportunities for feeding on specific organisms (Junk *et al.* 1989). Each of these functions varies by flow regime, climate, and the degree of river floodplain interaction. For example, in tropical rivers with long flood pulses, fishes adapt to feeding in food-rich floodplain areas during floods that may last for several months each year. By contrast, rivers with short and unpredictable flood pulses allow little opportunity for adaptation, and fishes may use floodplain habitats as short-term refuge areas during floods.

2.5.3 Fluvial processes: Sediment transport and storage

The rate of sediment transport relative to the rate of sediment supply to a reach (termed the relative transport rate or relative sediment supply) determines whether any individual reach is accumulating sediment, exporting sediment, or relatively stable. This supply-to-transport ratio is classically conceptualized by Lane's balance which illustrates that sediment supply exceeding transport capacity results in bed aggradation, whereas sediment supply less than transport capacity results in bed degradation (Figure 2.13). Shifts in relative transport capacity can result from changes in sediment supply or stream flow. Increases in sediment supply (e.g. from increased landsliding) shift reaches to the oversupplied or aggrading state, whereas decreased sediment supply shifts reaches to the undersupplied or degrading state. By contrast, increases in stream flow can result in relative sediment supply shifting to the undersupplied state, while decreases in stream flow can result in oversupply. Such shifts in sediment supply or stream flow are usually short-term effects under natural conditions; longer-term changes in sediment supply or stream flow usually result from land uses or installation of dams, and are discussed in Section 2.6.1.

At a low relative sediment supply (i.e. when sediment supply is low or sediment transport capacity is high), the bed surface will be relatively coarse and the fine sediment content of stream bed gravels will be low (Dietrich *et al.* 1989). As the ratio of sediment supply to transport capacity increases (i.e. the reach is increasingly oversupplied),

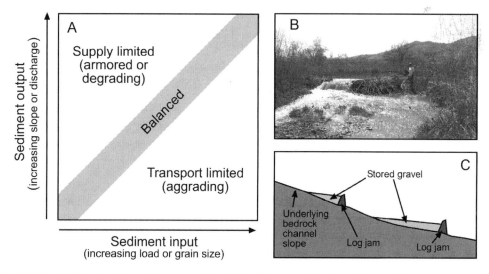

Figure 2.13 (A) The traditional sediment balance considers relative influences of sediment inputs and outputs. However, biological processes exert strong influences on the sediment balance in most natural river ecosystems, including (B) beaver dams that trap sediment and aggrade channels (Bridge Creek, western desert ecoregion, USA; photo Tim Beechie), and (C) wood debris in forested ecosystems that traps coarse gravels and creates alluvial reaches (Pacific Coastal Ecoregion, USA, based on Montgomery *et al.* 1996a).

increased fine sediment can be accommodated through fining of the bed surface, pool filling, channel aggradation, and finally channel widening (Lisle 1982; Madej 1982; Dietrich et al. 1989; Madej & Ozaki 1996). Channels may also become laterally unstable (Bergstrom 1982; Church 1983), which hinders revegetation of unvegetated bars and floodplains. When sediment supply is reduced, channels tend to incise and narrow to pre-aggradation widths, the bed surface becomes armored as fine sediments are winnowed away, and pool depths recover within a few years (e.g. Lisle 1982; Pitlick & Thorne 1987; Beechie et al. 2005b).

An important – but often overlooked – control on sediment transport and storage is the role of biota in regulating sediment transport and bank erosion. For example, in low-gradient channels dominated by fine sediment carried in suspension, suspended sediment may be trapped by vegetation on floodplains or in the channel (e.g. shrubs, herbs and grasses, wetland vegetation, and aquatic macrophytes) or beaver dams (Pollock et al. 2007; Beechie et al. 2008b; Gurnell et al. 2010). Both mechanisms reduce sediment transport capacity relative to sediment supply, and effectively shift the sediment balance to an oversupply state resulting in bed and floodplain aggradation (Figure 2.13). Similarly, dead wood in forest channels can reduce the transport of bedload sediment and shift channels to an oversupplied state, resulting in conversion of bedrock channels to alluvial channels or decreased grain size of the bed material (Montgomery et al. 1996a; Buffington & Montgomery 1999a, 1999b). Finally, fishes, crayfish, and aquatic macro-invertebrates can all increase sediment mobility or stability through various mechanisms. For example, salmon winnow away fine sediment during spawning which increases the grain size and reduces sediment transport capacity (Montgomery et al. 1996b), and crayfish activity increases the export of sand and finer sediments (Statzner et al. 2000). By contrast, some macro-invertebrate species such as net-spinning caddisflies (e.g. the families Hydropsychidae and Stenopsychidae) increase the velocity required to initialize sediment movement when their larvae construct filtration nets in pore spaces of stream beds (Cardinale et al. 2004; Takao et al. 2006).

Exchange of sediment between the floodplain and channel is also an important process regulating sediment transfer through reaches. Madej (1987) identified three zones of sediment storage in gravel bed channels: (1) active storage, or the channel bed; (2) semi-active storage, or bars; and (3) inactive storage, or the floodplain. Active sediment storage refers to sediment that is transported in most years, semi-active sediment is stored for years to decades, and inactive sediment is stored from decades to centuries or millennia (Madej 1987). Sediment stored in the bed and bars is activated by bedload transport during moderate to large floods, whereas floodplain sediment is activated by bank erosion during floods. Rates of transfer between these storage components varies as a function of channel size, as well as channel pattern and lateral migration rate (Beechie et al. 2006a). Smaller channels have much less capacity to erode their banks, and therefore have very low bank erosion rates (at most a few centimeters per year) and low rates of exchange between the channel and floodplain. By contrast, large channels have greater capacity to erode their banks (migrating as much as tens of meters per year), and rates of floodplain turnover vary as a function of channel pattern. Floodplain turnover rates (the amount of time it would take to erode the entire floodplain once) can be less than a century for mountain rivers with high bedload supply, high migration rates, and narrower floodplains (Beechie et al. 2006a). By contrast, turnover rates may be several centuries in large rivers dominated by fine sediment loads, in which lateral migration rates are low relative to the floodplain width and natural levees tend to form at the channel margins (Hughes 1997).

2.5.4 Channel and floodplain dynamics

Dynamic processes and continuous change are characteristics of intact natural river ecosystems (Jungwirth et al. 2002), and these dynamics create what has come to be known as the shifting habitat mosaic (Ward et al. 2002). In this shifting mosaic, some habitats are destroyed and others created each year, but the pattern and distribution of habitat types remains more or less constant through time (Ward et al. 2002; Beechie et al. 2006a). The most important processes that control these channel–floodplain dynamics include lateral channel migration (which includes bank erosion and bar deposition), avulsion, channel switching, floodplain building, variations in river discharge, and biological influences such as wood accumulation and beaver dam building.

Lateral-channel migration encompasses bank erosion at the outer bank of meander bends and bar and floodplain deposition on the inner bank. The rate of lateral migration (i.e. the average amount of bank retreat per year) controls the patch turnover rate, which has been quantified either as the floodplain turnover rate or the erosion return interval (the average number of years a floodplain patch persists before being eroded again by the river) (Hughes 1997; Beechie et al. 2006a). In addition to lateral migration, other physical processes such as channel avulsion or channel switching create new mainstem channels and leave abandoned mainstem channels that gradually evolve into smaller side channels. Floodplain building is the gradual aggradation of floodplain surfaces by overbank deposition of fine sediments, which operates more slowly but also contributes to diversity of floodplain forests and aquatic habitats (see also Section 2.5.1). Biological processes of wood recruitment and accumulation in jams create hard points that initiate island formation, or protect portions of floodplains from erosion. Animals such as beaver (*Caster* spp.) are also important ecosystem engineers that modify floodplain morphology and habitats, creating important slow-water habitats used by a wide range of aquatic species across North America, Europe and Asia.

These geomorphic dynamics play out over decadal timescales and longer, creating a complex suite of habitats ranging from lentic to lotic, from pool- to riffle-dominated, and from groundwater- to surface-water-fed. This habitat variability then drives variation in thermal and chemical habitat quality within floodplain reaches, which in turn contributes to high biological diversity (Poole *et al.* 2008). Set within this physically dynamic template is a hydrologic dynamic that operates on a timescale of days to months. The spatial extent, locations, and types of aquatic habitat vary as a function of river discharge, with the greatest spatial extent during floods and lowest spatial extent at low flows. Diversity and types of habitats vary substantially across the flow range, with flow depths and velocities generally increasing as discharge increases, and the diversity of habitats usually increasing to a mid-range flow and then decreasing as discharge approaches flood stage (Tockner *et al.* 2001; Stanford *et al.* 2005).

In the absence of human alterations, floodplain rivers are resilient to natural fluctuations in driving processes (stream flow, sediment supply, and riparian functions) because their form and function are an outgrowth of long-term variation in those processes. Moreover, floodplain rivers buffer effects of short-term variation because extreme events in headwater tributaries are muted in larger rivers (e.g. Naiman *et al.* 1992; Gomi *et al.* 2002). For example, large sediment inputs from landslides are attenuated in large rivers where timing of tributary sediment pulses are out of phase and sediment inputs are often small relative to transport capacity. Hence, river floodplain ecosystems that are unmodified by human land uses tend to maintain relatively constant form and condition from year to year (Beechie *et al.* 2006a), yet

these river–floodplain systems sustain a very high diversity of riparian forests and aquatic habitat types (Ward *et al.* 2002; Naiman *et al.* 2010).

2.5.5 Organic matter transport and storage

The dynamics of organic matter transport and storage are influenced by channel structure and floodplain interactions in much the same way as inorganic sediments. Particulate organic matter is a key basal resource in stream ecosystems, and its storage within a reach affects local ecosystem productivity. Larger particles can be trapped by large and small wood or between and under cobbles (Richardson *et al.* 2009), whereas smaller particles such as viable seeds settle in backwaters or are trapped by stands of aquatic and riparian vegetation (Gurnell *et al.* 2007, 2008). Generally, larger particles such as deciduous leaves and conifer needles are more easily trapped (Hoover *et al.* 2010). Increasing channel complexity decreases the rate of transport of organic matter, allowing it to be processed and consumed within the reach (Negishi & Richardson 2003). Finally, fine organic particulates are trapped by filter-feeding organisms and processed through the food web.

2.5.6 Instream biological processes

A wide range of instream biological processes influences the structure and function of stream ecosystems, including habitat selection, feeding, competition, and predation. These processes are not completely independent of each other as access to food and avoidance of predators are often key influences on habitat selection, and habitat selection by one species may influence habitat selection by other species. While these processes influence the behavior of individuals, when viewed at larger scales (e.g. a stream reach) the collective behaviors of many species and individuals interact to structure biological communities, including the spatial arrangement of species and configuration of food webs. These biological processes are particularly important for aquatic insects, crustaceans, and fishes, which are motile species that control their movement and choice of habitats. Because migration and habitat selection allow species to move to suitable habitats, areas of high habitat diversity (e.g. boulder headwater streams, island-braided reaches, or tributary confluences) tend to have high species diversity when there are species available to occupy the available range of habitat types (Jackson *et al.* 2001; Kiffney *et al.* 2006; Rice *et al.* 2006; Gooderham *et al.* 2007). The ability of organisms to freely exploit shifting habitats is therefore essential to the full expression of potential species distributions and diversity in river ecosystems (McGarvey & Hughes 2008).

Fishes, crustaceans, and invertebrates select habitats for a variety of needs including access to food resources, refuge from strong currents or predators, or reproduction. Stream fishes feeding on drifting invertebrates usually occupy an area that provides cover from predators and relief from current (e.g. behind and under large wood or boulders), while at the same time offering a good view of food moving towards them in the water (e.g. Hughes 1998). Similarly, many invertebrates require particular flow conditions to optimize food delivery in the water column, and rivers with high habitat diversity offer the largest range of functional flow conditions for invertebrates (e.g. Wetmore *et al.* 1990; Hart & Finelli 1999). In addition to habitat selection for resting and feeding requirements, many aquatic organisms also adjust their habitat selection to avoid predation. For instance, some aquatic insects such as black fly larvae (family *Simuliidae*) find a refuge from predation by flatworms (phylum *Platyhelminthes*) and other predators by selecting fast currents where their predators are unable to pursue them (Hart & Merz 1998). Fishes (especially small ones) often use complex hiding cover such as roots, wood jams, or aquatic macrophytes to avoid predators. By contrast, species such as crayfish move to shallow water as juveniles to avoid predation by fishes but, as they grow larger (and are too large for fish to eat them), these crayfish move to deeper water to avoid predation by birds and mammals (Englund & Krupa 2000). Even the egg-laying behavior of many aquatic insects depends on having appropriate flow conditions and bed morphology with rocks protruding from the surface to allow adults to crawl under them (e.g. Lancaster *et al.* 2010). These same types of habitat requirements influence the distribution and the abundance of many stream organisms that require particular arrangements of physical habitats to secure food and find refuge from predators.

These instream biological processes vary with riparian conditions, stream flow, and habitat diversity, each of which can alter the availability of habitats and food resources and thereby alter food webs and community structure. For example, food webs in streams are based on two key basal resources: materials that enter from the riparian area (e.g. leaf litter, terrestrial invertebrates, and seeds) and primary production within streams (e.g. algae, mosses, rooted aquatic plants) (Richardson *et al.* 2010). In streams with high inputs of leaf litter, invertebrate communities at the base of the food web may be domi- nated by shredders and collectors that feed on detritus. By contrast, invertebrate communities in streams with low leaf litter inputs and high light conditions may be dominated by grazers or scrapers that are adapted to feeding on benthic algae. In either of these environments, changes in amounts of litter or light reaching the stream have significant effects on instream processes. For example, we know through experiments that reductions in leaf litter inputs can restrict the productivity of stream food webs (Richardson 1991; Wallace *et al.* 1999), or reduced shading from riparian vegetation can increase the pro- duction of algae in streams (Kiffney *et al.* 2003; Fanta *et al.* 2010). However, a substantial reduction of shade can lead to very high light conditions that divert nutrients to species that are less edible, such as blue-green algae. Finally, wood also provides a food resource to a number of invertebrates (Bondar *et al.* 2005; Coe *et al.* 2009) and reduction in wood amounts might reduce productivity of stream reaches.

Seasonal flow conditions and habitat heterogeneity also influence biological processes that affect population dynamics of stream organisms. Flow disturbance such as flooding causes changes in habitats, reductions in basal resources, and direct mortality of organisms, providing habitat opportunities for organisms with rapid life cycles and good colonizing ability (e.g. Wootton *et al.* 1996). Low-flow periods can also result in mortality of species incapable of responding on the appropriate timescales and succumbing to desiccation (McAuliffe 1984). Similarly, increases in fine sediment supply can shift invertebrate taxa toward burrowing species, which reduces food supply and growth rates for juvenile salmonids (Suttle *et al.* 2004).

Interactions between instream processes and riparian, sediment, and hydrologic influences can be quite complex, and in some cases may result in unexpected changes to stream ecosystems. For example, one important conse- quence of complex interactions among instream processes is the trophic cascade, in which changes in productivity within one trophic level can influence trophic levels at least two steps away. For instance, in streams of coastal California, some winter floods can reduce or even elimi- nate top predators – in this case, steelhead (*Oncorhynchus mykiss*) – with the effect that the small fish they feed on increase in numbers and reduce their invertebrate prey, leading to increases in algae. When steelhead are present in sufficient numbers they can reduce the numbers of small fish through predation, which releases the inverte- brates who then consume most of the algae (Power 1990). In this example, changes at the top level (steelhead) can influence food webs down to the algae and biofilm level, several trophic steps away through cascading effects.

2.6 Common alterations to watershed processes and functions

Humans alter riverine ecosystems through a wide range of direct and indirect pathways (Karr 2006). Direct pathways include activities such as modification of channels and habitats, removal of riparian and aquatic vegetation, changes to flow regimes, or introduction of non-native species. Indirect pathways include changes in land use that alter peak flows or sediment supply, generate pollutants, or introduce fertilizers that alter nutrient loading to streams. Both types of pathways influence biological endpoints via changes in habitat structure, stream flow, water quality, energy sources, or biological interactions (Karr 2006). We provide a brief summary of human alterations to watershed- and reach-scale processes to form a foundation for later chapters on analysis of watershed processes and identification and prioritization of necessary restoration actions (Table 2.2). We do not intend an exhaustive review of human influences on riverine ecosystems. Rather, we summarize common alterations that lead to riverine ecosystem degradation and are frequently addressed through watershed or river restoration actions.

2.6.1 Alteration of watershed-scale processes

Most alterations to watershed-scale processes (runoff, erosion, or nutrient delivery) fall into the category of indirect influences on riverine ecosystems because they are changed by land uses that are either far from the stream or widely distributed in a watershed. Runoff processes are altered through four common effects: (1) vegetation change or removal; (2) interception of subsurface flow and routing to streams (usually in road ditches); (3) soil compaction and decreased infiltration; and (4) creation of impervious surfaces (e.g. pavement, roof tops). Vegetation removal is particularly relevant in forested areas, where tree removal (logging or land conversion) reduces interception and evapotranspiration, which increases annual water yields and flood magnitudes in areas with transient snow packs (Harr 1986; Beschta *et al.* 2000; Figure 2.14). Roads cut into hillslopes intercept subsurface flow in the soil layer, routing it into ditches and rapidly to streams (Jones & Grant 1996). This rapid runoff increases flood flow responses to storm events, particularly in small watersheds.

Soil compaction by grazing and impervious surfaces in developed areas both decreases infiltration into soils and creates more rapid runoff to streams (Trimble & Mendel 1995; Booth & Jackson 1997). Alterations to runoff processes that reduce stream flow are less common, but include afforestation (usually in areas that were previously deforested), creating terraces or berms to capture water on slopes and reduce runoff, and decreasing stream flows by dams and withdrawals (e.g. Richter *et al.* 1996; Donato 1998; Spinazola 1998).

Changes to sediment supply are caused by a variety of land uses including changes in mass wasting due to logging and road building (e.g. Sidle *et al.* 1985), or increased surface erosion after vegetation clearing, tilling and soil exposure in agricultural fields (e.g. Megahan *et al.* 1995; Trimble & Mendel 1995), increased surface erosion from unpaved road surfaces, and surface erosion and gullying after grazing (Trimble & Mendel 1995; Figure 2.15). Mass-wasting processes can be increased by three main mechanisms: (1) over-steepening of slopes by construction practices (e.g. forest roads, urban or residential development); (2) forest clearing and loss of root strength and soil cohesion; or (3) concentration of water through road ditches or impervious surfaces and subsequent increases in pore-water pressure in soils on steep slopes. Each of these mechanisms influences the three main hillslope attributes that influence landslide occurrence: steepness of the slope; soil cohesion; and amount of water in the soil. While each of these influences occurs at individual sites, measurable increases in sediment supply at the watershed scale result from the cumulative increase in number or size of slides in a watershed.

Surface erosion is increased primarily by clearing of vegetation or grazing practices. Many land uses contribute to vegetation clearing, including tilling of fields for agriculture, construction of unpaved roads for logging, and clearing of construction sites for urban or residential development (Megahan *et al.* 1995; Trimble & Mendel 1995). As with mass wasting each land use affects a small area within a watershed, but the cumulative effect of multiple actions at multiple sites results in increased fine sediment loading within a watershed. For example, a small area of logging roads within a watershed may not have a detectable effect on sediment supply at the watershed scale, but construction of many roads creates a large enough increase in sediment supply to have detectable effects on riverine habitats and biota (e.g. Reid *et al.* 1981). Similarly, tilling of large areas of agricultural land creates substantial increases in sediment

Table 2.2 Common alterations to watershed and river ecosystem processes.

Ecosystem feature	Process	Common impacts
Watershed-scale processes		
Sediment	Surface erosion	Increased surface erosion by tilling (cultivation), grazing, construction and use of unpaved roads, land development
	Mass wasting	Increased landslide rates by logging (reduced root strength) or road construction (failure of sidecast or crossing fills), impervious areas increase runoff and erosion rates
	Soil creep	Not commonly impacted by human activity
Water	Interception	Removal of vegetation (especially trees) reduces interception and increases amount of precipitation reaching the ground
	Snow accumulation and melt	Removal of conifer trees increases snow accumulation on the ground and increases melt rate through higher wind velocities
	Surface runoff	Soil compaction and creation of impervious surfaces (e.g. roofs, pavement) reduces infiltration, and increases runoff
	Subsurface flow	Excavation for roads or other construction intercepts water
Nutrients/ pollutants	Nutrient production	Application of fertilizers to crops or lawns increases nutrient supply to rivers
	Nutrient/pollutant delivery	Vegetation removal increases nutrient delivery via surface runoff or subsurface flow
	Trace metal production and delivery	Urbanization increases production of trace metals and other pollutants, and paved areas increase delivery to streams
	Pesticide delivery	Application of pesticides in urban, agriculture, and forested areas introduces alien substances to streams
Reach-scale processes		
Sediment	Transport and storage in channel	Gravel mining or dredging, construction of dams, removal of beaver dams
	Floodplain building	Elimination of floodplain sediment deposition by dikes or levees
Water	Routing and stream flow	Alteration of flow regime by dams, abstraction for human uses
	Flood storage	Increased flood storage by dam construction, decreased flood storage by dike or levee construction
Riparian vegetation	Shading	Removal of canopy trees
	Root reinforcement of banks	Mechanical removal of bank vegetation, grazing or trampling of bank vegetation by livestock
	Wood supply	Removal of large trees within the recruitment zone (equivalent to tree height)
	Sediment retention	Livestock grazing or trampling of vegetation that traps suspended sediment
	Litter fall	Removal of trees that provide leaf litter, needles and branches to streams
Channel and floodplain dynamics	Channel movement, floodplain turnover, floodplain channel formation	Bank armoring or levees prevent bank erosion and channel migration, dredging or channel incision may reduce connectivity between the channel and floodplain habitats
	Pool formation by wood	Wood removal from channels reduces frequency and depths of pools, dam construction can reduce supply of wood from upstream and reduce pool formation
	Habitat formation by beaver dams	Removal of beaver or beaver dams reduces area of pond habitats
Instream biological processes	Migration	Dams and other blockages to fish migration alter species ranges and local community composition
	Primary and secondary production	Pesticides can reduce algae or invertebrate production
Direct manipulation of ecosystem features		
Habitat diversity	Hydraulics and sediment transport	Dredging and wood removal from river channels alters hydraulics and habitat formation, resulting in simplified habitats and communities
Species composition	Species interactions	Addition or removal of species alters food webs and community composition

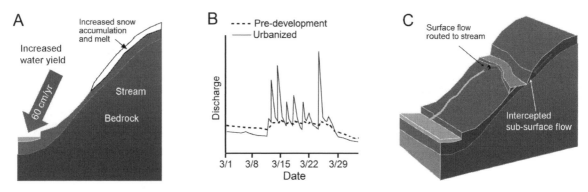

Figure 2.14 Runoff processes are altered by (A) vegetation removal, (B) increased impervious surfaces areas in urban watersheds (modified from Booth *et al.* 2002), and (C) interception of subsurface flows and ditch routing to streams.

Figure 2.15 Erosion processes are altered by (A) tilling practices that remove vegetation and leave bare soils, (B) logging that reduces root strength and increases landslide rates, (C) channel incision and increased bank erosion, and (D) increased surface erosion from unpaved roads (photos Tim Beechie).

supply, whereas small land clearings may only have localized effects if they are near a stream.

Nutrients (even limiting nutrients such as nitrogen and phosphorous) can effectively become pollutants at high concentrations. Agricultural runoff, waste processing plants, urbanization, industrial waste, and other sources can result in elevated concentrations of nutrients, which in turn often lead to nuisance blooms of algae or rooted, aquatic plants (e.g. Hudon & Carignan 2008). Supply of nutrients to streams can also be altered by forest harvesting, which reduces the uptake of nutrients in the short term and leads to higher fluxes to streams via groundwater and surface water (Likens et al. 1970; Feller 2005; Nitschke 2005). In agricultural or urban lands nutrients from fertilizers are transported to streams, especially during the non-growing season (Lowrance et al. 1985). Nutrients can be delivered to streams via surface runoff, subsurface flow, groundwater flow, and litter and soil inputs from certain nitrogen-fixing species, but use of vegetated buffer strips can reduce nutrient delivery to streams (Lowrance et al. 1984; Osborne & Kovacic 1993; Volk et al. 2003).

Most pollutants are simply alien to riverine environments, including pesticides, metals and other materials that are often increased by human activity. Pollutant sources can be either point source (e.g. urban wastewater treatment plants) or non-point sources (e.g. ditches, street runoff, groundwater). Moreover, pollutants can arrive from within a catchment (e.g. Kolpin et al. 2002; Kidd et al. 2007), travel long distances as airborne chemicals such as pesticides, mercury, and particulates from combustion (e.g. Temme et al. 2007), or by a combination of transport mechanisms. For example, mercury may be aerially transported to forests from distant sources of combustion, and forest fire can lead to subsequent mobilization of mercury from forest soils and delivery to streams as a non-point-source pollutant (Kelly et al. 2006). By contrast, point-source discharges of treated sewage or storm runoff can include a vast array of personal care and pharmaceutical compounds, from antibiotics and pesticides to livestock growth hormones, prescription drugs and even human birth control chemicals (Kolpin et al. 2002; Kidd et al. 2007). The addition of human-derived caffeine to surface water as an effluent from wastewater treatment plants has even been used as a tracer for following other organic contaminants entering fresh water (Standley et al. 2000). Mining impacts can greatly increase the rates of acidification of water through oxidation of sulfur-dense rock, and elevate concentrations of heavy metals with many known health

effects (e.g. Clements et al. 2010). Many individual pollutants are also bio-magnified up the food chain to higher-level predators, leading to high concentrations of pollutants in some stream fishes.

2.6.2 Alteration of reach-scale processes

Routing and storage of sediment and water through reaches are closely related, and are often affected by the same human actions. The most common and severe alteration is construction of dams, which interrupts downstream water, sediment, and organic matter transport, and alters the timing and magnitude of delivery of these resources to downstream reaches. In most cases, dams trap all but the finest of sediments (usually only clay-sized particles remain in suspension through a reservoir) and, if sediment-transporting flows are not significantly reduced by the dam, downstream bed materials coarsen as mobile gravels are transported away and the channel may incise (Kondolf 1997). Decreases in sediment-transporting flows can also result in reduced channel size and the floodplain may become heavily vegetated (Shafroth et al. 2002; Negishi et al. 2008). Reduction of low flows or intermediate flows alters seasonal habitat conditions, establishment or maintenance of riparian vegetation, habitat availability for aquatic biota, and survival of biota that migrate through heavily modified river systems (Kareiva et al. 2000; Poff et al. 2007; Richter & Thomas 2007). Stream flows are also altered by small-scale diversions for irrigation or other consumptive uses, and have many of the same effects on habitats, riparian vegetation, and instream biota as large dams. For example, water abstraction directly impacts aquatic systems by reducing the amount of available habitat through reduced stream flow or degraded water quality, or indirectly by reducing the ability of a system to transport or route sediment. Finally, gravel extraction directly alters habitat conditions in river and reduces downstream gravel transport, thereby affecting aquatic biota and ecosystem processes (Kondolf 1997).

Bank armoring and levee construction are common channel engineering practices that limit channel migration and formation of floodplain channels, which reduces habitat and biological diversity in the main channel, side channels, and riparian ecosystems (Fischenich 2003). Levees dramatically alter floodplain habitats by eliminating flood flows, bank erosion, and access to habitats for aquatic biota. Effects of these changes include reduced habitat capacity for commercially important species (Beechie et al. 1994, 2001), and reduced diversity of floodplain-dependent species. Bank armoring primarily affects bank erosion and formation of floodplain habitats. While bank armoring has relatively little effect on flood flows, it reduces bank erosion, channel migration, and connectivity between the main channel and floodplain habitats. Bank armoring also alters habitat quantity and quality in main river channels, which may also have significant effects on mainstem biota (Fischenich 2003).

Removal of riparian forests has a variety of effects on streams, including dramatically altering the amounts and sizes of wood in streams (e.g. Bilby & Ward 1991; Dahlström & Nilsson 2004). Decreases in wood loading have profound effects on habitat structure, including reduction in number and sizes of pools, decreased sediment retention and increased gravel size, and altered organic matter transport and storage (Bilby 1981; Montgomery et al. 1995; Beechie & Sibley 1997; Buffington & Montgomery 1999b). Grazing, vegetation management, and agricultural practices can lead to changes in channel width and form, as reductions in root strength can lead to increased bank erosion, channel widening, and conversion of single-thread channels to braided channels (e.g. Sweeney et al. 2004; Zanoni et al. 2008). Removing riparian vegetation, especially to the bank, can also result in higher inflows of nutrients and pollutants to stream reaches (e.g. Correll 2005), or in changes to the amount and composition of leaf litter delivered to streams (Kiffney & Richardson 2010). Both of these effects can ultimately lead to changes in food web dynamics in streams.

2.6.3 Direct manipulation of ecosystem features

Most direct manipulations of river ecosystems are either physical manipulations of channels and floodplains or alterations of biota. The most common physical manipulation of river ecosystem features includes channel dredging and clearing of wood from channels, usually combined with construction of levees or bank revetments that eliminate river–floodplain interactions (e.g. Collins et al. 2002; Hohensinner et al. 2003). This direct manipulation of habitat features shifts the relative proportions of specific habitat types and reduces habitat diversity through simplification of the channel cross-section, removal of channel roughness, and removal of complex hiding cover for fishes and other biota. Where habitat heterogeneity from boulders contributes to the trapping and retention of organic matter (Negishi & Richardson 2003), channel simplification and channelization can significantly diminish organic matter storage (Rosenfeld

et al. 2010). This habitat simplification eliminates or changes spawning and rearing habitats for fishes, as well as important wood or gravel substrates for primary and secondary production (e.g. algae and aquatic insect production) (Coe et al. 2009). Removal of beaver and beaver dams also simplifies channel structure, reduces sediment and organic matter retention, and may reduce floodplain habitat diversity for both riparian and aquatic species (Pollock et al. 2003). Finally, filling of wetlands and side-channels, usually during conversion of lands to agriculture or development, eliminates important habitats from the landscape and decreases habitat and biological capacity in river ecosystems (e.g. Beechie et al. 1994; Hohensinner et al. 2003).

The second common manipulation of river ecosystems is altering the spatial distribution of aquatic species by blocking fish migration or transplanting fishes, thereby altering species composition in watersheds or reaches (e.g. Pess et al. 2003). Removal of species from parts of the landscape through fishing, hatchery practices, or habitat modification (range reduction) can shift the abundance of species available in riverine ecosystems and the potential for dispersal to upstream and downstream reaches. Shifts in species abundances in turn can alter food webs and productivity of river ecosystems. For example, dams that block upriver migration of Pacific salmon reduce inputs of important nutrients and energy for aquatic and terrestrial food webs (e.g. Schindler et al. 2003). Other intentional alterations include transport of species to areas outside their natural range to create sport fisheries, inadvertent transport of organisms such as zebra mussels (*Dreissena polymorpha*), and expansion of species ranges through habitat manipulation (range expansion or invasion) (e.g. Carlton 1992; Sanderson et al. 2009).

2.7 Process-based restoration

The concept of process-based river restoration has gained momentum in recent years, with many authors pressing for more holistic restoration efforts that better address root causes of ecosystem degradation and more cost-effectively restore river ecosystems (Beechie & Bolton 1999; Brierley et al. 2002; Wohl et al. 2005; Palmer & Allan 2006; Kondolf et al. 2006). The aim of process-based restoration is to 're-establish normative rates and magnitudes of physical, chemical, and biological processes that create and sustain river and floodplain ecosystems' (Beechie et al. 2010). Underlying this aim are two primary needs: (1) to ensure that restoration plans and actions address root causes of degradation rather than symptoms; and (2) to support sustainable restoration that does not require repeated maintenance or intervention to achieve restoration objectives. While complete restoration of processes is not always possible, it is important to recognize this as a goal during restoration planning and implementation. A focus on restoring processes to the fullest extent possible will guide restoration actions toward those that are most likely to succeed, and ultimately reduce restoration costs over the long term. However, where socio-economic constraints limit restoration actions to habitat improvement or construction, incorporating a process understanding into the design of individual restoration plans is also critical to improving the effectiveness of river restoration efforts. That is, design of effective restoration requires a clear understanding processes such as sediment transport, channel migration, or riparian functions, even where those processes have been altered by human activities. We argue that process-based restoration will improve our ability to effectively restore important physical and ecological functions of rivers, and maintain or restore the important ecosystem services which they provide. In this section we summarize a set of process-based principles to guide restoration efforts, and briefly illustrate the application of the approach with an example.

2.7.1 Process-based principles for restoration

Process-based restoration is guided by four basic principles (Beechie *et al.* 2010):
1. target root causes of habitat and ecosystem change;
2. tailor restoration actions to local potential;
3. match the scale of restoration to the scale of physical and biological processes; and
4. clearly define expected outcomes, including recovery time.

The purposes of these principles are to ensure that restoration actions are effective over the long term, and to allow river ecosystems to respond to future climate change or other stochastic processes without continual human intervention. Such actions also accommodate natural spatial variation in habitat and ecosystem attributes and functions, as well as annual variation in storms and floods that are important to the creation and maintenance of habitat structure in river ecosystems.

Each of the four principles has a specific purpose in guiding effective restoration. The core principle is to address root causes of degradation so that restoration is

sustained and built structures are not overwhelmed by unrepaired processes (Beechie & Bolton 1999). Because channel and riparian conditions in each reach of a river network are expressions of processes operating at multiple scales, restoration actions will not lead to long-term improvements unless they are designed to correct the ultimate and proximate causes of degradation (i.e. the disrupted driving processes) and to redirect channel and habitat conditions toward the local physical and biological potential (Brierley *et al.* 2002). Analyses that help identify disruptions to driving processes will therefore help restoration planners identify appropriate restoration actions, and also help describe expected physical or biological conditions in a restored system (Chapter 3).

The second principle guides restoration designs and techniques to be consistent with local physical and biological potential, which is controlled by the landscape template as well as watershed processes (Kern 1992). Here we stress that 'local potential' does not necessarily mean 'natural potential'. Where aiming to return segments of rivers to natural or near-natural conditions is possible, adopting those conditions as targets will likely lead to the most effective restoration actions. However, where certain human constraints will not be removed (e.g. a large dam or a city on the floodplain), the definition of local potential must acknowledge those constraints. For example, floodplain restoration actions should consider sediment supply and lateral migration rates, which are functions of both watershed-scale and reach-scale processes. If floodplain restoration actions are designed for a river with relatively high sediment supply and high lateral migration rates, projects should be designed to accommodate channel movement and annual restructuring of some parts of the floodplain. By contrast, if a floodplain restoration action is downstream of a dam that is unlikely to be removed in the near future, then the local potential is one of very low sediment supply and low lateral migration rate. In this case, the restoration action should be designed with a stable channel in mind (the local potential for the foreseeable future), even though the natural potential might have been one of high sediment supply and a mobile channel.

The third principle (matching the scale of restoration to the scale of environmental problems) is intended to foster recognition of the scales of physical and biological processes that must be addressed (Brierley *et al.* 2002; Hughes *et al.* 2005). The conventional focus of watershed and river restoration has been on small actions in small streams (Bernhardt *et al.* 2005), but restoration efforts in the past decade have progressively increased in both scale and scope. Moreover, many years of emphasis on watershed analyses in river management planning have shown that watershed-scale problems require watershed-scale solutions, whereas reach-scale problems can be addressed locally. Hence, the third principle guides restoration planners to explicitly identify the correct scale for restoration, and to press for restoration actions that are of sufficient magnitude to successfully restore or rehabilitate river ecosystems.

The fourth principle guides restoration planners to set realistic expectations for both the restoration outcome and the time required to achieve that outcome (Beechie *et al.* 2000; Brierley *et al.* 2002; Pollock *et al.* 2007). In part this principle has the aim of forcing recognition that many restoration actions have limited benefit or long recovery time, whereas other projects have large benefits or short recovery time. Acknowledging realistic predicted outcomes for restoration actions serves the purpose of maintaining appropriate expectations for the magnitude and pace of ecosystem recovery. These predictions help avoid giving policy makers, funders, and managers a false expectation of dramatic and rapid ecosystem recovery in cases where that is unlikely. An important feature of these expected outcomes is that they are typically described as ranges of potential outcomes rather than fixed attributes (e.g. Beechie *et al.* 2008b). Quantifying the range of expected outcomes and the recovery time frame for restoration actions is critical for estimating how much restoration is needed for ecosystem recovery, for understanding lag times between action implementation and ecosystem recovery, and for designing monitoring programs that evaluate the effectiveness of actions relative to stated targets.

Together these four principles aim to restore the dynamics of rivers, guide the design of restoration actions to restore process regimes rather than states, and produce variability in conditions rather than uniformity (Brierley *et al.* 2002; Beechie *et al.* 2010). For example, most actions intended to restore processes will remove human influences on those processes (e.g. levee removal to allow channel migration, or removal of dams to restore movement of water, sediment, and biota), allowing natural variation in processes to drive habitat formation and ecosystem function. By contrast, actions that partially restore processes do not fully remove human influences, but attempt to restore key attributes of natural processes. For example, restoration of environmental flows in a regulated river aims to restore a range of flood flows to low flows that mimic components of the natural hydrologic regime. These diverse flow targets restore such

functions as formation and erosion of floodplain features (large floods), wood recruitment and habitat formation (channel-forming floods), and maintenance of aquatic organisms (low flows) (Poff *et al.* 2007), while still retaining water for irrigation or power generation. Ultimately such restoration approaches aim to restore dynamic and self-sustaining river ecosystems, thereby reducing restoration costs over the long term.

2.7.2 Applying the principles to restoration

Complete restoration of watershed and riverine processes is rarely possible, so river restoration employs a suite of strategies ranging from fully restoring processes to habitat creation efforts that artificially construct or modify habitat features as a substitute for natural functions. As discussed in Chapter 1, most restoration actions fall into one of three classes: (1) full restoration of processes (restoration); (2) partial restoration of processes (rehabilitation); and (3) habitat creation or improvement (enhancement) (Beechie *et al.* 2010). *Full restoration* actions restore habitat-forming processes, *partial restoration* actions restore selected ecosystem processes and functions, and *habitat creation* or improvement actions are focused on building habitat rather than addressing the root causes of degradation (Cairns 1988). Each of these types of actions has different benefits and longevity, yet in each case the process-based principles can be applied to help identify actions with highest likelihoods of success and sustainability given the surrounding constraints. Full and partial process restoration actions can use all four of the process-based principles to guide actions toward success and sustainability. By contrast, habitat creation, such as excavating a side channel, does not address the causal mechanism of degradation (Principle 1). Nevertheless, the remaining principles can be applied to ensure that actions are suited to the site potential (Principle 2), at an appropriate scale (Principle 3), and have clearly stated expected outcomes including recovery time and longevity (Principle 4).

To illustrate application of the principles to the three types of restoration actions, we describe a suite of restoration options to restore a 15-km lowland reach of the Kiso River, Central Japan (catchment area of 5275 km^2). In this reach, past gravel extraction and reduction of sediment supply by dam construction have caused an average incision of 3 m (up to 20 m at maximum). As a result, this historically braided and meandering part of the Kiso River has narrowed, floodplain surfaces are less frequently inundated and have become heavily vegetated,

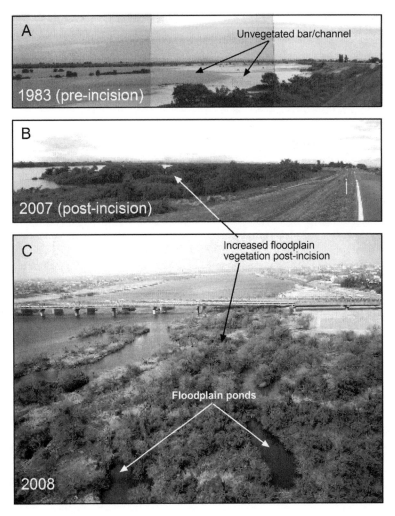

Figure 2.16 Ground-based photographs of (A) the incised reach of the Kiso River, Japan prior to incision in 1983 (photo Chubu Regional Bureau), and (B) after incision in 2007 (photo Junjiro Negishi). (C) Oblique aerial photograph showing the same area in 2008 with floodplain water bodies that experience frequent hypoxia due to increased organic matter inputs and reduced flushing (photo Chubu Regional Bureau). Note that what was once an extensive sand bar has become densely vegetated over the 25-year period.

and floodplain pond habitat for endangered fauna including freshwater unionoid mussels has been degraded (Negishi *et al.* 2008, 2010; Figure 2.16). Extensive growth of riparian vegetation provides large inputs of organic matter into floodplain ponds, and the decreased inundation frequency reduces organic matter flushing (Negishi *et al.* 2012). Consequently, the major habitat impairment is the frequent occurrence of hypoxia, especially during summer periods. It is now recognized that restoring processes that will reduce the occurrence of hypoxia in

floodplain ponds is key to the successful restoration of target floodplain-dependent species (Negishi *et al.* 2012).

Restoration actions that achieve this objective may fall into any of the three restoration categories (Figure 2.17). The simplest and quickest option is a habitat creation strategy that relies on the removal of riparian vegetation and dredging organic matter from ponds to prevent organic matter accumulation and hypoxia from occurring. This action does not address the loss of sediment supply but addresses the local mechanisms causing

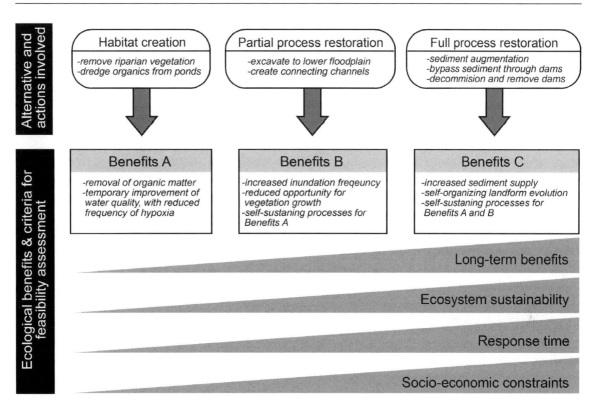

Figure 2.17 Illustration of three restoration alternatives for the lower Kiso River, Japan, showing shifts in expected benefits, sustainability, response time, and socio-economic constraints among the alternatives.

hypoxia, is at an appropriate scale, and recognizes that the expected benefits are temporary and will require continued intervention (following Principles 2–4). The second strategy would lower the floodplain elevation by excavation, thereby increasing inundation frequency to impede vegetation growth, reduce organic matter inputs, and increase flushing flows. While this option is akin to habitat creation in some respects (excavating the floodplain), the action restores floodplain connectivity and flooding processes, which should produce a more sustained reduction of organic matter inputs and increased flushing flows. Finally, the third strategy of restoring the sediment supply (the primary impaired process) should result in aggradation of the channel over the long-term, and ultimately restore floodplain connectivity and a more natural floodplain inundation frequency. This alternative follows all four principles by addressing the root cause of the problem, addressing the problem at an appropriate scale, matching the project to local potential, and recognizing that the response will be delayed but sustained over the longest time frame.

In many cases a combination of approaches may be used in restoration, especially in cases where short-term manipulations (habitat improvement) can be combined with long-term process restoration (full or partial process restoration). For example, in streams where wood abundance and habitat structure have been degraded by loss of riparian forests, wood placement can help achieve short-term improvements in habitat while riparian restoration aims to correct the root cause of the problem and eventually restore the source of wood to the system. The habitat improvement effort by itself cannot address the root cause of the problem, and the process restoration will take several decades to realize any change in wood recruitment (although root strength and shade functions are often achieved in a very short period of time). In other cases, process restoration achieves an immediate and long-lasting benefit (e.g. the removal of a migration blockage for anadromous fish or restoring stream flows during summer), and in those cases habitat improvement efforts will not help address either the symptom or the root cause of degradation. Therefore, a focus on restoring the key degraded process is usually necessary to achieve both short- and long-term benefits.

Where land-use constraints limit restoration opportunities, restoration actions are often limited in scope by infrastructure or altered processes that will not be corrected in the foreseeable future. Nevertheless, such actions can be designed in the context of the process-based principles to improve restoration effectiveness and longevity. While the first principle may not apply (i.e. it may not be possible to address the root cause of degradation), the remaining principles guide restoration or habitat creation actions to be: (1) consistent with local physical and biological potential (which is altered by human constraints); (2) at an appropriate and correct scale for the location and problem; and (3) reasonably limited in expectations for the restoration outcome (Beechie *et al.* 2010). For example, in the Trinity River in California, USA, the Lewiston dam has dramatically reduced sediment supply, stream flow, and river dynamics and restoration actions are planned and designed to match this constrained potential (US Fish and Wildlife Service & Hoopa Valley Tribe 1999; McBain & Trush, Inc. 2007). That is, the dam is treated as a constraint on restoration potential, and restoration designs are based on the recognition that the future Trinity River below the dam will be smaller and less mobile than the historical channel (e.g. Trush *et al.* 2000). Moreover, floodplain channels that might have existed prior to the dam are unlikely to form with reduced sediment supply and stream flow. The restoration plan therefore includes a combination of actions that build floodplain habitats (habitat creation) and partially restore sediment supply and stream flow (partial process restoration actions) (US Fish and Wildlife Service & Hoopa Valley Tribe 1999).

2.8 Summary

Processes that control river ecosystems are hierarchically nested, both spatially and temporally. Long-term geologic, tectonic, and climatic processes create the landscape template that controls the structure of river networks and valley forms, which are generally immutable over human time frames. The landscape template sets limits on the types of habitats and process rates that are expressed in river reaches, but watershed- and reach-scale processes control conditions at any point in time. Runoff and erosion processes control stream flow and sediment supply, while nutrient supply processes control primary productivity at the base of the food web. Reach-scale

processes, including routing of sediment and water, river–floodplain interactions, riparian processes, and instream biological processes control ecosystem conditions at the reach level. Human impacts to these processes include both indirect and direct effects on habitats and ecological systems. Indirect effects are those that affect processes away from the stream, such as land-use effects on erosion or runoff processes that ultimately affect stream flow or sediment supply, or riparian vegetation impacts that affect stream temperature or wood and nutrient supply. Direct manipulations of river channels and biota include impacts such as dredging or channel control structures, or directs effects on biota such as fishing, hatchery practices, and stocking of non-native species.

Successful river ecosystem restoration is rooted in a clear understanding of linkages between causes of habitat change and the resultant effects of habitat change on biota and ecosystem processes. These cause-effect linkages are the foundation of process-based restoration, which aims to restore watershed and river processes that drive ecosystem functions and features. Process-based restoration is guided by four fundamental principles, including targeting root causes of habitat and ecosystem change, tailoring restoration actions to local potential, matching the scale of restoration to the scale of physical and biological processes, and explicitly stating expected outcomes. The purpose of these principles is to guide river restoration toward actions that require minimal maintenance and create a resilient river system that adapts to future perturbations such as climate change. These principles also help define the needs of watershed assessments (Chapter 3), inform how restoration actions should be identified and prioritized (Chapters 3, 5 and 6), support development and design of restoration projects (Chapter 7), and guide the development of monitoring plans that track success or failure of projects (Chapter 8).

2.9 References

Abbe, T.B. & Montgomery, D.R. (2003) Patterns and processes of wood debris accumulation in the Queets River basin, Washington. *Geomorphology* **51**, 81–107.

Agee, J.K. (1988) Successional dynamics in forest riparian zones. In: Raedeke, K.J. (ed.) *Streamside Management: Riparian Wildlife and Forestry Interactions*. College of Forest Resources, University of Washington, Seattle, Washington, pp. 31–43.

Allan, J.D. (1995) *Stream Ecology: Structure and Function of Running Waters*. Chapman & Hall, New York, New York.

Allan, J.D. (2004) Landscapes and riverscapes: the influence of land use on stream ecosystems. *Annual Review of Ecology, Evolution and Systematics* **35**, 257–284.

Asaeda, T., Rajapakse, L. & Kanoh, M. (2010) Fine sediment retention as affected by annual shoot collapse: *Sparganium erectum* as an ecosystem engineer in a lowland stream. *River Research and Applications* **26**, 1153–1169.

Beechie, T.J. & Sibley, T.H. (1997) Relationships between channel characteristics, woody debris, and fish habitat in northwestern Washington streams. *Transactions of the American Fisheries Society* **126**, 217–229.

Beechie, T. & Bolton, S. (1999) An approach to restoring salmonid habitat-forming processes in Pacific Northwest watersheds. *Fisheries* **24**, 6–15.

Beechie, T., Beamer, E. & Wasserman, L. (1994) Estimating coho salmon rearing habitat and smolt production losses in a large river basin, and implications for habitat restoration. *North American Journal of Fisheries Management* **14**, 797–811.

Beechie, T.J., Pess, G., Kennard, P., Bilby, R.E. & Bolton, S. (2000) Modeling recovery rates and pathways for woody debris recruitment in northwestern Washington streams. *North American Journal of Fisheries Management* **20**, 436–452.

Beechie, T.J., Collins B.D. & Pess, G.R. (2001) Holocene and recent geomorphic processes, land use and salmonid habitat in two north Puget Sound river basins. In: Dorava, J.B., Montgomery, D.R., Fitzpatrick, F. & Palcsak, B. (eds) *Geomorphic Processes and Riverine Habitat*. Water Science and Application Volume 4, American Geophysical Union, Washington DC, pp. 37–54.

Beechie, T.J., Liermann, M., Beamer, E.M. & Henderson, R. (2005a) A classification of habitat types in a large river and their use by juvenile salmonids. *Transactions of the American Fisheries Society* **134**, 717–729.

Beechie, T.J., Veldhuisen, C.N., Schuett-Hames, D.E., DeVries, P., Conrad, R.H. & Beamer, E.M. (2005b) Monitoring treatments to reduce sediment and hydrologic effects from roads. In: Roni, P. (ed.) *Methods for Monitoring Stream and Watershed Restoration*. American Fisheries Society, Bethesda, Maryland, pp. 35–65.

Beechie, T.J., Liermann, M., Pollock, M.M., Baker, S. & Davies, J. (2006a) Channel pattern and river-floodplain dynamics in forested mountain river systems. *Geomorphology* **78**(1–2), 124–141.

Beechie, T.J., Ruckelshaus M., Buhle E., Fullerton A. & Holsinger, L. (2006b) Hydrologic regime and the conservation of salmon life history diversity. *Biological Conservation* **130**, 560–572.

Beechie, T.J., Moir, H. & Pess, G. (2008a) Hierarchical physical controls on salmonid spawning location and timing. In: Sear, D. & De Vries, P. (eds) *Salmonid Spawning Habitat in Rivers: Physical Controls, Biological Responses, and Approaches to Remediation*. American Fisheries Society, Symposium 65, Bethesda, Maryland, pp. 83–102.

Beechie, T.J., Pollock, M.M. & Baker, S. (2008b) Channel incision, evolution and potential recovery in the Walla Walla and Tucannon River basins, northwestern USA. *Earth Surface Processes and Landforms* **33**, 784–800.

Beechie, T.J., Pess, G.R., Pollock, M.M., Ruckelshaus, M.H. & Roni, P. (2009) Chapter 33: Restoring rivers in the 21st century: science challenges in a management context. In: Beamish, R.J. & Rothschild, B.J. (eds) *The Future of Fisheries Science in North America*, Springer, Heidelberg, pp. 695–716.

Beechie, T.J., Sear, D., Olden, J., Pess, G.R., Buffington, J., Moir, H., Roni, P. & Pollock, M.M. (2010) Process-based principles for restoring river ecosystems. *BioScience* **60**, 209–222.

Bekker, M.F. (2005) Positive feedback between tree establishment and patterns of subalpine forest advancement, Glacier National Park, Montana, USA. *Arctic, Antarctic, and Alpine Research* **37**, 97–107.

Benda, L. & Dunne, T. (1997a) Stochastic forcing of sediment supply to channel networks from landsliding and debris flow. *Water Resources Research* **33**, 2865–2880.

Benda, L. & Dunne, T. (1997b) Stochastic forcing of sediment routing and storage in channel networks. *Water Resources Research* **33**(12), 2849–2863.

Benda, L., Beechie, T.J., Wissmar, R.C. & Johnson, A. (1992) Morphology and evolution of salmonid habitats in a recently deglaciated river basin, Washington State, USA. *Canadian Journal of Fisheries and Aquatic Sciences* **49**, 1246–1256.

Benda, L., Poff, L., Miller, D. *et al.* (2004) The network dynamics hypothesis: how channel networks structure riverine habitats. *BioScience* **54**, 413–427.

Bennett, K.D. (1986) The rate of spread and population increase of forest trees during the postglacial. *Philosophical Transactions of the Royal Society B* **314**, 523–531.

Bennett, S.J., Wu, W., Alonso, C.V. & Wang, S.S.Y. (2008) Modeling fluvial response to in-stream woody vegetation: implications for stream corridor restoration. *Earth Surface Processes and Landforms* **33**, 890–909.

Bergstrom, F.W. (1982) Episodic behavior in badlands: its effects on channel morphology and sediment yields. In: Swanson, F.J., Janda, E.J., Dunne, T. & Swanson, D.N. (eds) *Sediment Budgets and Routing in Forested Drainage Basins*. General Technical Report PNW-141. United States Department of Agriculture Forest Service, Portland, Oregon, pp. 59–66.

Bernhardt, E.S., Palmer, M.A., Allan, J.D. *et al.* (2005) Synthesizing U.S. river restoration efforts. *Science* **308**, 636–637.

Bertoldi, W., Gurnell, A.M., Surian, N. *et al.* (2009) Understanding reference processes: linkages between river flows, sediment dynamics and vegetated landforms along the Tagliamento River, Italy. *River Research and Applications* **25**, 501–516.

Bertoldi, W., Drake, N.A. & Gurnell, A.M. (2011a) Interactions between river flows and colonizing vegetation on a braided river: exploring spatial and temporal dynamics in vegetation cover using satellite data. *Earth Surface Processes and Landforms* **36**, 1474–1486.

Bertoldi, W., Gurnell, A.M. & Drake, N.A. (2011b) The topographic signature of vegetation development along a braided river: results of a combined analysis of airborne LiDAR, colour air photographs and ground measurements. *Water Resources Research* **47**, W06525, doi:10.1029/2010WR010319.

Beschta, R.L., Pyles, M.R., Skaugset, A.E. & Surfleet, C.G. (2000) Peakflow response to forest practices in the western Cascades of Oregon, U.S.A. *Journal of Hydrology* **233**, 102–120.

Bilby, R.E. (1981) Role of organic debris dams in regulating the export of dissolved and particulate matter from a forested watershed. *Ecology* **62**, 1234–1243.

Bilby, R.E. & Ward, J.W. (1991). Characteristics and function of large woody debris in streams draining old-growth, clear-cut, and second-growth forests in southwestern Washington. *Canadian Journal of Fisheries and Aquatic Sciences* **48**, 2499–2508.

Bishop, P., Young, R.W. & McDougall, I. (1985) Stream profile change and longterm landscape evolution: early Miocene and modern rivers of the east Australian highland crest, central New South Wales, Australia. *The Journal of Geology* **93**, 455–474.

Bondar, C.A., Bottriell, K., Zeron, K. & Richardson, J.S. (2005) Does trophic position of the omnivorous signal crayfish (*Pacifastacus leniusculus*) in a stream food web vary with life history stage or density? *Canadian Journal of Fisheries and Aquatic Sciences* **62**, 2632–2639.

Booth, D.B. & Jackson, C.R. (1997) Urbanization of aquatic systems – degradation thresholds, stormwater detention, and the limits of mitigation. *Water Resources Bulletin* **33**, 1077–1090.

Booth, D.B., Hartley, D. & Jackson, C.R. (2002) Forest cover, impervious-surface area, and the mitigation of stormwater impacts. *Journal of the American Water Resources Association* **38**, 835–845.

Bradford, J.M. & Huang, C. (1994) Inter-rill erosion as affected by tillage and residue cover. *Soil and Tillage Research* **31**, 353–361.

Brierley, G.J. & Fryirs, K.A. (2005) *Geomorphology and River Management: Application of the River Styles Framework*. Blackwell Publishing, Oxford, United Kingdom.

Brierley, G.J., Fryirs, K., Outhet, D. & Massey, C. (2002) Application of the River Styles framework as a basis for river management in New South Wales, Australia. *Journal of Applied Geography* **22**, 91–122.

Brown, A.G., Cooper, L., Salisbury, C.R. & Smith, D.N. (2001) Late Holocene channel changes of the Middle Trent: channel response to a thousand-year flood record. *Geomorphology* **39**(1–2), 69–82.

Buffington, J.M. & Montgomery, D.R. (1999a) Effects of sediment supply on surface textures of gravel-bed rivers. *Water Resources Research* **35**, 3523–3530.

Buffington, J.M. & Montgomery, D.R. (1999b) Effects of hydraulic roughness on surface textures of gravel-bed rivers. *Water Resources Research* **35**, 3507–3521.

Cairns, J., Jr. (1988) Restoration and the alternative: a research strategy. *Restoration and Management Notes* **6**, 65–67.

Cardinale, B.J., Gelmann, E.R. & Palmer, M.A. (2004) Net spinning caddisflies as stream ecosystem engineers: the influence of *Hydropsyche* on benthic substrate stability. *Functional Ecology* **18**, 381–387.

Carlton, J.T. (1992) Introduced marine and estuarine mollusks of North America: an end-of-the-20th-century perspective. *Journal of Shellfish Research*, **11**, 489–505.

Church, M. (1983) Pattern of instability in a wandering gravel bed channel. *Special Publications of the International Association of Sedimentologists* **6**, 169–180.

Church, M. (2002) Geomorphic thresholds in riverine landscapes. *Freshwater Biology* **47**, 541–557.

Clarke, S.J. (2002) Vegetation growth in rivers: influences upon sediment and nutrient dynamics. *Progress in Physical Geography* **26**, 159–172.

Clements, W.H., Vieira, N.K.M. & Church, S.E. (2010) Quantifying restoration success and recovery in a metal-polluted stream: a 17-year assessment of physicochemical and biological responses. *Journal of Applied Ecology* **47**, 899–910.

Coe, H.J., Kiffney, P.M., Pess, G.R., Kloehn, K.K. & McHenry, M.L. (2009) Periphyton and invertebrate response to wood placement in large pacific coastal rivers. *River Research and Applications* **25**, 1025–1035.

Collins, B.D., Montgomery, D.R. & Haas, A.D. (2002) Historical changes in the distribution and functions of large wood in Puget Lowland rivers. *Canadian Journal of Fisheries and Aquatic Sciences* **59**, 66–76.

Compton, J.E., Church, M.R., Larned, S.T. & Hogsett, W.E. (2003) Nitrogen export from forested watersheds in the Oregon Coast Range: the role of nitrogen fixing Red Alder. *Ecosystems* **6**, 773–785.

Corenblit, D., Tabacchi, E., Steiger, J. & Gurnell, A.M. (2007) Reciprocal interactions and adjustments between fluvial landforms and vegetation dynamics in river corridors: A review of complementary approaches. *Earth-Science Reviews* **84**, 56–86.

Correll, D.L. (2005) Principles of planning and establishment of buffer zones. *Ecological Engineering* **24**, 433–439.

Dahlström, N. & Nilsson, C. (2004) Influence of woody debris on channel structure in old growth and managed forest streams in Central Sweden. *Environmental Management* **33**(3), 376–384.

Death, R.G. & Collier, K.J. (2010) Measuring stream macroinvertebrate responses to gradients of vegetation cover: when is enough enough? *Freshwater Biology* **55**, 1447–1464.

Dietrich, W.E., Kirchner, J.W., Ikeda, H. & Iseya, F. (1989) Sediment supply and the development of the coarse surface layer in gravel-bedded rivers. *Nature* **340**, 215–217.

Donato, M.M. (1998) *Surface-water/ground-water relations in the Lemhi River basin, east-central Idaho.* United States Geological Survey Water-resources Investigations Report 98-4185. United States Geological Survey, Washington, DC.

Doyle, M.W., Stanley, E.H., Strayer, D.L., Jacobson, R.B. & Schmidt, J.C. (2005) Effective discharge analysis of ecological processes in streams. *Water Resources Research* **41**, W11411.

Dunne, T. & Leopold, L.B. (1978) *Water in Environmental Planning.* Freeman, San Francisco.

Englund, G. & Krupa, J.J. (2000) Habitat use by crayfish in stream pools: influence of predators, depth and body size. *Freshwater Biology* **43**, 75–83.

Fanta, S.E., Hill, W.R., Smith, T.B. & Roberts, B.J. (2010) Applying the light: nutrient hypothesis to stream periphyton. *Freshwater Biology* **55**, 931–940.

Fastie, C.L. (1995) Causes and ecosystem consequences of multiple pathways of primary succession at Glacier Bay, Alaska. *Ecology* **76**, 1899–1916.

Fausch, K.D., Torgersen, C.E., Baxter, C.V. & Li, H.W. (2002) Landscapes to riverscapes: bridging the gap between research and conservation of stream fishes. *BioScience* **52**, 483–498.

Feller, M.C. (2005) Forest harvesting and streamwater inorganic chemistry in western North America: A review. *Journal of the American Water Resources Association* **41**, 785–811.

Fetherston, K.L., Naiman, R.J. & Bilby, R.E. (1995) Large woody debris, physical process, and riparian forest development in montane river networks of the Pacific Northwest. *Geomorphology* **13**, 133–144.

Fischenich, J.C. (2003) *Effects of Riprap on Riverine and Riparian Ecosystems.* Report number ERDC/EL TR-03-4. United States Army Corps of Engineers, Washington, DC.

Francis, R.A., Tibaldeschi, P. & McDougall, L. (2008) Fluvially-deposited large wood and riparian plant diversity. *Wetlands Ecology and Management* **16**, 371–382.

Frissell, C.A., Liss, W.J., Warren, C.E. & Hurley, M.D. (1986) A hierarchical framework for stream habitat classification: viewing streams in a watershed context. *Environmental Management* **10**, 199–214.

Gomi, T., Sidle, R.C. & Richardson, J.S. (2002) Headwater and channel network -understanding processes and downstream linkages of headwater systems. *BioScience* **52**, 905–916.

Gooderham, J.P.R., Barmuta, L.A. & Davies, P.E. (2007) Upstream heterogeneous zones: small stream systems structured by a lack of competence? *Journal of the North American Benthological Society* **26**, 365–374.

Gregory, S.V., Swanson, F.J., McKee, W.A. and Cummins, K.W. (1991) An ecosystem perspective of riparian zones. *BioScience* **41**, 540–551.

Gurnell, A. & Petts, G. (2006) Trees as riparian engineers: the Tagliamento River, Italy. *Earth Surface Processes and Landforms* **31**, 1558–1574.

Gurnell, A.M., Petts, G.E., Hannah, D.M. *et al.* (2001) Riparian vegetation and island formation along the gravel-bed Fiume Tagliamento, Italy. *Earth Surface Processes and Landforms* **26**, 31–62.

Gurnell, A., Tockner, K., Edwards, P.J. & Petts, G.E. (2005) Effects of deposited wood on biocomplexity of river corridors. *Frontiers in Ecology and Environment* **3**(7), 377–382.

Gurnell, A.M., Boitsidis, A.J., Thompson, K. & Clifford, N.J. (2006a) Seed bank, seed dispersal and vegetation cover: Colonization along a newly-created river channel. *Journal of Vegetation Science* **17**, 665–674.

Gurnell, A.M., Morrissey, I.P., Boitsidis, A.J. *et al.* (2006b) Initial adjustments within a new river channel: interactions between fluvial processes, colonizing vegetation and bank profile development. *Environmental Management* **38**, 580–596.

Gurnell, A.M., van Oosterhout, M.P., de Vlieger, B. & Goodson, J.M. (2006c). Reach-scale interactions between aquatic plants and physical habitat: River Frome, Dorset. *River Research and Applications* **22**, 667–680.

Gurnell, A.M., Goodson, J., Thompson, K., Clifford, N. & Armitage, P.D. (2007) The river bed: a dynamic store for viable plant propagules? *Earth Surface Processes and Landforms* **32**, 1257–1272.

Gurnell, A.M., Thompson, K., Goodson, J., Moggridge, H. (2008) Propagule deposition along river margins: linking hydrology and ecology. *Journal of Ecology* **96**, 553–565.

Gurnell, A.M., Surian, N. & Zanoni, L. (2009) Multi-thread river channels: a perspective on changing European alpine river systems. *Aquatic Sciences* **71**, 253–265.

Gurnell, A.M., O'Hare, J.M., O'Hare, M.T., Dunbar, M.J. & Scarlett, P.M. (2010) An exploration of associations between assemblages of aquatic plant morphotypes and channel geomorphological properties within British rivers. *Geomorphology* **116**, 135–144.

Harr, R.D. (1986) Effects of clearcutting on rain-on-snow runoff in western Oregon: A new look at old studies. *Water Resources Research* **22**, 1095–1100.

Hart, D.D. & Merz, R.A. (1998) Predator prey interactions in a benthic stream community: a field test of flow-mediated refuges. *Oecologia* **114**, 263–273.

Hart, D.D. & Finelli, C.M. (1999) Physical-biological coupling in streams: The pervasive effects of flow on benthic organisms. *Annual Review of Ecology and Systematics* **30**, 363–395.

Haslam, S.M. (2006) *River Plants*. Forrest Text, Tresaith, Ceredigion, UK.

Hill, W.R., Mulholland, P.J. & Marzolf, E.R. (2001) Stream ecosystem responses to forest leaf emergence in spring. *Ecology* **82**, 2306–2319.

Hohensinner S., Habersack, H., Jungwirth, M. & Zauner, G. (2003) Reconstruction of the characteristics of a natural alluvial river-floodplain system and hydromorphological changes following human modifications: the Danube River (1812–1991). *River Research and Applications* **20**, 25–41.

Hoover, T.M., Marczak, L.B., Richardson, J.S. & Yonemitsu, N. (2010) Transport and settlement of organic matter in small streams. *Freshwater Biology* **55**, 436–449.

Hovius, N., Stark, C.P. & Allen, P.A. (1997) Sediment flux from a mountain belt derived by landslide mapping. *Geology* **25**, 231–234.

Hudon, C. & Carignan, R. (2008) Cumulative impacts of hydrology and human activities on water quality in the St. Lawrence River (Lake Saint-Pierre, Quebec, Canada). *Canadian Journal of Fisheries and Aquatic Sciences* **65**, 1165–1180.

Huet, M. (1959) Profiles and biology of western European streams as related to fisheries management. *Transactions of the American Fisheries Society* **88**, 155–163.

Hughes, F.M.R. (1997) Floodplain biogeomorphology. *Progress in Physical Geography* **21**, 510–529.

Hughes, F.M.R., Colston, A. & Mountford, J.O. (2005) Restoring riparian ecosystems: the challenge of accommodating variability and designing restoration trajectories. *Ecology and Society* **10**(1), article 12. Available at: http://www.ecologyandsociety.org/vol10/iss1/art12/.

Hughes, N.F. (1998) A model of habitat selection by drift-feeding stream salmonids at different scales. *Ecology* **79**, 281–294.

Ibarra, A.A., Park, Y-S., Bosse, S., Reyjol, Y., Lim, P. & Lek, S. (2005) Nested patterns of spatial diversity revealed for fish assemblages in a west European river. *Ecology of Freshwater Fish* **14**, 233–242.

Imaizumi, F., Sidle, R.C. & Kamei, R. (2008) Effects of forest harvesting on the occurrence of landslides and debris flows in steep terrain of central Japan. *Earth Surface Processes and Landforms* **33**, 827–840.

Jackson, D.A., Peres-Neto, P.R. & Olden, J.D. (2001) What controls who is where in freshwater fish communities – the roles of biotic, abiotic, and spatial factors. *Canadian Journal of Fisheries and Aquatic Sciences* **58**, 157–170.

Jones, J.A. & Grant, G.E. (1996) Peakflow responses to clear-cutting and roads in small and large basins, western Cascades, Oregon. *Water Resources Research* **32**, 959–974.

Jungwirth, M., Muhar, S. & Schumtz, S. (2002) Re-establishing and assessing ecological integrity in riverine landscapes. *Freshwater Biology* **47**, 867–887.

Junk, W.J., Bayley, P.B. & Sparks, R.E. (1989) The flood pulse concept in river-floodplain systems. In: Dodge, D.P. (ed.) *Proceedings of the International Large River Symposium*. Canadian Special Publications Fisheries and Aquatic Sciences 106, pp. 110–127.

Kareiva, P., Marvier M. & McClure, M. (2000) Recovery and management options for spring/summer Chinook salmon in the Columbia River basin. *Science* **290**, 977–979

Karr, J.R. (1991) Biological integrity: A long-neglected aspect of water resource management. *Ecological Applications* **1**(1), 66–84.

Karr, J.R. (2006) Seven foundations of biological monitoring and assessment. *Biologia Ambientale* **20**(2), 7–18.

Karrenberg, S., Edwards, P.J. & Kollmann, J. (2002). The life history of Salicaceae living in the active zone of floodplains. *Freshwater Biology* **47**, 733–748.

Kawaguchi, Y. & Nakano, S. (2001) Contribution of terrestrial invertebrates to the annual resource budget for salmonids in forest and grassland reaches of a headwater stream. *Freshwater Biology* **46**, 303–316.

Kelly, E.N., Schindler, D.W., St. Louis, V.L., Donald, D.B. & Vladicka, K.E. (2006) Forest fire increases mercury accumulation by fishes via food web restructuring and increased mercury inputs. *Proceedings of the National Academy of Sciences* **103**, 19380–19385.

Kern, K. (1992) Rehabilitation of streams in South-West Germany. In: Boon, P.J., Calow P. & Petts, G.E. (eds), *River Conservation and Management*, John Wiley & Sons Ltd, Chichester, UK, pp. 321–336.

Kidd, K.A., Blanchfield, P.J., Mills, K.H. *et al.* (2007) Collapse of a fish population after exposure to a synthetic estrogen. *Proceedings of the National Academy of Sciences* **104**, 8897–8901.

Kiffney, P.M. & Richardson, J.S. (2010) Organic matter inputs into headwater streams of southwestern British Columbia as a function of riparian reserves and time since harvesting. *Forest Ecology and Management* **260**, 1931–1942.

Kiffney, P.M., Richardson, J.S. & Bull, J.P. (2003) Responses of periphyton and insects to experimental manipulation of riparian buffer width along forest streams. *Journal of Applied Ecology* **40**, 1060–1076.

Kiffney, P.M., Greene, C.M., Hall, J.E. & Davies, J.R. (2006) Tributary streams create spatial discontinuities in habitat, biological productivity, and diversity in mainstem rivers. *Canadian Journal of Fisheries and Aquatic Sciences* **63**, 2518–2530.

Kollmann, J., Vieli, M., Edwards, P.J., Tockner, K. & Ward, J.V. (1999) Interactions between vegetation development and island formation in the Alpine river Tagliamento. *Applied Vegetation Science* **2**, 25–36.

Kolpin, D.W., Furlong, E.T., Meyer, M.T. *et al.* (2002) Pharmaceuticals, hormones, and other organic wastewater contaminants in US streams, 1999–2000: A national reconnaissance. *Environmental Science & Technology* **36**, 1202–1211.

Kondolf, G.M. (1997) Hungry water: Effects of dams and gravel mining on river channels. *Environmental Management* **21**, 533–551.

Kondolf, G.M., Boulton, A.J., O'Daniel, S. *et al.* (2006) Process-based ecological river restoration: visualizing three-dimensional connectivity and dynamic vectors to recover lost linkages. *Ecology and Society* **11**(2), Article 5. Available at: http://www.ecologyandsociety.org/vol11/iss2/art5/

Lancaster, J., Downes, B.J. & Arnold, A. (2010) Oviposition site selectivity of some stream-dwelling caddisflies. *Hydrobiologia* **652**, 165–178.

Leopold, L.B., Wolman, M.G. & Miller, J.P. (1964) *Fluvial processes in geomorphology*. Freeman and Co., San Francisco.

Liffen, T., Gurnell, A.M., O'Hare, M.T., Pollen-Bankhead, N. & Simon, A. (2011) Biomechanical properties of the emergent aquatic macrophyte *Sparganium erectum*: implications for landform development in low energy rivers. *Ecological Engineering* **37**, 1925–1931.

Likens, G.E., Bormann, F.H., Johnson, M., Fisher, S.W. & Pierce, R.S. (1970) Effects of forest cutting and herbicide treatment on nutrient budgets in the Hubbard Brook Watershed ecosystem. *Ecological Monographs* **40**, 23–47.

Lisle, T.E. (1982) Effects of aggradation and degradation on riffle-pool morphology in natural gravel channels, northwestern California. *Water Resources Research* **18**, 1643–1651.

Lowrance, R., Todd, R., Fail, Jr., J., Hendrickson, Jr., O., Leonard, R. & Asmussen, L. (1984) Riparian forests as nutrient filters in agricultural watersheds. *BioScience* **34**, 374–377.

Lowrance, R.R., Leonard, R.A., Asmussen, L.E. & Todd, R.L. (1985) Nutrient budgets for agricultural watersheds in the southeastern coastal plain. *Ecology* **66**, 287–29.

Madej, M.A. (1982) Sediment transport and channel changes in an aggrading stream in the Puget Lowland, Washington. In: Swanson, F.J., Janda, E.J., Dunne, T. & Swanson, D.N. (eds) *Sediment Budgets and Routing in Forested Drainage Basins*. United States Forest Service General Technical Report PNW-141, Portland, Oregon, pp. 97–108.

Madej, M.A. (1987) Residence times of channel-stored sediment in Redwood Creek, northwestern California. In: Beschta, R.L., Blinn, T., Grant, G.E., Ice, G.G. &

Swanson, F.J. (eds) *Erosion and Sedimentation in the Pacific Rim.* IAHS Publication 165, Wallingford, United Kingdom, pp 429–438.

Madej, M.A. & Ozaki, V. (1996) Channel response to sediment wave propagation and movement, Redwood Creek, California, USA. *Earth Surface Processes and Landforms* **21**, 911–927.

Mahoney, J.M. & Rood, S.B. (1998) Streamflow requirements for cottonwood seedling recruitment: an integrative model. *Wetlands* **18**: 634–645.

McAuliffe, J.R. (1984) Competition for space, disturbance, and the structure of a benthic stream community. *Ecology* **65**, 894–908.

McBain & Trush, Inc. (2007) *Coarse Sediment Management Plan: Lewiston Dam to Douglas City, Trinity River, California.* Trinity River Restoration Program, Weaverville, CA.

McGarvey, D.J. & Hughes, R.M. (2008) Longitudinal zonation of Pacific Northwest (USA) fish assemblages and the species-discharge relationship. *Copeia* **2008**, 311–321.

Megahan, W.F., King, J.G. & Seyedbagheri, K.A. (1995) Hydrologic and erosional responses of a granitic watershed to helicopter logging and broadcast burning. *Forest Science* **41**, 777–795.

Micheli, E.R. & Kirchner, J.W. (2002) Effects of wet meadow riparian vegetation on stream bank erosion: Remote sensing measurements of streambank migration and erodibility. *Earth Surface Processes and Landforms* **27**, 627–639.

Millar, R.G. (2000) Effect of bank vegetation on alluvial channel patterns. *Water Resources Research* **36**, 1109–1118.

Minshall, G.W., Cummins, K.W., Petersen, R.C. *et al.* (1985) Developments in stream ecosystem theory. *Canadian Journal of Fisheries and Aquatic Sciences* **42**, 1045–1055.

Montgomery, D.R. (1999) Process domains and the river continuum. *Journal of the American Water Resources Association* **35**, 397–410.

Montgomery, D.R. (2002) Valley formation by fluvial and glacial erosion. *Geology* **30**, 1047–1050.

Montgomery, D.R. & Buffington, J.M. (1997) Channel-reach morphology in mountain drainage basins. *Geological Society of America Bulletin* **109**, 596–611.

Montgomery, D.R. & Bolton, S.M. (2003) Hydrogeomorphic variability and river restoration. In Montgomery, D., Bolton, S., Booth, D., & Wall, L. (eds) *Restoring Puget Sound Rivers.* University of Washington Press, Seattle, pp. 39–80.

Montgomery, D.R. & Abbe, T.B. (2006) Influence of logjam-formed hard points on the formation of valley-bottom landforms in an old-growth forest valley, Queets River, Washington, USA. *Quaternary Research* **65**, 147–155.

Montgomery, D.R., Buffington, J.M., Smith, R.D., Schmidt, K.M. & Pess, G. (1995) Pool spacing in forest channels. *Water Resources Research* **31**(4), 1097–1105.

Montgomery, D.R., Abbe, T.A., Buffington, J.M., Peterson, N.P., Schmidt, K.M. & Stock, J.D. (1996a) Distribution of bedrock and alluvial channels in forested mountain drainage basins. *Nature* **381**, 588–589.

Montgomery, D.R., Buffington, J.M., Peterson, P., Scheutt-Hames, D. & Quinn T.P. (1996b) Streambed scour, egg burial depths and the influence of salmonid spawning on bed surface mobility and embryo survival. *Canadian Journal of Fisheries and Aquatic Sciences* **53**, 1061–1070.

Moore, R.D., Spittlehouse, D.L. & Story, A. (2005) Riparian microclimate and stream temperature response to forest harvesting: a review. *Journal of the American Water Resources Association* **41**(4), 813–834.

Mount, J.F. (1995) *California Rivers and Streams – The Conflict Between Fluvial Process and Land Use.* University of California Press, Berkeley, California.

Naiman, R.J., Melillo, J.M., Lock, M.A., Ford, T.E. & Reice, S.E. (1987) Longitudinal patterns of ecosystem processes and community structure in a subarctic river continuum. *Ecology* **68**, 1139–1156.

Naiman, R.J., Lonzarich, D.G., Beechie, T.J. & Ralph, S.C. (1992) General principles of classification and the assessment of conservation potential in rivers. In: Boon, P.J., Calow, P. & Petts, G.E. (eds) *River Conservation and Management.* John Wiley and Sons, New York, pp. 93–124.

Naiman, R.J., Decamps, H. & McClain, M.E. (2005) *Riparia: Ecology, Conservation, and Management of Streamside Communities.* Elsevier, New York.

Naiman, R.J., Bechtold, J.S., Beechie, T., Latterell, J.J. & Van Pelt, R. (2010) A process-based view of floodplain forest dynamics in coastal river valleys of the Pacific Northwest. *Ecosystems* **13**, 1–31.

Negishi, J.N. & Richardson, J.S. (2003) Responses of organic matter and macroinvertebrates to placements of boulder clusters in a small stream of southwestern British Columbia, Canada. *Canadian Journal of Fisheries and Aquatic Sciences* **60**, 247–258.

Negishi, J.N., Kayaba, Y. & Sagawa, S. (2008) Environmental degradation of floodplains and endangered freshwater mussels. *Civil Engineering Journal*, **50**, 44–45. (In Japanese)

Negishi, J.N., Sagawa, S., Kayaba, Y. *et al.* (2010) Using airborne scanning laser altimetry (LiDAR) to estimate surface connectivity of floodplain water bodies. *River Research and Applications* **28**, 258–267.

Negishi, J.N., Sagawa, S., Kayaba, Y., Sanada, S., Kume, M. & Miyashita, T. (2012). Mussel responses to flood pulse frequency: the importance of local habitat. *Freshwater Biology* **57**, 1500–1511.

Nitschke, C.R. (2005) Does forest harvesting emulate fire disturbance? A comparison of effects on selected attributes in coniferous-dominated headwater systems. *Forest Ecology and Management* **214**, 305–319.

Osborne, L.L. & Kovacic, D.A. (1993) Riparian vegetated buffer strips in water-quality restoration and stream management. *Freshwater Biology* **29**, 243–258.

Osterkamp, W.R. & Hupp, C.R. (2010) Fluvial processes and vegetation: Glimpses of the past, the present, and perhaps the future. *Geomorphology* **116**, 274–285.

Palmer, M.A. & Allan, J.D. (2006) Restoring rivers. *Issues in Science and Technology*, **Winter**, 40–48.

Perucca, E., Camporeale, C. & Ridolfi, L. (2007) Significance of the riparian vegetation dynamics on meandering river morphodynamics. *Water Resources Research* **43**, W03430.

Pess, G.R., Montgomery, D.R., Beechie, T.J. & Holsinger, L. (2003) Anthropogenic alterations to the biogeography of salmon in Puget Sound. In: Montgomery, D.R., Bolton, S., Booth, D.B. & Wall, L. (eds) *Restoration of Puget Sound Rivers*. University of Washington Press, Seattle, Washington, pp. 129–154.

Pitlick, J.C. & Thorne, C.R. (1987) Sediment supply, movement, and storage in an unstable gravel-bed river. In: Thorne, C.R., Bathurst, J.C. & Hey, R.D. (eds) *Sediment Transport in Gravel-Bed Rivers*. John Wiley and Sons, London, pp. 121–150.

Poff, N.L., Allan, J.D., Bain, M.B. *et al.* (1997) The natural flow regime: a paradigm for river conservation and restoration. *BioScience* **47**, 769–784.

Poff, N.L., Olden, J.D., Merritt, M.D. & Pepin, D.M. (2007) Homogenization of regional river dynamics by dams and global biodiversity implications. *Proceedings of the National Academy of Sciences* **104**(14), 5732–5737.

Pollen-Bankhead, N., Thomas, R.E., Gurnell, A.M., Liffen, T., Simon, A. & O'Hare, M.T. (2011) Quantifying the potential for flow to remove the emergent aquatic macrophyte Sparganium erectum from the margins of low-energy rivers. *Ecological Engineering* **37**, 1779–1788.

Pollock, M.M., Naiman, R.J., & Hanley, T.A. (1998) Plant species richness in riparian wetlands – a test of biodiversity theory. *Ecology* **79**, 94–10.

Pollock, M.M., Heim, M. & Werner, D. (2003) Hydrologic and geomorphic effects of beaver dams and their influence on fishes. *American Fisheries Society Symposium* **37**, 213–233.

Pollock, M.M., Beechie, T.J. & Jordan, C.E. (2007) Geomorphic changes upstream of beaver dams in Bridge Creek, in incised stream channel in the interior Columbia River basin, eastern Oregon. *Earth Surface Processes and Landforms* **32**, 1174–1185.

Pollock, M.M., Beechie, T.J. & Liermann, M. (2009) Stream temperature relationships to forest harvest in western Washington. *Journal of the American Water Resources Association* **45**, 141–156.

Poole, G.C. (2002) Fluvial landscape ecology: addressing uniqueness within the river discontinuum. *Freshwater Biology* **47**, 641–660.

Poole, G.C., O'Daniel, S.J., Jones, K.L. *et al.* (2008) Hydrologic spiraling: the role of multiple interactive flow paths in stream ecosystems. *River Research and Applications* **24**, 1018–1031.

Power, M.E. (1990) Effects of fish in river food webs. *Science* **250**, 811–814.

Prosser, I.P. & Williams, L. (1998) The effect of wildfire on runoff and erosion in native *Eucalyptus* forest. *Hydrological Processes* **12**, 251–265.

Reid, L.M., Dunne, T. & Cederholm, C.J. (1981) Application of sediment budget studies to the evaluation of logging road impact. *Journal of Hydrology (New Zealand)* **20**, 49–62.

Rice, R.A. (1982) Sedimentation in the chaparral: how do you handle unusual events? In F. J. Swanson, E., Janda, J., Dunne, T. & Swanson, D.N. (eds) *Sediment Budgets and Routing in Forested Drainage Basins*, US Forest Service General Technical Report PNW-141, Portland, Oregon, pp. 39–49.

Rice, S.P., Ferguson, R.I. & Hoey, T.B. (2006) Tributary control of physical heterogeneity and biological diversity at river confluences. *Canadian Journal of Fisheries and Aquatic Sciences* **63**, 2553–2566.

Richardson, J.S. (1991) Seasonal food limitation of detritivores in a montane stream: an experimental test. *Ecology* **72**, 873–887.

Richardson, J.S., Hoover, T.M. & Lecerf, A. (2009) Coarse particulate organic matter dynamics in small streams:

towards linking function to physical structure. *Freshwater Biology* **54**, 2116–2126.

Richardson, J.S., Zhang, Y. & Marczak, L.B. (2010) Resource subsidies across the land-freshwater interface and responses in recipient communities. *River Research and Applications* **26**, 55–66.

Richter, B.D. & Thomas, G.A. (2007) Restoring environmental flows by modifying dam operations. *Ecology and Society* **12**(1), 12.

Richter, B.D., Baumgartner, J.V., Powell, J. & Braun, D.P. (1996) A method for assessing hydrologic alteration within ecosystems. *Conservation Biology* **10**, 1163–1174.

Richter, B.D., Mathews, R., Harrison, D.L. & Wigington, R. (2003) Ecologically sustainable water management: managing river flows for ecological integrity. *Ecological Applications* **13**, 206–224.

Rood, S.B., Samuelson, G.M., Braatne, J.H., Gourley, C.R., Hughes, F.M.R. & Mahoney, J.M. (2005). Managing river flows to restore floodplain forests. *Frontiers in Ecology and the Environment* **3**: 193–201.

Rosenfeld, J., Hogan, D., Palm, D., Lundqvist, H., Nilsson, C. & Beechie, T.J. (2010) Contrasting landscape influences on sediment supply and stream restoration priorities in northern Fennoscandia (Sweden and Norway) and coastal British Columbia. *Environmental Management* **47**(1), 28–39.

Rust, B.R. (1981) Sedimentation in an arid-zone anastomosing fluvial system: Cooper Creek, central Australia. *Journal of Sedimentary Petrology* **51**, 745–755.

Sand-Jensen, K. (1998) Influence of submerged macrophytes on sediment composition and near-bed flow in lowland streams. *Freshwater Biology* **39**, 663–679.

Sanderson, B.L., Barnas, K.A. & Rub, A.M.W. (2009) Nonindigenous species of the Pacific Northwest: an overlooked risk to endangered salmon? *BioScience* **59**, 245–256.

Schindler, D.E., Scheuerell, M.D., Moore, J.W., Gende, S.M., Francis, T.B., & Palen, W.J. (2003) Pacific salmon and ecology of coastal ecosystems. *Frontiers in Ecology and the Environment* **1**, 31–37.

Schlosser, I.J. (1985) Flow regime, juvenile abundance, and the assemblage structure of stream fishes. *Ecology* **66**, 1484–1490.

Schumm, S.A. (1985) Patterns of alluvial rivers. *Annual review of Earth and Planetary Sciences* **13**, 5–27.

Sear, D.A. (1994) River restoration and geomorphology. *Aquatic Conservation* **4**, 169–177.

Sear, D.A., Armitage, P.D. & Dawson, F.H. (1999) Groundwater dominated rivers. *Hydrological Processes* **13**, 255–276.

Shafroth, P.B., Stromberg, J.C., & Patten, D.T. (2002). Riparian vegetation response to altered disturbance and stress regimes. *Ecological Applications* **12**(1), 107–123.

Sheldon, A.L. (1968) Species diversity and longitudinal succession in stream fishes. *Ecology* **49**, 193–198.

Sidle, R.C., Pierce, A.J. & O'Loughlin, C.L. (1985) *Hillslope Stability and Land Use*. Water Resources monograph Series Volume 11. American Geophysical Union, Washington, D.C.

Skidmore, P.B., Thorne, C.R., Cluer, B., Pess, G.R., Castro, J., Beechie, T.J. & Shea, C.C. (2011) *Science Base and Tools for Evaluating Stream Engineering, Management and Restoration Proposals*. US Dept. Commerce, NOAA Tech. Memo. NMFS-NWFSC-112.

Smith, J.D. (2004) The role of riparian shrubs in preventing floodplain unraveling along the Clark Fork of the Columbia River in the Deer lodge Valley, Montana. In: Bennett, S.J. & Simon, A. (eds) *Riparian Vegetation and Fluvial Geomorphology*. American Geophysical Union, Washington, DC, pp. 71–85.

Spinazola, J. (1998) *A spreadsheet notebook method to calculate rate and volume of stream depletion by wells in the Lemhi River Valley upstream from Lemhi, Idaho*. United States Bureau of Reclamation, Boise, Idaho.

Standley, L.J., Kaplan, L.A. & Smith, D. (2000) Molecular tracers of organic matter sources to surface water resources. *Environmental Science and Technology* **34**, 3124–3130.

Stanford, J.A., Lorang, M.S. & Hauer, F.R. (2005) The shifting habitat mosaic of river ecosystems. *Verhandlungen des Internationalen Verein Limnologie* **29**(1), 123–136.

Statzner, B., Fievet, E., Champagne, J., Morel, R. & Herouin, E. (2000) Crayfish as geomorphic agents and ecosystem engineers: biological behavior affects sand and gravel erosion in experimental streams. *Limnology and Oceanography* **45**, 1030–1040.

Stromberg, J.C. (2001) Restoration of riparian vegetation in the southwestern United States: importance of flow regimes and fluvial dynamism. *Journal of Arid Environments* **49**, 17–34.

Suttle, K.B., Power, M.E., Levine, J.M. & McNeely, C. (2004) How fine sediment in riverbeds impairs growth and survival of juvenile salmonids. *Ecological Applications* **14**, 969–974.

Sweeney, B.W., Bott, T.L., Jackson, J.K. *et al.* (2004) Riparian deforestation, stream narrowing, and loss of stream ecosystem services. *Proceedings of the National Academy of Sciences* **101**, 14132–14137.

Taberlet, P., Fumagalli, L., Wust-Saucy, A.G. & Cosson, J.F. (1998) Comparative phylogeography and postglacial colonization routes in Europe. *Molecular Ecology* **7**, 453–464.

Tague, C. & Grant, G.E. (2004) A geological framework for interpreting the low-flow regimes of Cascade streams, Willamette River Basin, Oregon, *Water Recourses Research* **40**, W04303.

Tague, C., Grant, G.E., Farrell, M., Choate, J. & Jefferson, A. (2008) Deep groundwater mediates streamflow response to climate change. *Climatic Change* **86**, 189–210.

Takao, A., Negishi, J.N. Nunokawa, M., Gomi, T. & Nakahara, O. (2006) Potential influences of a net-spinning caddisfly (Trichoptera: *Stenopsyche marmorata*) on stream substratum stability in heterogeneous field environments. *Journal of the North American Benthological Society* **25**, 545–555.

Temme, C., Blanchard, P., Steffen, A. *et al.* (2007) Trend, seasonal and multivariate analysis study of total gaseous mercury data from the Canadian atmospheric mercury measurement network (CAMNet). *Atmospheric Environment* **41**, 5423–5441.

Tockner, K., Scheimer, F., Baumgartner, C. *et al.* (2001) The Danube restoration project: species diversity patterns across connectivity gradients in the floodplain system. *Regulated Rivers: Research and Management* **15**, 245–258.

Trimble, S.W. & Mendel, A.C. (1995) The cow as a geomorphic agent – a critical review. *Geomorphology* **13**, 233–253.

Trush, W.J., McBain, S.M. & Leopold, L.B. (2000) Attributes of an alluvial river and their relation to water policy and management. *Proceedings of the National Academy of Sciences* **97**(22), 11858–11863.

US Fish and Wildlife Service & Hoopa Valley Tribe (1999) *Trinity River Flow Evaluation*. US Department of the Interior, Washington, DC.

Vannote, R.L., Minshall, G.W., Cummins, K.W., Sedell, J.R., & Cushing, C.E. (1980) The river continuum concept. *Canadian Journal of Fisheries and Aquatic Sciences* **37**, 130–37.

Volk, C.J., Kiffney, P.M. & Edmonds, R.L. (2003) Role of riparian red alder in the nutrient dynamics of the Olympic peninsula, Washington, USA. In: Stockner, J.G. (ed.) *Nutrients in Salmonid Ecosystems: Sustaining Production and Biodiversity*. American Fisheries Society, Symposium 34, Bethesda, Maryland, pp. 213–228.

Wallace, J.B., Eggert, S.L., Meyer, J.L. & Webster, J.R. (1999) Effects of resource limitation on a detrital-based ecosystem. *Ecological Monographs* **69**, 409–442.

Waples, R.S., Teel, D.J., Myers, J.M. & Marshall, A.R. (2004) Life-history divergence in Chinook salmon: Historic contingency and parallel evolution. *Evolution* **58**, 386–403.

Waples, R.S., Pess, G.R. & Beechie, T. (2008) Evolutionary history of Pacific salmon in dynamic environments. *Evolutionary Applications* **1**, 189–206.

Ward, J.V., Tockner, K., Arscott, D.B. & Claret, C. (2002) Riverine landscape diversity. *Freshwater Biology* **47**, 517–539.

Wetmore, S.H., Mackay, R.J. & Newbury, R.W. (1990) Characterization of the hydraulic habitat of *Brachycentrus occidentalis*, a filter-feeding caddisfly. *Journal of the North American Benthological Society* **9**, 157–169.

Wohl, E. (2000) *Mountain Rivers*. Water Resources Monograph 14. American Geophysical Union, Washington DC.

Wohl, E., Angermeier, P.L., Bledsoe, B. *et al.* (2005) River restoration. *Water Resources Research* **41**, W10301.

Wondzell, S.M. & King, J.G. (2003) Postfire erosional processes in the Pacific Northwest and Rocky Mountain regions. *Forest Ecology and Management* **178**, 75–87.

Wootton, J.T., Parker, M.S. & Power, M.E. (1996) Effects of disturbance on river food webs. *Science* **273**, 1558–1561.

Zanoni, L., Gurnell, A.M., Drake, N. & Surian, N. (2008) Island dynamics in a braided river from an analysis of historical maps and air photographs. *River Research and Applications* **24**, 1141–1159.

3 Watershed Assessments and Identification of Restoration Needs

Tim Beechie[1], George Pess[1], Sarah Morley[1], Lucy Butler[2], Peter Downs[3], Alistair Maltby[4], Peter Skidmore[5], Steve Clayton[6], Clint Muhlfeld[7] & Karrie Hanson[1]

[1]Northwest Fisheries Science Center, National Oceanic and Atmospheric Administration, USA
[2]Eden Rivers Trust, Penrith, UK
[3]University of Plymouth, UK
[4]The Rivers Trust, Stoke Climsland, UK
[5]Skidmore Restoration Consulting, Montana, USA
[6]CH2M HILL, Idaho, USA
[7]United States Geological Survey, Montana, USA

3.1 Introduction

The term 'watershed assessment' refers to a general methodology for evaluating both the condition of riverine ecosystems and the landscape and land-use factors that influence those conditions (Beechie *et al.* 2003a). The primary purposes of watershed assessments are to identify causes of habitat degradation, list necessary restoration actions, and determine which of those actions are most likely to help achieve restoration goals. From these assessments a restoration strategy can be developed, which helps focus limited restoration dollars on actions and locations that will most improve the status of biota. To accomplish these aims, watershed assessments proceed through five key steps: clearly stating the restoration goal; determining which watershed assessment components are needed to identify necessary restoration actions; conducting the assessments; interpreting and summarizing the assessments; and identifying the restoration needs (Figure 3.1). Once restoration needs are identified, the most appropriate restoration techniques can be selected (Chapter 5), restoration actions can be pri-oritized (Chapter 6), and individual projects designed (Chapter 7).

In this chapter we first describe the role of restoration goals in guiding watershed assessments, focusing on how the goal drives the selection of specific methods (Steps 1 and 2). We describe key assessment methods, grouping them into assessments of: (1) causes of habitat and biological degradation; (2) habitat alteration; and (3) changes in biota. Assessments of causes of changes to watershed processes are further organized by scale as in Chapter 2 (landscape, watershed, and reach). For each component and scale, we review a range of common assessment methods, suggest general criteria for choosing between them, and identify key products that an assessment should produce to help identify restoration actions (Step 3). We then describe basic approaches to summarizing and interpreting the results, as well as key principles that guide the identification of restoration opportunities based on the integrated results from assessments of watershed processes, habitat alteration, and biological changes (Steps 4 and 5). Finally, we present two case studies to illustrate alternative presentations of watershed assessment results and development of restoration strategies and actions.

Stream and Watershed Restoration: A Guide to Restoring Riverine Processes and Habitats, First Edition. Philip Roni and Tim Beechie.
© 2013 John Wiley & Sons, Ltd. Published 2013 by John Wiley & Sons, Ltd.

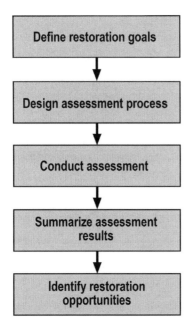

Figure 3.1 Illustration of five key steps in developing and implementing a watershed assessment to identify restoration actions.

3.2 The role of restoration goals in guiding watershed assessments

Restoration goals are statements of broad aims to be achieved, whereas restoration objectives are specific and measureable steps that must be completed to attain the goal (Barber & Taylor 1990; Tear *et al.* 2005; Ryder *et al.* 2008). Restoration goals for conservation organizations often focus on single species (e.g. Trout Unlimited, Atlantic Salmon Trust), or on broader ecosystem restoration efforts (e.g. The Nature Conservancy). However, restoration goals are perhaps most commonly driven by legislation that spurs restoration activity and funding, including laws or acts that focus on conservation or recovery of individual species, restoration or improvement of ecosystems or landscapes, or restoration of ecosystem services such as recreation or clean drinking water (Parker 1997; Beechie & Bolton 1999; Ehrenfeld 2000; Nakamura & Tockner 2004) (Table 3.1). Because these legislative mandates have divergent aims and restoration goals, criteria that determine restoration 'success' will vary depending on the legislation that drives river restoration

Table 3.1 Examples of legislation supporting river restoration.

	Legislation	Purpose
Species focused		
United States	Endangered Species Act (1973)	Provide a means whereby the ecosystems upon which endangered species and threatened species depend may be conserved
European Union	Habitats Directive (1992)	Contribute towards ensuring biodiversity through the conservation of natural habitats and of wild fauna and flora
Japan	Act on Conservation of Endangered Species of Wild Fauna and Flora (1992)	Contribute to a healthy and culturally rich life for current and future Japanese citizens by preserving endangered species of wild fauna and flora as well as the natural environment those species depend upon
Australia	Endangered Species Protection Act (1992)	Promote the recovery of species and ecological communities that are endangered or vulnerable
Ecosystem focused		
United States	Clean Water Act (1972)	Restore and maintain the chemical, physical, and biological integrity of the nation's waters
European Union	Water Framework Directive (2000)	Prevent further deterioration and protect and enhance the status of aquatic ecosystems
Japan	River Law (1964, amended 1997)	Contribute to land conservation, the development of the country and thereby maintain public security and promote public welfare, by administering rivers comprehensively . . . , and maintain the normal functions of the river water by maintaining and conserving the fluvial environment
Australia	Environment Protection and Biodiversity Conservation Act (1999)	To provide for the protection of the environment . . . and promote the conservation of biodiversity

efforts. Subsequently, the types of assessments selected for identifying causes of degradation and necessary restoration actions should be tailored to match the restoration goals, maintaining a clear linkage between legislative mandates, restoration goals, and watershed assessments. This ensures that the restoration actions identified through the watershed assessments are the key actions needed to achieve restoration goals.

3.2.1 Stating restoration goals

Restoration goal statements should include at least three main components: (1) clearly identified ecological aims; (2) a focus on addressing underlying causes of degradation; and (3) acknowledgement of social and economic constraints on restoration (Slocombe 1998; Beechie *et al.* 2008a). These elements of a restoration goal reflect not only legislative or organizational drivers for restoration (the ecological or biological aims), but also stakeholder values and the need to address root causes of degradation rather than symptoms (Table 3.2). Restoration goals that

are driven by species-focused legislation or non-profit organizations often have very narrowly stated outcomes for the species, but they usually acknowledge that restoration of ecosystems is the appropriate means to achieving that goal. By contrast, broader restoration goals might target recovery of biological integrity or diversity, and also include ecosystem restoration as a means of achieving the goal. Such goals guide restoration practitioners in choosing how to identify necessary restoration actions, how to prioritize restoration efforts, and help to inhibit drift in management objectives through time (Barber & Taylor 1990).

It is worth noting that the terms *goals* and *objectives* are often used interchangeably, but they do in fact have distinct definitions (Box 3.1). For the purposes of restoration planning, we define *goals* as broadly stated aims or desired outcomes of a restoration effort, including the main biological outcome to be achieved. *Objectives*, by contrast, are specific and measureable achievements that are necessary to reach a restoration goal. In this

Table 3.2 Examples of ecological restoration goals, indicating a variety of desired endpoints such as improved salmonid populations, increased biological diversity, and ameliorating impacts to watershed processes.

Location and legal drivers	Goal statement
Lower Columbia River, USA; for salmonids listed under the Endangered Species Act	Salmon, steelhead and bull trout are recovered to healthy, harvestable levels that will sustain productive sport, commercial, and tribal fisheries through the restoration and protection of the ecosystems upon which they depend and the implementation of supportive hatchery and harvest practices (Lower Columbia Fish Recovery Board 2010)
Tyne River, UK; for restoration of numerous species and habitat types, including species listed under the EU Habitats Directive	To conserve, protect, rehabilitate, and improve the rivers, streams, watercourses, and water impoundments of the River Tyne catchment including its estuary and adjacent coastal area (Tyne Rivers Trust, http://www.tyneriverstrust.org/home/about-the-trust/our-vision)
Northern Ireland, National Atlantic Salmon Management Strategy	To conserve, enhance, restore, and rationally manage salmon stocks in catchments throughout Northern Ireland through two Salmon Management Plans (Milner *et al.* 2008)
Kissimmee River, Florida, USA; to restore portions of the Everglades ecosystem	Reestablish an ecosystem that is capable of supporting and maintaining a balanced, integrated, and adaptive community of organisms having a species composition, diversity, and functional organization comparable to the natural habitat of the region (Kissimmee River, Florida, USA; Toth 1995)
Anacostia River, Maryland, USA; to restore ecological integrity of the Anacostia River	Protect and restore the ecological integrity of the Anacostia River and its streams to enhance aquatic diversity, increase recreational use, and provide for a quality urban fishery (Anacostia Watershed, Maryland, USA; Anacostia Watershed Restoration Partnership, http://www.anacostia.net/restoration.html)
Upper Truckee River, Nevada, USA; to reduce sediment inputs to Lake Tahoe and improve habitats for fishes	'Reduce erosion, fine sediment and nutrient loads…, restore natural river flows, floodplains and meadows . . . , expand the habitat corridor to strengthen the natural ecosystem . . . , and maintain recreational and economic benefits.' (Upper Truckee River Restoration, http:// www.restoreuppertruckee.net/)

Box 3.1

Goal – broadly stated aims or desired outcomes of a restoration effort, including the main biological outcome to be achieved.

Objectives – specific and measureable achievements that are necessary to reach a restoration goal.

context, objectives are specific restoration targets that must be attained in order to achieve the broadly stated restoration goal. For example, a simplified restoration goal might be: 'Restore local fish populations by restoring watershed processes and habitats that sustain them.' Objectives needed to achieve this goal are identified by the analyses of watershed processes, habitats, and biota, and might include:

1. increase fish rearing habitat capacity by reconnecting x% of historical floodplain habitats;

2. reduce stream temperatures in x km of key rearing habitats by restoring riparian vegetation; and

3. reduce erosion and sediment delivery from agricultural lands by x%.

The watershed assessment also identifies specific restoration actions (or at least potential restoration actions) that are necessary to achieve the objectives, and therefore to achieve the restoration goal.

3.2.2 Designing the watershed assessment to reflect restoration goals and local geography

Once the restoration goal is stated, the assessment process can be constructed to identify specific objectives (i.e. restoration actions) necessary to achieve that goal. While legislative mandates vary in environmental purposes, most are based on concepts such as ecosystem conservation, biodiversity, or physical, chemical, and biological integrity. Moreover, the same watershed processes drive riverine ecosystems regardless of geographic setting. A common set of watershed-scale and reach-scale processes should therefore be assessed regardless of whether the goal is species-focused, water-quality-focused or ecosystem-focused (Table 3.3). However, specific assessment methods for each process vary depending on the geographic setting of the watershed (i.e. physiography and common land uses). For example, erosion processes are included in most watershed assessments, but assessments to identify land-use effects on sediment supply in low-relief agricultural areas might focus on agriculture effects on surface erosion processes, whereas assessments in steep forested

mountains might focus on forestry effects on landslides. Similarly, runoff processes and stream flows are a part of most assessments, but areas with extensive irrigation needs might focus assessments on the effects of water withdrawals on low flows, whereas urban watersheds might focus analyses on the effects of impervious surfaces on flood flows. Hence, both the focus and methodology for each process assessment should be tailored to the local geography.

While legal and organizational drivers of restoration efforts have relatively little influence on the suite of watershed-scale and reach-scale processes included in assessments, the drivers strongly influence habitat and biological evaluations. Where restoration efforts are driven by species-focused goals, assessments commonly focus on changes in habitats that are important to the species, as well as on status of the species and key restoration actions that are needed to improve its status (e.g. Harwell *et al.* 1999; McElhany *et al.* 2000; Tear *et al.* 2005). By contrast, restoration efforts that focus more broadly on ecological status or ecosystem health might include assessments that target key water quality attributes, as well as on multimetric indicators of ecological health (Karr 1991; Wright *et al.* 2000). Beyond legislative or organizational needs, assessment methods for habitats and biota are also partly driven by geographic setting, as species and habitats vary considerably even over short distances. Hence, the focus and methods for assessing habitats and biota should be tailored to both restoration goals and local geography.

In constructing the watershed assessment, it is important that each component is focused on one or more cause–effect linkages between watershed processes, habitat conditions, and biota in riverine ecosystems (Karr 2006; Beechie *et al.* 2008a). We organize these cause–effect linkages by the two main questions that a watershed assessment is intended to answer (Figure 3.2). The first question is focused on identifying causes of habitat change (the altered processes that have led to changes in habitat) (Beechie *et al.* 2003a, b), and the second question addresses how habitat changes have affected biota. Answering the second question includes assessing habitat loss or degradation and estimating the effect of those changes on one or more species. Answering these two basic questions through watershed assessment is key to identifying which kinds of restoration actions are most needed in a watershed, and to understanding the potential biological benefits of various restoration options (Beechie *et al.* 2008a). Finally, a third question focuses on identifying constraints on restoration options, as

Table 3.3 Common assessment components for identifying restoration actions.

Process group	Description
Watershed-scale processes	
Runoff and stream flow	Assess effects of changing land cover on runoff and stream flow, such as effects of increased impervious surface area or forest removal on flood flows
Erosion and sediment supply	Assess effects of land uses on erosion processes, such as logging effects on landslides or tilling and grazing effects on surface erosion
Nutrient delivery	Assess sources of increased or decreased nutrient supply to streams, such as fertilizer applications in agricultural lands or reduction of salmon populations and loss of marine-derived nutrients
Reach-scale processes	
Riparian processes	Assess riparian vegetation conditions and effects on stream shading, wood supply to streams, sediment retention, root reinforcement of banks and detritus inputs to riverine ecosystems
Stream flow and flood storage	Assess direct alteration of stream flow by dams or water diversions
Sediment transport and storage	Assess direct alteration of sediment transport and storage by dams, dredging, mining, or other in-channel activities
Channel, floodplain, and habitat dynamics	Assess loss of floodplain habitats by levees, dikes, and revetments (bank armoring)
Habitat alteration	
Habitat type and quantity	Evaluate condition of habitat features relative to expected natural conditions, or relative to best attainable reference condition if full restoration is not possible; map blockages to fish migration
Water quality	Evaluate human impacts to important water-quality attributes such as temperature, pollutants, and nutrients
Changes to biota	
Single species	Evaluate status of important populations or species, especially for restoration efforts driven by species-focused legislation
Multiple species	Evaluate condition of communities or assemblages, especially for restoration efforts driven by ecosystem-focused legislation
Non-native species	Evaluate presence, abundance and potential impacts of non-native species

many restoration efforts are located in heavily managed watersheds or parts of watersheds (such as urban or agricultural zones) and it may not be possible to restore stream or watershed functions to their natural potential (Geist & Galatowitsch 1999). Together, the answers to these questions guide identification of the kinds of restoration actions that are needed to achieve a restoration goal, as well as constraints that limit the degree to which watershed processes and habitats can be restored (Poff & Ward 1990; Ebersole & Liss 1997; Frissell *et al.* 1997; Pess *et al.* 2003a).

To illustrate how geography, legislation and key questions guide the design of watershed assessments, we briefly summarize the development of assessment methods for two contrasting watersheds: one from the USA (Skagit River) and one from the UK (River Eden). The geography of the Skagit River basin of Washington State, USA (drainage area of 8270 km^2) is dominated by steep forested mountains with wide mainstem river valleys and a large delta in the lower reaches of the basin. Upland erosion is dominated by landsliding, and forestry activities are the primary influence on landsliding. A sediment budget approach was therefore used to analyze changes in landslide rates and estimate increases in sediment supply (Table 3.4). By contrast, the lower river floodplains and delta are dominated by agriculture, and the primary land-use impact has been loss of floodplain and delta habitats due to levees and conversion to agriculture or urban land uses. Hence, reach-scale process assessments were focused on losses of floodplain habitats

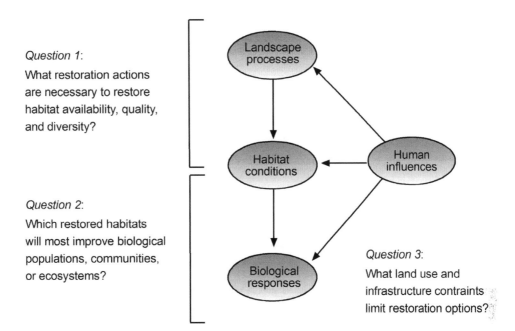

Figure 3.2 Conceptual diagram of process linkages between landscape processes, stream habitats, and biota, and the key questions to be answered by restoration assessments. Modified from Beechie *et al.* 2003, 2008.

and changes in riparian condition. Because there are several salmon species of economic interest in the basin, the Skagit Watershed Council recognized that a restoration goal of 'restoring and maintaining landscape processes that formed and sustained the habitats to which salmonid stocks are adapted' would help guide restoration actions toward those that support habitats for all species (Beechie & Bolton 1999). Habitat assessments were therefore designed to quantify changes in habitats used by all of the local salmonid species. However, Chinook salmon (*Oncorhynchus tshawytscha*) and steelhead (*O. mykiss*) are currently listed under the United States Endangered Species Act (Federal Register 1999, 2007), and most restoration funding is currently for restoring Chinook salmon. Therefore, the biological analysis focused on identifying restoration actions that will most benefit Chinook salmon.

By contrast, the 2300 km² Eden River basin in the UK is predominantly low rolling hills, and more than 90% of the landscape is used for mixed livestock and arable farming, dairy farming, or upland sheep farming. The Eden Rivers Trust recognized that root causes of environmental degradation included both distributed watershed processes (e.g. diffuse pollution from agriculture) and reach-scale riparian processes (e.g. impacts of grazing livestock), and that identifying sustainable solutions to these root causes required a range of innovative assessments based upon remote sensing, aerial photography, Geographical Information Systems (GIS), environmental modeling, and ecological surveying. The erosion and nutrient assessments focus on identifying erosion-prone fields and pastures that are hydrologically connected to streams (i.e. eroded sediment was likely delivered to the stream) (Table 3.4), as agricultural runoff is considered a likely cause of siltation, nutrient enrichment, and fish mortality events across the watershed. Riparian assessments are focused on impacts of grazing livestock on bank erosion, as well as on identifying areas where riparian restoration might reduce delivery of sediment or nutrients to streams. The European Union Habitats Directive protects the habitats of numerous species in the River Eden, and the Trust chose to focus on age 0+ Atlantic salmon (*Salmo salar*) and brown trout (*S. Trutta*) as the best indicators of local habitat conditions and water quality for the broad range of species present. Hence, the habitat and biological assessments were focused mainly on sediment characteristics that influence Atlantic salmon and brown trout spawning and survival of eggs, as well as on the effects of water quality on survival of young juveniles. Finally, statistical relationships

Table 3.4 Examples of selecting appropriate assessment components and methods for the Skagit River basin, USA and River Eden, UK.

	Skagit River basin, USA	River Eden, UK
Watershed-scale processes		
Runoff and stream flow	Assess effects of increased impervious surface area or forest removal on flood flows	Assess connectivity of potential sources of fine sediment to pinpoint key areas for restoration
Erosion and sediment supply	Assess effects of logging and road construction on landslides	Assess effects of land use on surface erosion and sediment supply
Nutrient delivery	Not assessed; streams are largely oligotrophic	Assess risk of increased nutrient supply to streams via connectivity analysis
Reach-scale processes		
Riparian processes	Assess effects of logging and land conversion on riparian forest structure using satellite data and aerial photography	Assess effects of grazing on riparian vegetation conditions using high-resolution (20 cm) photography
Channel, floodplain, and habitat dynamics	Assess loss of floodplain habitats by levee construction using historical maps and field data	Not assessed; focus was on salmonid spawning areas (i.e. headwaters and sub-basins where channel is mostly confined with few floodplains)
Habitat alteration		
Habitat type and quantity	Assess changes in habitat types relevant to key salmon species (e.g. changes in tributary pools, beaver pond areas, mainstem habitat areas)	Assess channel width and substrate characteristics in relation to grazing and bank erosion from high-resolution aerial photography; assess channel gradient and associated physical biotope from 5 m DTM
Water quality	Not addressed at the basin scale; water quality issues are relatively localized in the lower basin.	Not directly assessed; cumulative risk of high nutrient load was estimated by reach from the connectivity analysis above
Changes to biota		
Single species	Use life-cycle models and a variety of habitat specific studies to identify habitat changes most important to Chinook salmon	Use rapid electrofishing method to evaluate status of age 0+ Atlantic salmon and brown trout by reach, as an index of local habitat quality

between land uses, reach-scale conditions, and species abundances were used to identify restoration areas and actions that were most important to the focal species.

The contrasting assessment methods for watershed- and reach-scale processes between these two basins are primarily a function of how local geography determines the dominant processes active in a watershed (e.g. surface erosion versus landsliding), and also how varying land-use patterns influence those underlying processes. However, contrasting assessment methods for habitats and biota are primarily a function of key species in the watershed and legislative mandates for protection or restoration of those species and their habitats. Nevertheless, the same suite of processes is considered for both river

basins, and all processes that are relevant to the local geography and species are evaluated in the watershed assessment. Later in this chapter we revisit these two river basins to illustrate assessment methods for erosion processes and riparian conditions in more detail (Section 3.3), and how to summarize assessments and use those results to develop restoration strategies (Section 3.8).

3.3 Assessing causes of habitat and biological degradation

We summarize assessment methods for each of the key watershed processes, focusing on those processes that: (1)

drive riverine ecosystems; (2) are commonly influenced by land and water uses; and (3) are targets of restoration activity (Table 3.3). For each process we briefly review common assessment methods, discuss general criteria for choosing appropriate methods depending on the geographic setting and restoration goals, and describe the products each assessment should produce to help identify necessary restoration actions. More detailed descriptions of assessment procedures are described in other texts and references, which we identify within each section (e.g. Hendry & Cragg-Hine 1996; Kondolf & Piegay 2003; Hauer & Lamberti 2006).

We note here that we address longitudinal, lateral and vertical connectivity under each process assessment as appropriate. For example, longitudinal connectivity can refer to continuity of upstream movements of fishes or downstream fluxes of sediment and water, and each of these functions is addressed under sections on assessments of habitats, sediment routing, and stream flow. Lateral connectivity is addressed primarily under assessment of floodplain processes, and vertical connectivity is addressed under fluvial process assessments.

3.3.1 Use of landscape and river classification to understand the watershed template

A watershed's geology, topography and climate (the watershed template) control the limits of restoration potential, whereas watershed- and reach-scale processes control current habitat and biological conditions within those limits (Chapter 2). Large-scale features such as topography, network structure and valley types tend to be relatively stable over human time frames, and these features define the range of potential conditions that may be expressed within reaches (Naiman *et al.* 1992; Newson *et al.* 1998; Cullum *et al.* 2008; Figure 3.3). By contrast, small-scale features such as reach types or habitat unit types can change as a result of either natural or human disturbances, and they reflect the current state of habitat in a particular location (Frissell *et al.* 1986). Hierarchical landscape and river classification systems are often used to describe these limits and map current conditions, and these classification systems are particularly well suited to identifying: (1) landscape or valley features that define the range of possible conditions; and (2) reach-scale features that describe the current condition within that range (Naiman *et al.* 1992; Brierley *et al.* 2008).

Classification of large-scale features defines suites of potential channel forms based on geology, topography, or valley form (Montgomery 1999), or based on a combination of physical and ecological attributes (e.g. Omernik 1986). These landscape units are commonly used to identify portions of landscapes in which channel characteristics and processes are generally similar (Montgomery & Bolton 2003). Nested within landscape units are valley types, which are primarily a function of formative processes (e.g. glacial or fluvial), floodplain width, and valley slope (Cupp 1989; Naiman *et al.* 1992). Valley types set fundamental limits on channel patterns and habitats that can form within valley segments. Each of these classification systems defines an aspect of restoration potential that is independent of human impacts (or natural disturbances), and helps to describe potential geomorphic, aquatic habitat or biological conditions that may exist within each reach.

Reach-scale classification systems are generally more sensitive to natural and human disturbances, and therefore the classification of a reach can change through time if there are major changes in stream flow, sediment supply, or riparian vegetation. Common classification schemes range from simple (a few process-based classes) to complex (more than 20 reach types). Simple systems include the Montgomery & Buffington (1997) system for small streams and the large river channel types proposed by Leopold & Wolman (1957) (see Figure 2.5 in Chapter 2), while complex systems include those such as the Rosgen (1994) system and the river styles framework in Australia (Brierley *et al.* 2002; Table 3.5). Most classification systems for small streams distinguish channel types according to sinuosity, bed material, slope, width or depth, and bedform or planform (e.g. Naiman *et al.* 1992; Rosgen 1994; Montgomery & Buffington 1997). These systems commonly identify reach types that are identified in the field – such as pool-riffle, cascade or plane-bed channels – that reflect not only channel slope and size, but also sediment and wood supply to the reach (Montgomery & Buffington 1997). Classification of large rivers is usually based on channel pattern, which reflects variations in slope, discharge, sediment supply, and floodplain vegetation (Nanson & Hickin 1986; Millar 2000; Church 2002; Eaton *et al.* 2004). The simplest classification system for floodplain rivers identified three channel patterns: straight, meandering and braided (Leopold & Wolman 1957). Since then researchers have developed a wide variety of classification systems for alluvial channel patterns (e.g. Kellerhals *et al.* 1976; Schumm 1985; Knighton & Nanson 1993; Ward *et al.* 2001), resulting in overlapping and conflicting terminology (Beechie *et al.* 2006). Nevertheless, the underlying drivers of channel pattern

Immutable

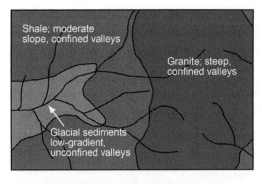

Landscape template

The landscape template defines topographic and erosional controls on suites of valley segments and stream reaches.

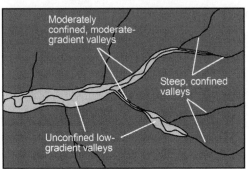

Valley form

Valley segments define the range of potential channel types that can be expressed within each valley segment, based largely on slope and confinement.

Mutable

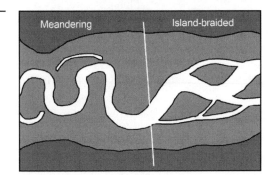

Reach type

Reach types (or channel patterns) reflect local slope, sediment supply, and hydrology. They can change as land or water uses alter sediment supply or stream flow regime.

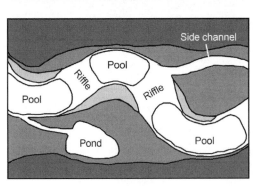

Habitat units

Reach types (or channel patterns) reflect local slope, sediment supply, and hydrology. They can change as land or water uses alter sediment supply or stream flow regime.

Figure 3.3 Hierarchical classification includes higher-level classes that are unchanged by human impacts (e.g. litho-topographic units or valley forms) whereas lower-level classes respond to human impacts (e.g. reach types or habitat units).

Watershed Assessments and Identification of Restoration Needs 59

Table 3.5 Examples of common classification systems for varying spatial scales of assessment.

Scale	Classification system	Description
Landscape	Litho-topographic units	Based on geology, topography and valley form. Identifies areas of common physiography (Montgomery 1999)
	Ecoregions	Based on a combination of physical and ecological attributes. Identifies areas with common physical and ecological attributes (Omernik 1986)
Valley segment	Valley types	Based on valley slope, shape (e.g. U-shaped or V-shaped) and confinement (floodplain width relative to channel width) (Cupp 1989; Naiman *et al.* 1992)
Floodplain	River channel pattern	Based on arrangement of channels on the floodplain (e.g. braided, meandering, straight), and includes a wide range of terminologies and levels of detail (Leopold & Wolman 1957; Schumm 1985; Knighton & Nanson 1993;Ward *et al.* 2001; Beechie *et al.* 2006)
Stream reach	Stream type	Based on channel slope, sinuosity, entrenchment, width-depth ratio, bed material size, and number of channels (Rosgen 1994)
	Channel type	Based on bed forms (cascade, step-pool, pools, riffles, dunes), and wood forcing of pools (Montgomery & Buffington 1997)
Hierarchical	Hierarchical	Classifies streams and habitats at five scales: watershed, segment, reach, habitat unit, and microhabitat unit (Frissell *et al.* 1986)
	River styles	Based on valley setting (ranging from confined to alluvial), number and form of channels, and geomorphic units within a valley segment (Brierley *et al.* 2002)

are well supported (Nanson & Hickin 1986; Millar 2000; Church 2002), and a simple classification of channel patterns is a useful tool for understanding the range of habitat types that each reach can support as well as for predicting responses to changes in discharge, sediment supply, and riparian vegetation.

Choosing a classification system that is useful for a particular watershed assessment can be guided by five key principles (Brierley & Fryirs 2005; Cullum *et al.* 2008; Table 3.6). First, the selected system should be firmly grounded in hierarchy theory, recognizing that higher-level controls influence lower levels (see also Chapter 2). Second, it should reflect an understanding that processes outside a reach strongly influence conditions within that reach (Montgomery & Buffington 1997; Montgomery 1999). Third, the system should be tailored to local restoration goals and suited to the geomorphic, ecological, and historical context of the region (Cullum *et al.* 2008; Walter & Merritts 2008). Fourth, it should be ecologically relevant and stratify important environmental and biological variables. Finally, a classification system should explicitly recognize uncertainties associated with both the classification system and predictions of ecosystem responses to changes in driving processes (Kondolf *et al.*

Table 3.6 Guiding criteria for selecting or designing an effective classification system for analyzing geological and topographic controls on stream and river habitats (adapted from Kondolf *et al.* 2003).

Criterion	Description
Hierarchical	Recognizes that lower-level conditions are controlled by higher-level processes
Process-based	Based on, and predictive of, responses to changes in watershed or reach-level processes
Locally tailored	Suited to the local geomorphic, ecological, and human historical context
Ecologically relevant	Relevant to local ecological or environmental restoration goals and objectives
Recognizes uncertainty	Explicitly recognizes uncertainties in both classification of sites and predicted responses to process changes

2003). These principles will help ensure that the chosen classification system fits the restoration assessment needs, recognizes regionally specific physical and biological processes, and understands that human history imparts legacy effects that may confound our perception of reference conditions.

Ultimately, the selected stream classification system should help restoration planners understand and describe the underlying restoration potential of each reach in a river network (Brierley *et al.* 2002). To do this, a stream classification system should normally have at least two scales of classification: a landscape scale and a valley segment scale. The landscape scale of classification identifies areas of the landscape within which watershed processes operate relatively similarly, and the valley scale more specifically defines the range of potential channel and habitat types that each reach can support based on valley form, slope, and floodplain width. Note that the

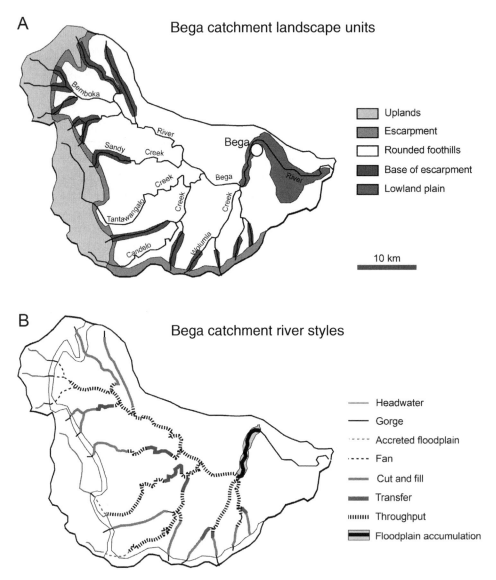

Figure 3.4 Example of (A) landscape- and (B) valley-scale classification in the Bega catchment, Australia. Based on Brierley *et al.* (2002).

classification system need not describe the current condition of the channel or habitats in detail, as the fluvial geomorphology and habitat assessments will address these later (see Sections 3.3.3 and 3.4). With the landscape scale of classification, areas are identified to have broadly similar rates and types of watershed-scale processes, whereas the valley scale of classification should at a minimum help to illustrate the restoration potential of each reach.

For example, landscape classification of the Skagit River basin identified four lithologic and topographic regions with differing erosion processes and background (i.e. natural) rates of sediment delivery. Each region is prone to different modes and rates of landsliding, has its own background erosion rate, and requires unique suites of restoration actions to ameliorate land-use effects on sediment supply. Similarly, landscape-scale classification in the Bega Catchment, Australia shows key landscape features related to erosion rates and areas of floodplain formation (Figure 3.4). For valley-scale classification, Brierley *et al.* (2002) used the River Styles framework to classify stream reaches into specific geomorphic settings that indicate potential responses of each reach, so that restoration planners can begin to develop and clearly articulate a restoration vision for each part of a river basin (Figure 3.4).

3.3.2 Assessing watershed-scale (non-point) processes

Assessing watershed processes or functions that cause habitat or ecosystem degradation focuses on understanding how those processes have been altered from their background rates to identify where and what kinds of restoration actions are needed to reduce land-use effects (Beechie *et al.* 2008a). Watershed-scale processes include erosion and sediment supply, runoff and stream flow, and inputs of nutrients or pesticides. Quantifying how these processes have been altered and identifying restoration actions that are required for their recovery requires two different kinds of assessments. First, process assessments commonly use remotely sensed data to identify the degree to which a process has been altered by land use, and where in the watershed these changes have occurred. Second, field inventories locate specific restoration actions that are needed for recovery, focusing on key causes of disrupted processes identified in the process assessment.

Changes to watershed processes are typically assessed using a budgeting approach, which can be stated in equation form as

$$\Delta S = I - O$$

where ΔS is change in storage, I is input and O is output (Reid & Dunne 1995). In general, S is the stream condition for any parameter (e.g. the amount of sediment, nutrients, or a pollutant in a stream reach), and quantifying changes in inputs or outputs indicates how land uses have altered the stream ecosystem. Because the watershed-scale assessments are most often concerned with identifying changes in inputs (I), the budget often focuses only on quantifying how inputs have been altered by land uses, which is called a partial sediment budget. Methods and terminology for calculating changes in inputs vary depending on which processes are being evaluated. Notably, the term 'budget' is not always explicitly stated, especially with runoff or erosion models that simulate basic processes contributing to stream flow volumes or sediment loads. Nevertheless, the budgeting concept underlies most of these models. The following three sections describe assessment methods for sediment supply, stream flow and nutrients or pollutants.

3.3.2.1 Sediment supply: Erosion and delivery to streams

The partial sediment budget aims to quantify changes in sediment supply due to land use, and commonly focuses on two main processes: surface erosion and landsliding. The analysis is conducted by: (1) quantifying erosion rates from aerial photography or maps (mainly used for landsliding); and (2) extrapolation of limited empirical data or modeling (mainly used for surface erosion) (Reid & Dunne 1995; Beechie *et al.* 2003a). Both approaches focus on identifying where sediment supplies to streams have been significantly altered, and can help focus restoration efforts on areas that contribute large amounts of sediment. The primary output of these analyses should be a map of changes in sediment supply due to land use, where map polygons express a percent increase in sediment supply over a background rate or an absolute increase over background rate (e.g. tons of sediment delivered) (Reid & Dunne 1995). Estimating change from background rate is an important part of the calculation, as background rates of sediment supply vary with landform, slope, soil type, vegetation cover, and other factors (Reid 1998).

In mountainous areas, a partial sediment budget is often constructed by conducting landslide inventories from historical aerial photographs and estimating contributions of fine sediments from road surface erosion (e.g. Reid & Dunne 1995; Paulson 1997). In these inventories, landslides are enumerated and measured on each aerial photograph, and the volume of each landslide is calculated based on a relationship of photo-measured area to field-measured volume for a subset of the recent landslides (Reid & Dunne 1995). Land use is also recorded for each landslide (e.g. clear-cut, road, or mature forest), allowing estimation of the aggregate impact of land use on the sediment input, as well as identification of the land uses most responsible for changes in sediment supply (Figure 3.5). Estimates of surface erosion from unpaved roads can be based on characteristics of road surfaces, cut and fill slopes, and precipitation (e.g. Ketcheson *et al.* 1999). The two assessment components (landslides and surface erosion) can be summarized by sub-basins to identify areas where sediment supplies have been most altered and where modified timber harvest practices or road modifications can have the greatest impact on ecosystem recovery (e.g. Beechie *et al.* 2003a; Benda *et al.* 2007). For example, landslide data for the Skagit River basin, USA indicated that natural erosion rates varied with lithology and logging roads had the largest influence on sediment supply rates for all parts of the basin (Figure 3.5). Extrapolating these data across the basin based on geology, forest ages, and road locations showed that erosion rates were likely more than double the natural background rate in 13 sub-basins of the Skagit River, and

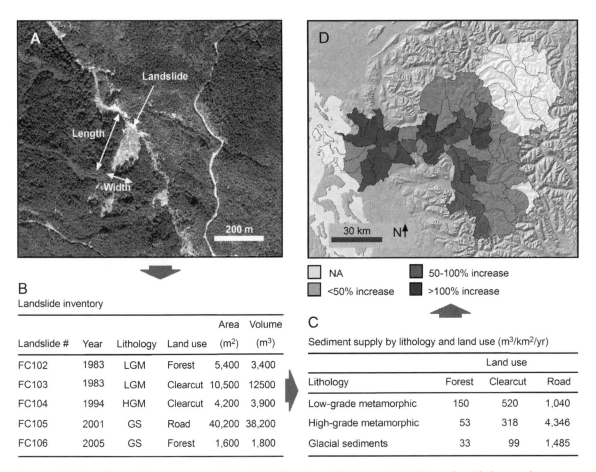

Figure 3.5 (A) A sediment budget constructed for the Skagit River basin, Washington State, USA used aerial photography to identify and measure landslides in 10 sub-basins with varying lithology and land use. (B) Landslide data including lithology, land use and year of photograph were used to calculate (C) sediment production rates from all combinations of lithology and land use. (D) These data were then used in GIS to map sub-basins with significant increases in sediment supply relative to the forested background rate. Modified from Beechie *et al.* 2003a.

that restoration efforts should be focused on road removal or reconstruction in those sub-basins.

While the sediment budget for forest lands indicates where sediment supply has increased and where restoration or rehabilitation may be necessary, it does not identify specific restoration actions needed to reduce sediment supplies in impaired sub-basins. Three types of inventory can be used to identify specific restoration actions. First, mapping of landslide hazard areas identifies areas that are particularly prone to landsliding, and high-hazard areas can be removed from timber harvest or development plans to allow recovery of sediment supply rates (Beechie et al. 2003a). Second, inventory of road landslide hazards in forested mountain areas identifies specific road segments that can be rehabilitated to reduce landslide rates. Road inventories should identify segments of road that are at risk of failure (e.g. Rennison 1998), as well as specific stream crossings, cross drains, or fills that are likely to fail. Each potential failure site can be listed and prioritized based on potential impact to the stream ecosystem. Third, the road surface erosion assessment can identify segments of road that produce large amounts of fine sediment, as well as important mitigation actions such as improving road surfacing or redirecting runoff away from streams.

Analysis of surface erosion from croplands or grazing lands commonly relies on measured erosion rates from soils with varying vegetation cover, which are incorporated into sediment budgets or surface erosion models to estimate changes in sediment supply (Dunne & Leopold 1978; Renard et al. 1991, 1997). Bare soils erode at rates as much as 10 times higher than soils with cover crops, and erosion rate varies with soil type, rainfall, slope, cover type and other factors (see overview of processes and rates in Dunne & Leopold 1978, Toy et al. 2002). Erosion rates are commonly estimated using one of several erosion models, many of which are based on the Universal Soil Loss Equation (Wischmeier & Smith 1965, 1978) or Revised Universal Soil Loss Equation (Renard et al. 1991, 1997). Dunne & Leopold (1978) provides a good overview of the equation and its application along with charts and tables for estimating certain parameters in the equation, and the original handbooks can be consulted for greater detail on the methods. Examples of models commonly used include Water Erosion Prediction Project (WEPP; Nearing et al. 1989) and European Soil Erosion Model (EUROSEM; Morgan et al. 1998). However, the spatial resolution of models (i.e. the grid cell size of digital elevation data representing topography) varies considerably, and some studies have found that grid reso-

lution between 30 m and 90 m yields the most accurate erosion estimates (e.g. Rojas et al. 2008).

Other models, such as the Sensitive Catchment Integrated Modeling and Analysis Platform (SCIMAP; www.scimap.org.uk), focus on mapping the relative likelihood of sediment delivery to streams based on the concept of hydrological connectivity (Reaney et al. 2011). It identifies fields and locations where land uses with high erosion rates have a high likelihood of being hydrologically connected to the river and therefore a high likelihood of sediment delivery (Figure 3.6). Each of these methods allows more effective targeting of mitigation measures to reduce fine sediment delivery by identifying areas or sub-watersheds that have the largest increases in erosion rates. To identify and list specific restoration actions, these assessments may need follow-up field or aerial photograph inventories to identify specific fields that can be managed with cover crops or no-till agriculture, or areas where intensive livestock activity can be modified or removed to reduce erosion and delivery of sediment to streams (see also Chapter 5).

3.3.2.2 Hydrology: Runoff and stream flow

At the watershed scale, hydrologic analyses focus on understanding how land uses have altered stream flows via alterations in runoff processes such as interception, evapotranspiration, infiltration or routing through soils, or how dams have altered stream flows through flow regulation or abstraction (also termed withdrawal). These analyses are distinctly different from reach-scale *hydraulic* analyses, which focus on local flow characteristics such as channel roughness, velocity profiles, and turbulence (see Chapter 7). As with erosion rates, changes in runoff and stream flow due to land uses can be assessed using both hydrologic models and empirical methods (Table 3.7). Most hydrologic models are spatially explicit, meaning they use physical features and land covers for each grid cell in a watershed to estimate runoff from each cell and to integrate those grid-cell estimates into stream flow estimates. Model outputs may include maps indicating where runoff rates are highest, as well as hourly or daily stream flow data. Empirical relationships between land cover attributes and stream flow characteristics can also be used to estimate changes in stream flow due to land cover change, but with less ability to identify detailed flow response to land cover change or mitigation actions. Assessments of changes in stream flow regimes as a result of flow regulation or water abstraction most commonly focus on key flow attributes calculated from measured

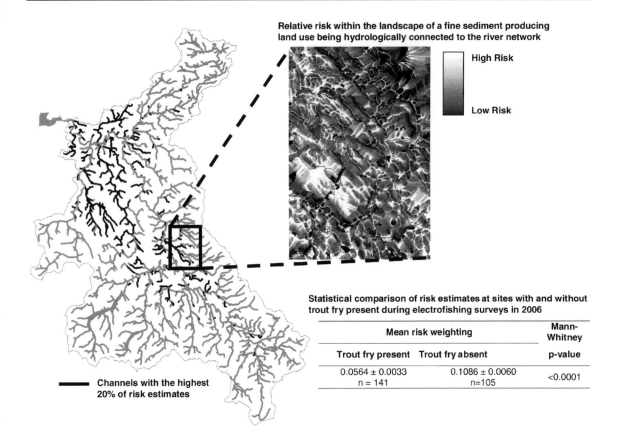

Relative risk within the landscape of a fine sediment producing
land use being hydrologically connected to the river network

High Risk

Low Risk

━━━━ Channels with the highest
20% of risk estimates

Statistical comparison of risk estimates at sites with and without
trout fry present during electrofishing surveys in 2006

	Mean risk weighting		Mann-Whitney
	Trout fry present	Trout fry absent	p-value
	0.0564 ± 0.0033 n = 141	0.1086 ± 0.0060 n=105	<0.0001

Figure 3.6 A risk-based modeling framework, SCIMAP (Sensitive Catchment Integrated Modeling and Analysis Platform) (www.scimap.org.uk), was used to assess the risk of fine sediment from agricultural sources being delivered to the Eden River system, Cumbria, UK. Sites with trout fry present were found to have a significantly lower risk estimate than those where trout were absent.

Table 3.7 Examples of methods for analyzing effects of land use and dams on stream flows.

Method	Description and citations
Hydrologic models	
Water Flow and Balance Simulation Model (WaSiM-ETH), Variable Infiltration Capacity (VIC) Model	Process-based runoff models for analyzing effects of land-use change and flow regulation by dams in a river network; grid cell resolution of 1 km² (Liang *et al.* 1994; Lohmann *et al.* 1998; Mattheussen *et al.* 2000; Verbunt *et al.* 2005; Kraus *et al.* 2007)
Distributed Soil Hydrology and Vegetation Model (DHSVM)	Process-based runoff model for analyzing effects of land-use change in a river network; grid cell resolution of 30 × 30 m (Wigmosta *et al.* 1994; Bowling *et al.* 2000; Cuo *et al.* 2009)
Analysis of streamflow data	
Comparison of individual flow metrics	Correlation of land-use metrics to peak flows provide a means of assessing forestry or urbanization effects on peak flows (Booth & Jackson 1997; Bartz *et al.* 2006)
Index of Hydrologic Alteration (IHA)	IHA software analyzes streamflow data for a wide range of flow metrics, and can compare effects of varying streamflow management options on these metrics (Richter *et al.* 1996; Poff *et al.* 2010); designed for analysis of effects of dams on flow regimes

stream flow data, and several empirical methods can be used to evaluate the degree of alteration. These methods compare stream flow data under the current management regime to historical or reference stream flow data, and indicate the degree to which water management has affected stream flows and ecological functions of those flows.

Hydrologic models simulate basic hydrologic processes including infiltration, interception, snow accumulation and melt, and routing through soils, and can therefore simulate how changes in forest cover, impervious surfaces, road densities, or climate change might affect stream flows. Models are typically process-based, and use both the water and heat energy balances to estimate changes in runoff and stream flow as a function of land-use change (Figure 3.7). The heat–energy balance is important for determining when and where precipitation falls as snow, and when snowmelt becomes stream runoff. Two coarse-resolution hydrologic models that are suitable for analyzing land-use effects on stream flows are the

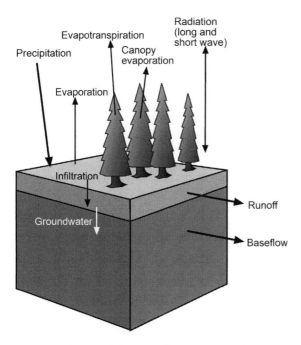

Figure 3.7 Conceptual diagram of key components in many hydrologic models, including the water and energy balances required to predict storage and routing of water to streams and rivers. The energy balance is especially important for accurate prediction of the form of precipitation (rain or snow), and storage and melting of snow that contribute to runoff and stream flow.

Variable Infiltration Capacity model (VIC; Liang *et al.* 1994; Lohmann *et al.* 1998; Mattheussen *et al.* 2000) and the Water Flow and Balance Simulation Model (WaSim-ETH; Verbunt *et al.* 2005; Krause *et al.* 2007; Bormann & Elfert 2010). Each model is grid based, with runoff processes modeled in each cell (1 km × 1 km) to estimate runoff and subsurface flows that are routed downslope and into stream channels. Because these models use large grid cells, they are most suitable for analyzing large watersheds or regions (i.e. >10,000 km²) where lack of fine resolution on physical features or land covers does not strongly influence the model result. The Distributed Hydrology Soils and Vegetation Model (DHSVM) is similar in its process-based structure, but DHSVM models stream flow and routing at a grid cell resolution of 30 m × 30 m and can be used to more accurately model land-use effects on stream flows in watersheds less than 10,000 km² in area (Wigmosta *et al.* 1994; Bowling *et al.* 2000; Cuo *et al.* 2009). Other basin-scale hydrologic models such as HSPF (Hydrology Simulation Program Fortran) and HEC-1 (Hydrologic Engineering Center, US Army Corps of Engineers) have similar resolutions and capabilities to DHSVM. While these detailed models can be used to identify areas or sub-watersheds in which land uses have substantially altered stream flow, they are usually more complicated and costly than necessary and empirical methods will suffice. Nevertheless, these detailed models may be well suited for evaluating the potential effects of various restoration alternatives in smaller basins. For example, HSPF was used to analyze changes in peak flow associated with urbanization in Des Moines Creek, Washington State, USA as well as to evaluate potential mitigation options for stormwater runoff (Figure 3.8).

Empirical methods may use landscape indicators that are known to alter peak flows as an index of peak flow change (e.g. Booth & Jackson 1997; Beschta *et al.* 2000), or they may use empirical relationships between those indicators and peak flows to project changes in peak flow hydrographs as a result of land cover changes (e.g. Booth & Jackson 1997; Bartz *et al.* 2006). Such methods are much easier to use than complex models, and provide a good general characterization of which sub-basins may have greater or lesser degrees of hydrological alteration by land uses. However, they are less suitable for analyzing detailed flow changes within sub-basins and the potential effects of mitigation actions. Finally, perhaps the simplest method of estimating stream flows is the Rational Method, which is recommended only for use in small watersheds <3 km² (Dunne & Leopold 1978). While this

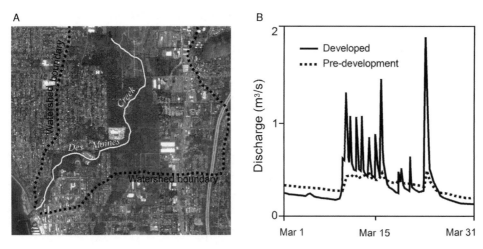

Figure 3.8 Example of use of a hydrologic model to evaluate effects of impervious surface areas on storm flows in Des Moines Creek near Seattle, Washington State, USA. (A) The watershed is largely urbanized, and (B) impervious surface areas associated with urban development dramatically increase storm flow responses to rainfall. (Based on Booth *et al.* 2002. Background image in (A) copyright Google, source USGS.)

method characterizes some of the same landscape features as process-based models (i.e. soil type and land use), it does not simulate runoff or routing explicitly. Its use is therefore generally limited to small watersheds with relatively simple landforms and uniform soil characteristics, and it is not suitable for most watershed analyses designed to estimate land-use effects on stream flows over large areas.

For alteration of flows by water storage and diversions, flow data availability varies between large dams and small private irrigation or water supply diversions, and analysis methods are tailored to the types of data available (Spence *et al.* 1996; Quigley & Arbelbide 1997). For large dams, analyses are usually empirical assessments of flow changes based on stream gage data, and indicators of hydrologic alteration (IHA) can be used to assess the degree of hydrologic change at one or more dams within a watershed (Richter *et al.* 1996; Stanford *et al.* 1996; Poff *et al.* 2010). This method uses an array of stream flow parameters that are both biologically relevant and sensitive indicators of flow change (Richter *et al.* 1996). Early analyses using this basic approach used relatively few flow parameters to evaluate the severity of stream flow alteration by dams (Figure 3.9; Richter *et al.* 1998), but more recent analyses use a larger range of flow metrics to evaluate stream flow alteration (Table 3.8). Software for analyzing 33 flow metrics and flow duration curves using IHA is available at http://conserveonline.org/workspaces/iha.

Where total reduction in stream flow is of interest, cumulative withdrawals from large dams can also be calculated based on stream flow data (e.g. Quigley & Arbelbide 1997). Data for smaller diversions and their effect on stream flows are usually less readily available (Spence *et al.* 1996), and new inventories of abstraction points and volumes may be needed to systematically identify stream reaches with impaired low flows. Assessments of changes to low flows typically include inventories of total abstraction and calculation of the proportion of stream flow removed (e.g. Donato 1998), as well as indirect estimates based on power consumption at pumping stations (e.g. Maupin 1999).

Alternatively, measured stream flows may be compared to modeled natural stream flows to estimate the effect of water abstraction (Benejam *et al.* 2010). Where measured or estimated daily stream flows are available, the IHA metrics can also be used to describe how stream flows have been altered by abstraction or flow regulation (Richter *et al.* 1996). Moreover, stream flows can be combined with reach-scale hydrodynamic models and fish life-history information to estimate how flow changes alter habitat suitability for fishes. For example, recent investigations downstream of Hungry Horse (South Fork Flathead River) and Libby (Kootenai River) reservoirs in the headwater reaches of the Columbia River in Montana, USA have assessed the effects of flow alterations on threatened bull trout (*Salvelinus confluentus*) popula-

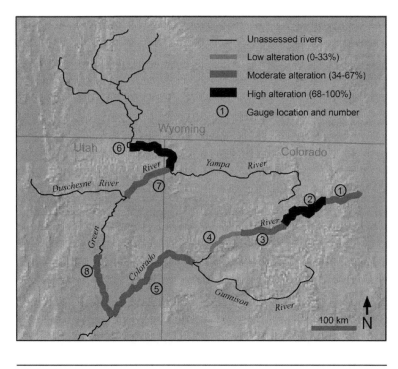

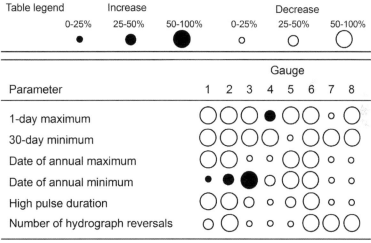

Figure 3.9 Evaluation of streamflow alteration in the upper Colorado River basin, illustrating percent change in six key flow parameters (lower panel) and overall change in hydrographs in eight reaches of the Colorado and Green Rivers, USA (upper panel). Based on Richter *et al.* (1998).

tions. Flow data were combined with two-dimensional hydrodynamic habitat models to assess discharge effects on useable habitats, and telemetry data revealed that bull trout move to shallow, low-velocity, shoreline areas at night, which are most sensitive to flow fluctuations (Muhlfeld *et al.* 2003, 2011). These two datasets combined showed that habitat availability under the natural flow regime (pre-dam, 1929–1952) was more suitable for bull trout than five post-dam flow management strategies (1953–2008), and that the current strategy best resembles natural flow conditions of all post-dam periods and helps restore channel margin habitats (Figure 3.10).

Table 3.8 Summary of indicators of hydrological alteration (based on Richter *et al.* 1996).

IHA statistics group	Flow parameters
Magnitude of monthly flows	Mean or median flow for each month (12 parameters)
Magnitude and duration of extreme flows	Annual minimum (1-day mean)
	Annual minimum (3-day mean)
	Annual minimum (7-day mean)
	Annual minimum (30-day mean)
	Annual minimum (90-day mean)
	Annual maximum (1-day mean)
	Annual maximum (3-day mean)
	Annual maximum (7-day mean)
	Annual maximum (30-day mean)
	Annual maximum (90-day mean)
	Number of zero flow days
	Base flow index (7-day minimum/mean annual flow)
Timing of extreme flows	Julian date of 1-day annual minimum flow
	Julian date of 1-day annual maximum flow
Frequency and duration of high/low pulses	Number of low pulses per year
	Mean or median duration of low pulses (days)
	Number of high pulses per year
	Mean or median duration of high pulses (days)
Rate and frequency of flow change	Means of all positive differences between daily means
	Means of all negative differences between daily means
	Number of reversals (or number of rises and number of falls)

3.3.2.3 Nutrients and pollutants

As with erosion and runoff assessments, both budgets and models that account for sources and routing are commonly used for nutrient-loading assessments (Arheimer & Olsson 2003). Budgets have been used to evaluate sources and fate of nutrients as a function of land-use practices including logging, agricultural practices, and urbanization (Likens *et al.* 1970; Feller & Kimmins 1984; Lowrance *et al.* 1985; Groffman *et al.* 2004). However, in a restoration planning context, such budgets are generally not used unless water quality or biological assessments indicate that nutrient loads are higher than expected. Once a biological or water quality assessment has identified a pollution or nutrient problem, a budget can be used to identify sources of the problem and potential restoration actions. The budget should focus on quantifying sources of nutrients or pollutants entering the river network from various land uses, which allows restoration planners to identify important sources that might need remediation or changes to management practices (e.g. Grimm *et al.* 2008). For example, a phosphorous budget for the Lake Mendota watershed in Wisconsin, USA showed that most phosphorous inputs were from fertiliz-

ers and feed supplements (Figure 3.11), and that net export of phosphorous to the lake – while large enough to cause eutrophication – was relatively small compared to that accumulating in soils and exported in agricultural products (Bennett *et al.* 1999). Therefore, management options to reduce phosphorous inputs to the lake must account for future release of phosphorous stored in soils as well as reducing short-term net export to Lake Mendota.

A variety of models can also be used to aid in watershed-scale assessments of nutrient and pollutant sources. For example, the Integrated Catchment Model (INCA) can be used to assess sources and metals discharged from mines and evaluate potential restoration strategies (Whitehead *et al.* 2009). Similarly, the Système Hydrologique Europeén Transport model (SHETRAN) is a physically based model for flow, solute and sediment transport in river basins (Ewen *et al.* 2000), although it is more limited in that it can model no more than 36 × 36 grid cells of any scale. A more versatile model for large catchments is Hydrologiska Byråns Vattenbalansavdelning – Nitrogen/Phosphorous (HBV-NP; Arheimer & Brandt 1998), which models nitrogen and phosphorous using a mass-balance framework. As with sediment supply or

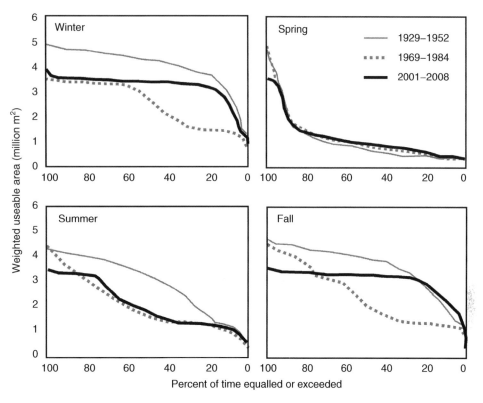

Figure 3.10 Alteration of rearing habitat suitability for bull trout (*Salvelinus confluentus*) by flow regulation during four flow seasons in Montana, USA. Time periods represent the pre-dam period (1929–1952), an early post-dam period (1969–1984), and a recent flow management period attempting to restore rearing habitat areas (2001–2008). Weighted usable area is the area of habitat that is considered suitable rearing habitat based on depth and velocity criteria (based on Muhlfeld *et al.* 2011).

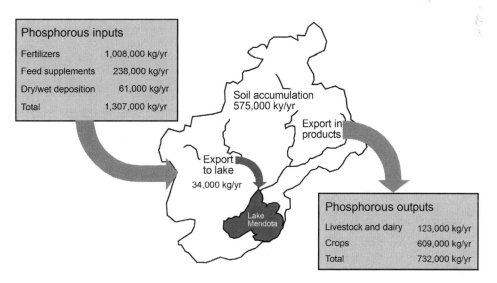

Figure 3.11 Phosphorous budget for the Lake Mendota watershed in Wisconsin, USA illustrating calculation of phosphorous inputs, outputs and accumulation in soils (Bennett *et al.* 1999). Budget analyses such as these provide a clear picture of sources and fate of nutrients or pollutants and suggest where remediation actions might be needed.

hydrologic assessments, selection of the appropriate method for any nutrient or pollution assessment depends on the geographic setting, local nutrient and pollutant issues, and biological endpoints stated in the restoration goal. Most importantly, the assessment method should identify causes of degradation (sources and delivery pathways) and clearly identify sites and types of restoration actions that can ameliorate the problem.

3.3.3 Assessing reach-scale processes

Reach-level processes are those processes that directly affect the adjacent reach, including riparian processes, floodplain–channel interactions, and local fluvial processes. Riparian functions include supply of wood and leaf litter to streams, stream shading, root reinforcement of stream banks, sediment retention, and filtration of nutrients or pesticides (Beschta 1987; Elmore 1992; Naiman et al. 2005). Dominant functions vary by climate and riparian species composition (Platts 1991), and human alteration of riparian zones has led to significant riverine ecosystem changes worldwide. Channel and floodplain interactions form a wide array of habitats that

are important to stream ecosystems (Sedell & Froggatt 1984; Peterson & Reid 1984; Collins et al. 2002; Ward et al. 2002), and human occupation of floodplains, levee construction, and channel incision have been among the greatest impacts to river ecosystems (Sedell & Froggatt 1984; Collins et al. 2002; Hohensinner et al. 2003). Finally, alteration of sediment transport and stream flows by dams and water diversions impacts channel morphology and habitat availability and diversity (Poff et al. 2010). Assessments for each of these groups of processes focus on identifying how and where land and water uses have altered them, and on identifying specific restoration actions needed to restore those processes and functions.

3.3.3.1 Riparian processes

The purpose of riparian assessments is to identify where and to what degree riparian areas have been degraded within a watershed, and to identify areas where restoration may be needed. As with watershed-level processes, remote sensing data can be used to identify where riparian processes and functions have been disrupted, but

Table 3.9 Examples of common methods of assessing riparian condition.

Type	Description	Citations
Satellite imagery		
Landsat TM	Multi-spectral imagery with resolution of 25–30 m; suitable for coarse-resolution land cover/vegetation classification	Lunetta et al. 1997; Fullerton et al. 2006; Brooks et al. 2008
Quickbird II	Multi-spectral imagery with resolution of c. 4 m; suitable for classification of riparian forest types	Gergel et al. 2007
IKONOS	Photograph imagery with resolution of 1 m; suitable for moderate resolution cover type classification	Goetz et al. 2003
Aerial imagery		
Aerial photography	May be black and white or color; range from low-resolution to moderate-resolution; suitable for coarse-resolution land-cover classification	Hyatt et al. 2004; Fullerton et al. 2006
Hyperspectral imagery	Usually moderate- to high-resolution; improved ability to distinguish species composition within cover classes	Hamada et al. 2007
Field assessment		
Rapid assessment	Visual classification of riparian condition; commonly includes vegetation community type and size class; may include disturbance classifications (e.g. bank condition, livestock access)	Munné et al. 2003; Dixon et al. 2006
Low-elevation photography	Acquired using tethered balloons or remote-controlled aircraft; high-resolution; suitable for describing vegetation cover types and species	Booth et al. 2007
Vegetation sampling	Detailed data on species and size or age using plots or transects	Brooks et al. 2008

field inventories of riparian sites must be used to design specific restoration or conservation actions (Clary & Leninger 2000; Beechie *et al.* 2003a). In general, riparian vegetation condition is evaluated with respect to a natural reference condition, which is usually based on historical information or reference sites (e.g. Harris 1999; Collins & Montgomery 2001; Hyatt *et al.* 2004). Where historical data or reference sites are not available, riparian conditions may be evaluated relative to a local desired condition or by assessing instream biota across a gradient of riparian conditions to hindcast reference conditions (Kilgour & Stanfield 2006). The products of a riparian assessment (maps and data tables of riparian conditions) indicate where riparian functions deviate from expected natural conditions (e.g. Lunetta *et al.* 1997; Hyatt *et al.* 2004;

Brooks *et al.* 2008), how land-use practices differ in their impacts on riparian functions (Amy & Robertson 2001), and where restoration efforts are most needed.

Remote sensing data for riparian classification include satellite data, aerial photography and hyperspectral imagery (Goetz *et al.* 2003; Hyatt *et al.* 2004; Brooks *et al.* 2008) (Table 3.9). Satellite data is generally the coarsest-resolution remote sensing data for classification of riparian forests, and its accuracy should be assessed with higher-resolution imagery or field sampling of riparian vegetation (Goetz *et al.* 2003; Brooks *et al.* 2008). Aerial photography is higher resolution than most satellite imagery, and riparian conditions can be classified by buffer width, stand type or vegetation type, and age or size of vegetation (Figure 3.12). Hyperspectral imagery

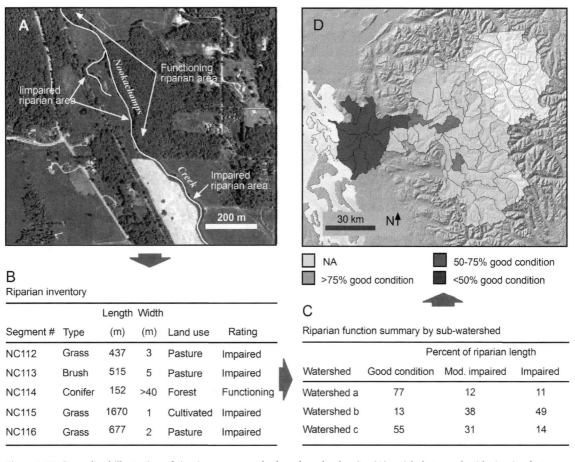

Figure 3.12 Generalized illustration of riparian survey methods and results showing (A) aerial photograph with riparian features identified (background image copyright Google, source USGS); (B) excerpt from a riparian inventory database; (C) summarized data by sub-watershed; and (D) final map of riparian conditions by sub-watershed (map based on unpublished data, Skagit Watershed Council, Mount Vernon, Washington State, USA).

provides the highest-resolution information, and can often be used to identify species composition in addition to buffer width and age or size of vegetation. Choosing between these data sources depends on cost of the imagery and the restoration planning needs. For example, satellite imagery is commonly used for assessment of large watersheds because it is relatively inexpensive and data storage requirements are low, but it can also be combined with higher-resolution data to better understand specific riparian conditions within each riparian condition class (e.g. Beechie *et al.* 2003a; Brooks *et al.* 2008). By contrast, high-resolution hyperspectral imagery typically has very large data-storage requirements, and is better suited to assessing smaller areas in great detail.

Low-elevation aerial photography or rapid field surveys may also be used to assess riparian conditions (Raven *et al.* 1998; Booth *et al.* 2007). While such methods are generally more costly than other remote sensing methods, some types of impacts such as bank erosion or grazing of non-woody vegetation are generally not discernible from coarser-resolution remotely sensed data. For example, where the primary disturbance is livestock grazing, field measures of riparian conditions such as stubble height or length of banks trampled by livestock can be used to measure disturbance (Clary & Leninger 2000; Turner & Clary 2001). Alternatively, high-resolution aerial photography (e.g. taken using a tethered balloon or remote controlled aircraft) can be used to assess riparian condition (Booth *et al.* 2007; Figure 3.13). Each of these methods is usually too costly to use for mapping riparian conditions in all reaches of a watershed, but coupling these methods with complete mapping of riparian classes can be used to stratify the riparian landscape (remote sensing data) and interpret detailed conditions within

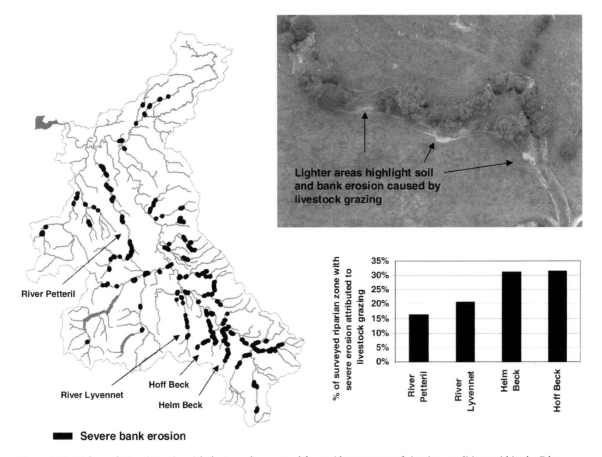

Figure 3.13 High-resolution (20 cm) aerial photography was used for rapid assessment of riparian conditions within the Eden River basin, Cumbria, UK. The mapping identified reaches where severe bank erosion due to livestock grazing and trampling was damaging riparian habitats, and which tributaries were in greatest need of riparian restoration.

each class (field or low-elevation photography data) (Beechie *et al.* 2003a; Harris & Olson 1997; Brooks *et al.* 2008).

3.3.3.2 Floodplain processes

Floodplain processes can be disrupted by a variety of land and water uses, including altered flow regimes or sediment and wood supplies downstream of dams, installation of dikes and riprap to control flooding or channel movement, and channel incision that isolates a channel from its floodplain (Beechie *et al.* 1994, 2008b; Hohensinner *et al.* 2003). However, assessment methods for identifying altered floodplain conditions (e.g. forest age structure and abundance of various habitat types) or processes (e.g. channel migration rates) are often the same regardless of the cause of change. For example, aerial photograph inventories of channel and vegetation changes are commonly used to assess changes in channel pattern and riparian conditions, and airborne laser mapping (using LiDAR, or Light Detection and Ranging) can be used to identify changes in connectivity of floodplain habitats (Shafroth 1999; Kloehn *et al.* 2008; Negishi *et al.* 2010). These assessments focus on mapping vegetation types through time, or may use grid-based sampling designs to estimate the proportion of floodplains occupied by various age classes or species of riparian vegetation (e.g. Collins *et al.* 2002; Kloehn *et al.* 2008).

Detailed assessments of floodplain vegetation and habitats based on historical surveys provide the most comprehensive picture of floodplain changes and perhaps provide the best guidance on restoration potential (Collins *et al.* 2002; Hohensinner *et al.* 2005). These assessments can be used regardless of causes of altered river–floodplain dynamics and, in cases where causes of floodplain alteration are obvious (e.g. floodplain land uses or levees), these assessments can also identify the cause of alteration. In cases where the cause of altered river–floodplain dynamics is indirect or less visible (e.g. a change in flow or sediment, or bank armoring), causes of change will be identified in other assessments. For example, flow and sediment regimes downstream of dams are assessed in either the hydrology or sediment supply assessments, bank armoring may be assessed in habitat surveys, and channel incision is analyzed in the fluvial process assessment. Each of these types of assessments produce maps of impacts to floodplain processes and riparian or habitat conditions.

Where floodplains have been isolated from rivers by levees, changes to river and floodplain habitats can be identified through measurement of channels no longer connected to the river (Sedell & Froggatt 1984; Beechie *et al.* 1994) or mapping of historical and current habitats or vegetation (Collins *et al.* 2002; Hohensinner *et al.* 2003). The least time-consuming and versatile of these approaches is measurement or estimation of habitat areas that are no longer connected to a main river channel (Beechie *et al.* 1994). These estimates can be made by measuring channels on historical maps, and provide a comprehensive estimate of total floodplain habitat losses in a river basin. Detailed mapping of historical and current channels in GIS provides more specific quantification of habitat losses, and also supports more quantitative estimates of restoration benefits based on flow regimes, surface elevations, or other criteria (Hohensinner *et al.* 2003; Figure 3.14). Products of the floodplain process assessments may include maps of changes in channel migration rates, changes in forest age structure that indicate increasing or decreasing channel migration rates, or maps of changes to floodplain habitat area (e.g. Hohensinner *et al.* 2003; Kloehn *et al* 2008; Draut *et al.* 2010). Each of these products can help identify where restoration actions might improve diversity and abundance of floodplain habitats.

3.3.3.3 Fluvial processes and conditions

Fluvial processes include water and sediment transport, as well as processes such as lateral migration or pool and riffle formation (Chapter 2). Evaluating how fluvial processes have been altered may involve direct assessment of processes (e.g. sediment transport or channel migration), but more commonly focuses on assessment of channel conditions that are indicators of altered processes (e.g. sediment size or pool depths). Assessment of channel conditions indicates where river morphology deviates significantly from expected morphology, and therefore points to changes in watershed processes, riparian functions or direct manipulations of the river. By contrast, directly diagnosing changes to fluvial processes is focused on understanding how rates of reach-scale fluvial processes have been altered by local actions such as gravel mining or impacts of dams. Assessments of both processes and conditions can be used in conjunction with assessments of other watershed-scale and reach-scale processes (e.g. erosion and sediment supply or riparian processes) to understand causes of channel change and to identify restoration actions that might resolve or mitigate those impacts on channel conditions. We note here that direct assessments or modeling of reach-scale processes such as sediment transport, bank erosion or meander migration are most often conducted during the design of individual restoration projects or plans,

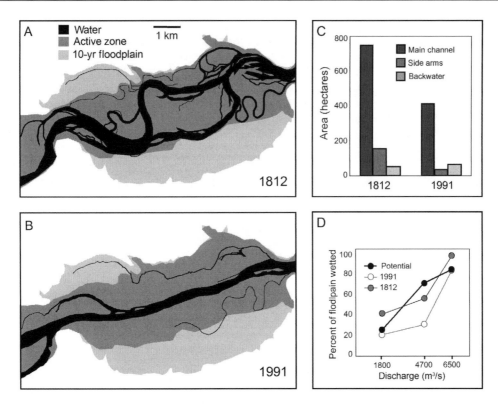

Figure 3.14 Example of historical analysis of river channel form, extent of floodplain connectivity and floodplain channel formation in the River Danube. Panels (A) and (B) show mapped channels in 1812 and 1991, which were used to estimate (C) loss of habitat area in the main channel and floodplain habitats. Modeling of floodplain inundation at various discharges was used to indicate (D) change in habitat availability with changes in discharge and under historical and potential restored configurations. Adapted from Hohensinner *et al.* (2003).

and we discuss these assessment procedures in detail in Chapter 7.

In general, it is far easier to detect changes in channel conditions than to evaluate fluvial processes, so changes to fluvial processes are often first diagnosed based on assessment of altered channel conditions. Diagnostics of channel condition include channel pattern, width-depth ratio, pool depths and frequency, bar size and height, grain size and degree of incision or bank erosion (Montgomery & MacDonald 2002; Table 3.10). For the majority of channel condition assessments, channel alteration can be measured either as deviations from natural conditions (where restoration can conceivably achieve near-natural endpoints), or as deviations from conditions expected under contemporary process regimes (e.g. where dams preclude restoration to near-natural conditions). While all of these can be measured in the field, some diagnostics can also be measured from aerial photography or LiDAR. Field reconnaissance surveys must be rapid and efficient

to be useful for a watershed assessment, because they must examine a large number of reaches to portray fluvial conditions across a watershed (e.g. Downs & Thorne 1996; Thorne 1998). Aerial photography or LiDAR also provide a rapid means of measuring certain conditions, with the added advantage of allowing a historical perspective that field measures typically do not (unless there have been past field measurements). We discuss each of these approaches in more detail below.

Each diagnostic parameter used in a river channel assessment should be an indicator of changes in watershed or fluvial processes. For example, the diagnostics summarized in Table 3.10 can indicate reductions in flood magnitude or coarse sediment supply following flow regulation by large dams, increases in intermediate flood frequency, increases in sediment supply due to land uses, or changes in wood supply. They can also indicate changes in local fluvial processes, such as reductions in bank strength and increases in water temperature that

Table 3.10 Common diagnostic measures of channel conditions (based on Dietrich *et al.* 1989; Kondolf 1997; Montgomery & MacDonald 2002).

Attribute	Description
Channel pattern	Changes in channel pattern can be diagnostic of changes in sediment supply, flood flows or reinforcement of stream banks. For example, increased braiding can indicate increases in sediment supply or reduced bank strength, whereas conversion to a single straight channel can indicate decreased sediment supply
Width:depth ratio	Increased width:depth ratio often indicates increased sediment supply, but may also be a short-term response to a recent large flood
Pool depths and frequency	Few and shallow pools can indicate loss of large wood, increased sediment supply, or direct channel modification such as dredging
Bar size and height	Large, high bars indicate high sediment supply, whereas few or no bars indicates low sediment supply relative to transport capacity
Channel incision	Channel incision may indicate downstream channel modifications that initiated head-cutting, loss of sediment retention mechanisms such as wood jams, floodplain, or aquatic vegetation (e.g. grasses or macrophytes in small streams), or decreased sediment supply due to upstream dams
Bank erosion	Unusually severe bank erosion commonly indicates loss of root strength in banks, increased sediment supply, or recent large floods
Grain size	Fining of the bed surface indicates high sediment supply relative to transport capacity, whereas a heavily armored bed (coarse surface layer over finer material) indicates relatively low sediment supply

Table 3.11 Examples of common methods of assessing channel conditions or modifications to rivers such as levees or bank armoring.

Approach	Description	Citations
Channel classification	Comparison of current conditions to *a priori* expected conditions based on channel classification	Montgomery & Buffington (1997)
Statistical methods	Uses statistical ordination of physical measurements (e.g. Principal Components Analysis) to identify reaches that deviate from 'reference' conditions	Raven *et al.* (1998)
Interpretation of reconnaissance surveys	Regional reconnaissance survey data are interpreted by experts to identify reaches impacted by land uses	Brierley & Fryirs (2005, 2008); Sear *et al.* (2009)
Regional reference equations	Regional equations define expected conditions, which can be compared to current conditions to identify deviations and possible altered fluvial processes	Thorne *et al.* (1996)

follow riparian vegetation clearing, knick points generated through instream gravel mining (or following capture of floodplain pits) and, most directly, deliberate channel modifications by channelization or bank armoring. Importantly, changes in channel conditions are frequently initiated from downstream, so it is also necessary to examine evidence of management actions such as channel straightening that might have initiated upstream-migrating knick points. Note that each diagnostic can indicate several different potential causes of change,

reinforcing the need for other watershed process assessments to identify restoration needs with greater certainty.

A number of field protocols have been developed for evaluating channel condition, including simple classification methods as well as procedures for evaluating a number of diagnostic features (Table 3.11). One popular approach is to classify or characterize a suite of channel conditions as part of a regional survey and to identify reaches that are out-of-character in some way. Classification procedures (see Section 3.3.1) typically use a top-down structuring of channel classes to identify expected conditions and altered fluvial processes (e.g. Rosgen 1994). Through the multi-parameter structure of such a classification scheme it is possible to identify which parameter is 'odd' and thus start to look for the causes of fluvial process alteration. A second approach is channel characterization schemes such as the River Habitat Survey in Europe (Raven *et al.* 1998), which relies on measuring a suite of variables and using statistical ordination techniques to identify 'altered' conditions. With this approach, deviations in channel conditions can be identified without the need for *a priori* expectation of conditions, thus avoiding the sometimes circular reasoning that accompanies classification approaches. A third approach is to use expert interpretation of reconnaissance surveys as the basis for classifying river typology (Downs & Thorne 1996; Thorne 1998; Montgomery & MacDonald 2002). For example, the River Styles technique (Brierley & Fryirs 2005, 2008; see Figure 3.4) is – on the surface – highly qualitative and requires significant training, but has the advantage of delivering a process-reasoned summary of conditions in their spatial and temporal context. This provides a solid foundation for planning process-based restoration. Similarly, the Fluvial Audit technique used in the UK (Sear *et al.* 2009) provides an integrated summary of conditions linked to potentially destabilizing phenomena at the watershed scale. Finally, regional equations describing hydraulic geometry relationships can be used to identify expected conditions for width, depth, meander wavelength, and other parameters, which can then be compared to current conditions to identify reaches that deviate from those expected values (Thorne *et al.* 1996). For example, in urbanized watersheds channels are often 'over-sized' relative to expected dimensions estimated from regional regime equations, indicating significant increases in runoff and flood magnitudes compared to a more natural river for that region.

Widespread channel incision is often considered a special case for geomorphological assessment, as its causes and remedies are often difficult to ascertain and describe. In many cases, channel incision partly reflects a historical legacy of severe erosion and aggradation associated with past river damming, channelization, and other forms of upriver sediment trapping (Fitzpatrick *et al.* 2009; Walter & Merritts 2008). Such legacy effects are not always readily apparent with common channel measurements, and detailed field studies are often required to determine historical aggradation and degradation sequences and to identify causes of apparent channel incision. In the absence of legacy effects, field mapping and measurement of incised channel locations and volumes provides data needed to illustrate where restoration is needed and the magnitude of the problem at various locations (Beechie *et al.* 2008b; Figure 3.15). Key diagnostic measures include bankfull channel dimensions, inset floodplain dimensions and dimensions of the incised channel. Identifying restoration options may also require analysis of sediment transport and storage mechanisms that may lead to channel recovery, and estimating the length of time required to meet a specified restoration objective may involve assessments of sediment supply and typical aggradation rates under varying restoration scenarios (Pollock *et al.* 2007; Beechie *et al.* 2008b). Finally, the potential impacts of an upstream migrating knick point can be predicted using the channel evolution model (Simon 1989), which requires combining basic morphological parameters (e.g. width, depth) with an interpretation of the condition of incised channel banks. An advantage of this approach is that incised channel conditions are evaluated in a spatio-temporal context that allows the model to be used to predict future incision, channel widening, or inset floodplain development.

In many cases, a few river reaches within a watershed have experienced obvious changes to fluvial processes (e.g. reaches downstream of dams) and merit closer examination to diagnose changes to processes and to identify suitable habitat restoration actions. For example, in the Trinity River in California, USA the Lewiston dam has dramatically reduced sediment supply, stream flow, and river dynamics, and fluvial process analyses have focused on understanding impacts of the altered flow and sediment regimes on channel and riparian conditions (US Fish and Wildlife Service & Hoopa Valley Tribe 1999; McBain & Trush Inc. 2007). Such detailed evaluations of fluvial processes or channel conditions might also examine time series of maps or field data to measure changes in process rates (e.g. channel migration). For example, overlays of historical topographical maps and aerial photographs have long been used to indicate the timing and impact of past management schemes and the dynamics of river meander processes (Gurnell *et al.* 1994). More recently, these techniques have been used within a GIS to determine the time-integrated probability of channel planform change (e.g. Graf 2000; Tiegs & Pohl 2005). Repeat surveys of channel cross-sections and bed elevation perhaps provide the clearest indication of channel response to prevailing fluvial processes, but unfortunately such data are rare. Nevertheless, where

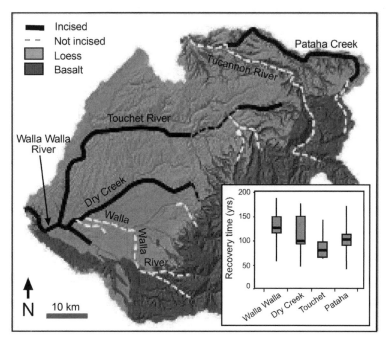

	Incised channel volume (m³)	Annual storage with beaver dams (m³)	Percent of annual sediment yield
Walla Walla River	8,897,000	78,600	3.7%
Dry Creek	5,559,000	44,900	7.1%
Touchet River	11,262,000	113,900	7.5%
Pataha Creek	6,765,000	56,500	17.7%

Figure 3.15 Mapping and analysis of channel incision in the Walla Walla and Tucannon River basins, Washington State, USA. Mapping specifies where incision has occurred and altered habitat conditions. The sediment budget (summarized in the table) indicates how much sediment has been lost during incision, how much sediment is moved through the system annually, and how long it may take for incised reaches to fully aggrade to the historical floodplain (box and whiskers plot in lower right corner of the map). Adapted from Beechie *et al.* 2008b.

they exist, the combination of bed elevation surveys and planform surveys can be used to develop a three-dimensional understanding of channel morphology and dynamics, including as trends in channel incision and active bed width (Downs *et al.* 2006). Yearly surveys of bed sediment size can also be assembled without great expense using the 'pebble count' technique (Wolman 1954) and can help determine whether the channel bed is coarsening or fining over time. Repeat bulk sediment samples can indicate whether the channel surface is armoring or whether fine sediments are infilling the interstices of coarse grains. Fine sediment accumulation in pools (e.g. the V* method, Lisle & Hilton 1992) can indicate the relative fine sediment load of the channel over time if the surveys are repeated.

Finally, the budgeting approach introduced earlier (Section 3.3.2) can help restoration planners to understand localized – but critical – restoration needs in a watershed. For example, as part of a long-term remediation plan to clean up impacts from a century of mining for silver, lead, and other metals in northern Idaho, USA the US Environmental Protection Agency (EPA) developed water, sediment, and lead budgets for the lower basin of the Coeur d'Alene River (CH2M HILL 2010). Discharge and suspended sediment measurements at the upper and lower boundaries of a 37-mile reach were used to quantify the volumes of water and sediment delivered to the lower basin. Over a 20-year period (1989–2008), the annual average volume of water and sediment leaving the lower basin was slightly more than the volume entering, a difference attributed primarily to tributaries in the lower basin (Figure 3.16). In contrast, the annual average mass of lead leaving the lower basin (estimated from relationships between suspended sediment and lead concentrations) was almost an order of magnitude greater than the mass entering. Based on these calculations, the

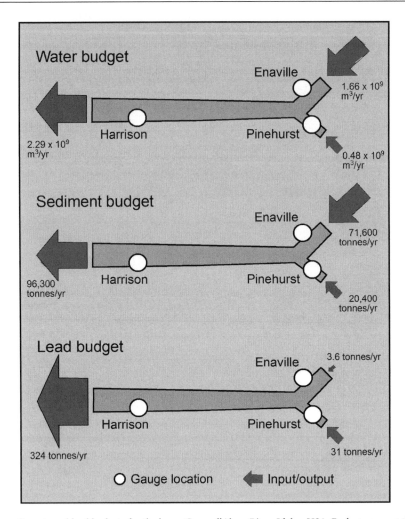

Figure 3.16 Water, sediment, and lead budgets for the lower Coeur d'Alene River, Idaho, USA. Budgets suggest that most lead entering Coeur d'Alene Lake originates from lead-laden sediments in the lower Coeur d'Alene River. Remediation actions to remove lead from the river system must therefore consider stored lead in river bed sediments in addition to lead sources in the upper watershed. Adapted from CH2M HILL 2010.

large net export of lead from the study reach suggests that most lead leaving the basin originates from lead-laden sediments stored in the river channel (rather than from the upper basin where mining once took place). These conclusions are consistent with earlier field measurements collected by the USGS (Clark & Woods 2001; Berenbrock & Tranmer 2008), and together these findings indicate that remediation of mining impacts will need to address lead originating from previously stored sediment in the lower river, not just the mine-related sources in the upper basin.

Important products of fluvial assessments include maps and tables of (1) reaches in which fluvial processes or channel conditions are impaired and in need of restoration or (2) reaches that are key response reaches for restoration of watershed-scale processes. An understanding of reach-scale deviations of fluvial processes and changes in processes provides an indication of the likely impairment of fluvial morphology and dynamics, which helps identify which restoration actions might most benefit river ecosystems. When set in a watershed context, this information focuses restoration efforts on likely causal mechanisms that should be addressed by restoration actions.

3.4 Assessing habitat alteration

The purpose of evaluating habitat change is to identify where habitats have been degraded or lost, and to broadly indicate where and what types of habitat restoration efforts are possible. The selection of habitat assessment methodologies for restoration planning depends largely on the legislation or goals driving river restoration, as those goals define the desired outcomes of restoration and the types of habitat information required for restoration planning. For example, restoration planning to address endangered species often requires information on habitat characteristics and status of a species, whereas restoration planning focused on water quality may require suites of water quality and biological metrics to evaluate restoration needs. Restoration planning to address broadly defined 'ecological status' or 'health' of rivers will likely include the broadest array of habitat assessments, including measures of habitat quantity and quality, as well as relevant biological metrics. Diagnosis of habitat degradation includes assessments of habitat structure and quality (i.e. habitat surveys and water quality sampling), and a wide variety of parameters can indicate deviations from expected habitat conditions or regulatory thresholds (e.g. characteristics of stream substrate, pool area or frequency, abundance or size of wood debris, and concentrations of nutrients or pollutants). In this section we first describe assessments of habitat type and quantity, and then assessments of water quality. In Section 3.5 we describe approaches to understanding the biological importance of habitat changes, including models that indicate where habitat restoration will most benefit species, and empirical sampling of suites of biota that can be correlated with landscape or habitat changes.

3.4.1 Habitat type and quantity

Habitat survey methods generally fall into two main categories: continuous habitat surveys and transect surveys (Figure 3.17). Continuous habitat surveys measure or tally a variety of habitat features continuously along the channel length (e.g. Bisson *et al.* 2006), whereas transect methods record habitat attributes at specified points

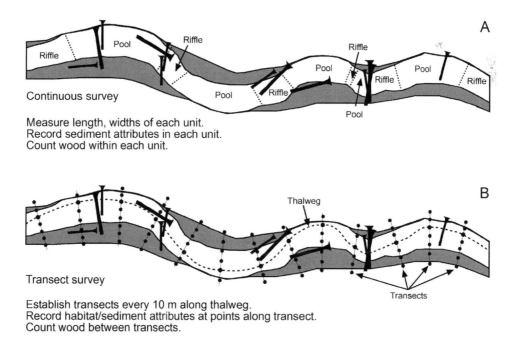

Figure 3.17 Habitat survey designs include (A) continuous surveys of physical habitat features or (B) transect methods that measure attributes at points across equally spaced cross-sections.

along cross-channel transects (e.g. Fitzpatrick *et al.* 1998). Continuous habitat surveys are usually based on measurement of habitat units (e.g. pools and riffles), with length and widths of all units estimated or measured and other attributes such as substrate size and wood abundance recorded within each unit (Bisson *et al.* 2006; Harding *et al.* 2009). Transect methods are usually based on a systematic sampling design, with transects placed at equal distances along the stream length (Fitzpatrick *et al.* 1998; Raven *et al.* 1998; Environment Agency 2003). Transect methods typically record a wider range of habitat attributes within a reach, and some transect-based methods also include continuous data collection of certain attributes between transects (Kaufmann *et al.* 1999). Continuous survey methods produce simple data describing areas of habitat unit types and sums or averages of other features (e.g. total areas of pools and riffles, or total wood in a reach or wood/meter). Transect surveys produce point data that can be summarized as proportions (e.g. percent pool or riffle), but habitat areas are not measured directly.

The choice of basic survey design depends largely on the kinds of analyses that are needed for restoration planning. For example, where survey data are intended to estimate habitat losses and changes in habitat capacity (e.g. for specific species such as Pacific or Atlantic salmon), continuous habitat surveys more directly produce habitat areas needed to estimate habitat losses and changes in productive capacity (e.g. Reeves *et al.* 1989; Beechie *et al.* 1994). Moreover, continuous rapid surveys give a more accurate portrayal of basin-wide habitat abundance and distribution despite reduced precision at the site scale, primarily because between-reach variation in habitat abundance is high and sampling more reaches with less detail produces a more accurate estimate of total habitat availability than sampling fewer reaches with more detail (Hankin 1984; Hankin & Reeves 1988). By contrast, transect methods are more precise at individual sites, and are more appropriate where the primary intent is to detect change in habitat attributes through time or to have precise measures of habitat within a reach for rating habitat quality (Fitzpatrick *et al.* 1998; Kaufmann *et al.* 1999).

Most habitat survey methods are designed for small, wadeable streams, and relatively few are available for larger rivers. Small-stream survey methods typically quantify some combination of habitat type, substrate size, cover characteristics, wood abundance and size, or aquatic vegetation type and density (Table 3.12). Physical habitat attributes measured in a survey should be useful

Table 3.12 Common habitat survey metrics.

Attribute	Purpose	Metrics
Habitat structure	Identify habitat changes relative to natural or reference condition	Channel units, depth/velocity distributions, hiding cover, wood counts, aquatic vegetation coverage
Sediment condition	Identify spawning areas for important species, indicator of increased fine sediment supply	Gravel size, percent fines, embeddedness
Bank condition	Identify areas with greater than natural bank erosion, or areas with artificially stabilized banks	Length of eroding bank, length of armoring

in interpreting where and what kinds of restoration are needed to achieve a restoration goal, and should also be practical and statistically reliable. General criteria for inclusion of habitat parameters include: (1) relevance to restoration goals; (2) relevance to common causes of habitat degradation in the region; (3) repeatability of the measure; and (4) practicality (level of detail required, amount of area to cover, resources and skills available) (Beechie *et al.* 2003a). For example, habitat types that meet the first criterion (i.e. relevance to restoration goals) should not only be useful predictors of abundance for species of interest or useful indicators of 'healthy' stream structure, but also sensitive to land use and restoration actions (i.e. they should change as a result of degradation or restoration). Most of the commonly used habitat types meet these criteria, and there is a broad similarity among typing systems regardless of environment and species of interest (Figure 3.18). Differences among habitat typing systems are usually in level of detail and the number of habitat types included.

Habitat attributes included in a survey should also reflect relevant indicators of local causes of habitat degradation (the second criterion). For example, tallies of wood debris are commonly included in forest streams where instream wood is an important habitat feature for

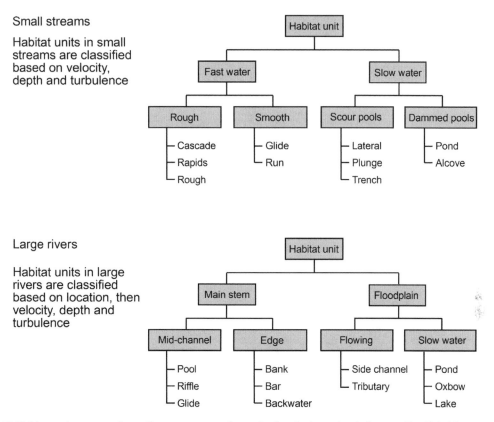

Small streams

Habitat units in small streams are classified based on velocity, depth and turbulence

Large rivers

Habitat units in large rivers are classified based on location, then velocity, depth and turbulence

Figure 3.18 Habitat typing systems in small streams commonly use depth, velocity, and turbulence to identify habitat types, whereas large river systems commonly also incorporate location on the floodplain to identify habitat types. Adapted from Wilcox (1993), Beechie *et al.* (2005), and Bisson *et al.* (2006).

many species, and wood abundance is sensitive to land-use change or restoration actions. Wood abundance is therefore both a sensitive indicator of environmental change and a relevant predictor of effects of those changes on biota. By contrast, lowland or grassland streams may naturally contain very little wood debris, and habitat surveys might instead include aquatic macrophytes as a key habitat attribute that is also a sensitive indicator of changes in land use. The third criterion considers repeatability of a measure, which is focused on selecting attributes that can be consistently identified by multiple observers and has relatively low measurement error. In general, attributes that are measureable are more repeatable than qualitative attributes (e.g. distances or grain sizes rather than 'good', 'fair', or 'poor' ratings of habitat condition), providing more reliable data for assessing habitat change (Kaufmann *et al.* 1999). Finally, the fourth criterion (practicality of a survey method) considers time

and funding relative to the assessment needs. In essence, this criterion focuses on how many and which parameters to include in a survey, which should consider the important tradeoff between knowing more about individual sites versus knowing more about the watershed. Restoration planning at the watershed scale generally benefits from simpler surveys with few parameters because the survey can then cover a sufficient area for the assessment. More detailed surveys with a large number of parameters may be better for assessment of reach-scale conditions in the project design phase (Chapter 7), or for monitoring situations (Chapter 8).

Habitat assessment methods for large rivers differ from those of small streams in that surveys must be conducted from a boat or from aerial photography because rivers are too deep to wade (e.g. Wilcox 1993; Beechie *et al.* 2005). However, as with small-stream methods, the choice of sampling method (continuous survey versus transect) and habitat attributes to measure depends on the restoration goals, the needs of the assessment and practical considerations. In general, large-river assessments are more likely to be continuous surveys rather than transect surveys simply because rivers cannot be crossed on foot. Habitat typing systems for large rivers are conceptually similar to those for small streams, but they often include both mainstem habitat types and floodplain habitat types. Habitat types for mainstem channels are generally relatively simple, focusing on a few mid-channel and edge habitat types, while floodplain habitats commonly include both lotic and lentic habitat types (Figure 3.18). Some of these units can be mapped from remote sensing data, but edge habitat units and floodplain units are often only discernible in the field due to vegetation cover.

Most habitat assessments cannot survey all – or even most – reaches in a watershed. Therefore, streams are often stratified by reach type and land-use factors using remotely sensed coarse-resolution data, and a sample of reaches within each stratum are field surveyed to characterize habitat conditions at the watershed scale. The reach type classification stratifies the potential range of habitat conditions, and the land-use component facilitates analysis of where and how much degradation has occurred within each reach type. Reach classification systems may be selected from the many systems available (e.g. see Section 3.1), or a simpler system may be developed based on relatively few landscape and land-use factors (e.g. channel slope and land use adjacent to the reach). Land-use or land-cover classes derived from satellite data are often publically available, and a relatively simple set of classes are usually sufficient for land-use classification (e.g. developed, agriculture, forest, grassland, and wetland). Regardless of the classification system chosen, a relatively small number of strata is desirable to reduce the number of reaches that need to be sampled across the watershed, as well as to simplify the presentation of results. Once the classification has been completed, reaches are sampled either randomly or systematically from each stratum, and habitat conditions are surveyed as described above to generate estimates of habitat abundance or condition at the watershed scale.

An additional and often important habitat assessment procedure is evaluation of fish migration blockages, often described by the term 'longitudinal connectivity'. Portions of tributaries that are blocked from fish access can be mapped using estimates or inventories of habitat upstream of migration barriers (Beechie et al. 2003a; Pess et al. 2003b; Sheer & Steel 2006). Natural barriers to migration must first be identified to delineate the assess-

ment area, and then all structures crossing streams within the assessment area (culverts, bridges, small dams) should be inventoried to determine if they meet passage criteria. Finally, habitat areas upstream of each man-made barrier must be surveyed to determine how much habitat is inaccessible (i.e. the length or area upstream of the identified barrier and downstream of the natural barrier to fish migration; Figure 3.19). Use of a standardized method for determining blockages will streamline identification and prioritization of isolated habitats and provide a standard set of criteria for monitoring progress toward reopening these habitats.

For most watershed assessments, quantifying a historical or reference condition provides a useful baseline for assessing degradation of habitat, and for determining where certain types of habitat restoration might be needed to achieve the biological aims of a restoration program. Even where some areas cannot be fully restored, the baseline condition helps to understand what kinds of habitats were historically present and what kinds of created habitats are compatible with local driving processes (Kondolf 2006). Methods for estimating historical or reference habitat conditions include historical analysis, reference site data or model predictions (Table 3.13), and the choice of methods depends on habitat type and availability of information (Buijse et al. 2002; Sear & Arnell 2006). Historical analyses of large river habitats commonly rely on pre-1900s maps of river channels and vegetation patterns to estimate historical areas of main river channels, floodplain channels, and lakes (e.g. Beechie et al. 1994; Collins & Montgomery 2001; Hohensinner et al. 2005; see also Figure 3.14). For smaller features in tributaries, such as channel units or beaver ponds, reference site data and model predictions are usually more appropriate because historical data and maps are not detailed enough to estimate habitat conditions (Beechie et al. 1994; Pollock et al. 2004). Where present-day reference sites can be surveyed to represent historical habitat types and areas, field survey methods should be the same as those described above for current conditions. When neither historical data nor reference site data are available, models may be used to estimate the historical abundance of certain habitats. For example, historical beaver pond areas can be estimated based on number and area of beaver ponds in a few relatively undisturbed areas combined with literature values (e.g. Pollock et al. 2004). In most cases, some combination of these methods is required to compile a comprehensive picture of salmon habitat changes in a watershed. An example of combining these methods of

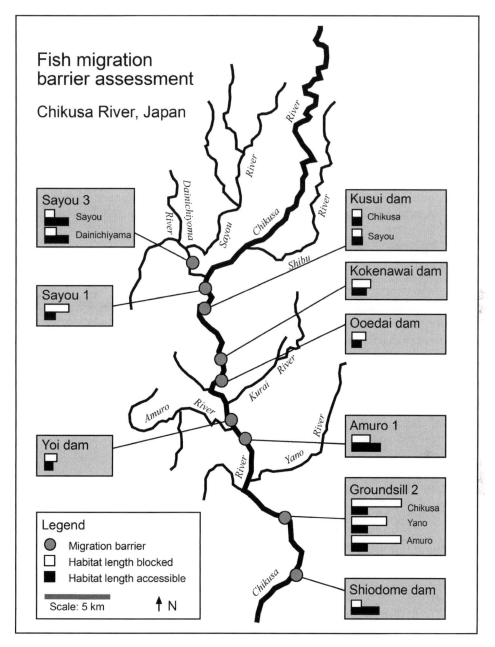

Figure 3.19 Inventories of fish migration barriers in the Chikusa River, Japan include barrier locations as well as surveys of amounts of habitat that can be opened with each barrier removal (Mitsuhashi *et al.*, unpublished data 2011). Both locations and habitat areas are needed to inform restoration planning.

Table 3.13 Summary of three general approaches to estimating reference condition.

Method	Description	Citations
Historical data	Reference condition estimated from historical maps, notes or photos, and often field verified by evidence of their prior locations	Beechie *et al*. 1994; Collins & Montgomery 2001; Hohensinner *et al*. 2005
Reference site data	Collect data from relatively natural or undisturbed sites to estimate reference condition	Hughes *et al*. 1986; Harris 1999; Stoddard *et al*. 2006
Models	Model the reference condition based on first principles or data from a variety of natural sites	Pollock *et al*. 2004; Kilgour & Stanfield 2006

Table 3.14 Examples of methods for estimating historical or reference abundances of various habitat types used in analysis of habitat change in the Skagit River basin, USA (modified from Beechie *et al*. 2003b).

Habitat type	Analysis methods	References
Reduced off-channel or wetland areas	Historical or reference habitat areas estimated from historical maps, notes or photos, and often field verified by evidence of their prior locations	Beechie *et al*. 1994; Collins & Montgomery 2001; Hohensinner *et al*. 2005
Lakes	Changes to lake areas measured directly from historical and current maps	Beechie *et al*. 1994
Beaver ponds	Pre-settlement or reference beaver pond areas estimated based on frequencies of beaver ponds in relatively pristine areas, or predictive methods using stream and valley characteristics	Pollock *et al*. 2004; Beechie *et al*. 2001
Tributary and mainstem blockages	Barriers to fish movement mapped using inventories of anthropogenic and natural barriers, as well as surveys of estimates of habitat upstream of migration barriers	Beechie *et al*. 1994; Sheer & Steel 2006
Altered pool abundance in tributaries	Based on measured pool areas in reference sites; may also use historical information where available	Beechie *et al*. 1994; Nickelson & Lawson 1998

estimating reference conditions for salmon habitats in northwestern Washington State, USA is shown in Table 3.14. The combined outputs of historical and current assessments include summaries of changes in habitat condition throughout the watershed, as well as general assessments of the degree of degradation associated with various causes (e.g. channel modification, beaver trapping, hydropower, forestry activities, etc.) (Figure 3.20).

3.4.2 Water quality

Water-quality assessments generally rely on direct measurements of individual water quality parameters such as water temperature, nutrient concentrations, turbidity, or pollutant concentrations. Products of water quality

assessments indicate reaches in a watershed that are impaired for one or more water quality parameters, including temperature, turbidity, high nutrient loads, or specific pollutants, based on comparison to regional reference conditions or specific legal standards. For stream temperature, for example, reference conditions are often difficult to acquire due to a lack of suitable reference sites, so temperatures may be compared to temperature standards that aim to avoid exceeding maximum temperature thresholds for aquatic biota. However, where reference conditions can be acquired, comparisons to reference data give a better indication of where and how temperatures deviate from the natural regime, indicating both the degree of degradation and potential restored conditions. In addition to maximum temperature thresholds, analy-

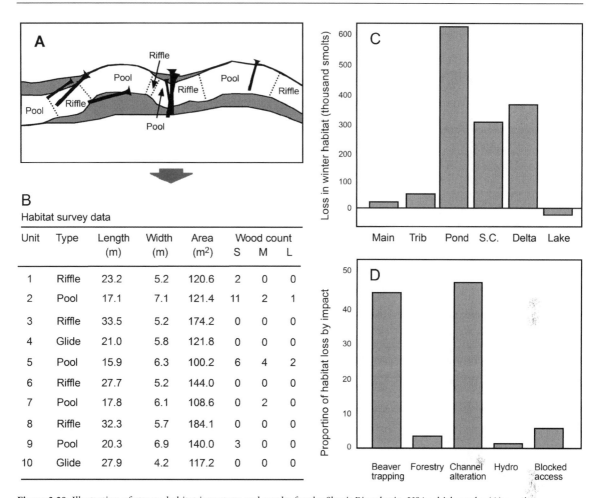

Figure 3.20 Illustration of stream habitat inventory and results for the Skagit River basin, USA which used a (A) continuous habitat survey design. (B) Survey data are recorded in simple data records, and changes in habitat capacity can be summarized by (C) habitat type and by (D) cause of loss. Based on data in Beechie *et al.* (1994, 2001).

ses can examine a broader range of temperature attributes to identify where temperature regimes have been altered by increasing or decreasing the magnitude or duration of key temperatures (e.g. assessing diel or monthly variation, or duration of exceedence of certain thresholds) (Steel & Lange 2007). As with other habitat assessments, the key products of these assessments are maps of locations and degree of habitat degradation. This information identifies reaches that have been affected by changes in sediment, nutrient or pollutant sources or delivery, but watershed-scale assessments are usually needed to identify appropriate restoration actions (i.e. the cause of the problem and likely remediation actions) (see Section 3.3.2).

Methods for measuring stream temperature include both field measurements and airborne remote sensing methods. Field measurements commonly employ inexpensive temperature recorders that record stream temperature at specified time intervals (usually between 15 minutes and hourly) over several months to years (Dunham *et al.* 2005). Remote sensing methods such as the airborne Thermal Infrared (TIR) system can also be used to identify temperature patterns in streams, which indicate areas of increased temperature or cool-water refugia (Torgerson *et al.* 2001). Choosing between these two methods generally depends upon the purpose of the assessment, and the relative costs of acquiring temperature data. Field measurements have the advantage of characterizing *temporal* variation at specific sites, whereas TIR has the advantage of describing *spatial* variation in stream temperature at a single point in time. Where analyses aim to characterize certain thermal habitat features and their locations in a watershed (e.g. cool-water refugia), TIR provides the most suitable data for identifying those areas as potential habitats to target for protection. By contrast, if the assessment requires a better understanding of when and for how long certain temperature conditions exist, then temperature recorders provide more appropriate data.

Measurement of nutrient, pollutant or turbidity concentrations requires laboratory analysis of water samples from selected sites. As with stream temperatures, water-quality metrics are commonly compared to regulatory standards to identify impaired sites, and to indicate whether water quality is 'safe' for a wide range of biota. However, water-quality sampling for a wide range of nutrients and pollutants is often cost-prohibitive, and use of biological sampling methods as indicators of water quality degradation are often used to identify areas with water quality impairments based on analysis of biological communities (see Section 3.5). Nevertheless, measurement of specific water-quality attributes may be necessary to better understand causes of biological change and to identify restoration needs. Some parameters, such as turbidity, are monitored regularly at gauging stations, and data can be acquired for those few sites at a minimum. Other parameters may not be monitored continuously at gauges, and sampling water quality at key locations may be required to identify specific causes of biological change.

Selection of sample sites should consider stratification by reach type and land use, ensuring that each stratum is represented and there is sufficient sample density to adequately characterize each stratum (Madrid & Zayas 2007). In the context of watershed assessments for restoration planning, sample sites will most commonly be located in areas of relatively heavy land use in order to focus sampling effort on those areas most likely in need of restoration (e.g. agriculture or urban areas). Water-quality samples at each site may then be collected at a specified depth or by using a depth-integrated sampler to characterize the entire water column (Lane *et al.* 2003). Determining whether sites exceed regulatory standards or deviate from reference conditions requires analyses that account for temporal variation in concentrations, as concentrations of many compounds may spike briefly during runoff events. Moreover, some regulations (e.g. the total maximum daily load or TMDL rules of the US EPA) now require consideration of the amount of a particular pollutant that may be present in a river and allocation of allowable pollutant loads among sources. Hence, modeling of nutrient or pollutant loads in river systems is an increasingly common assessment tool for water quality (e.g. Borsuk *et al.* 2002).

3.5 Assessing changes in biota

Selecting assessment methods for changes in biota parallels selection of habitat assessment methods in that the choice is largely driven by the restoration goals. The purpose of assessing changes in biota is to identify where an individual species, life stage, or species assemblage has been altered or lost. Such measurements are also often used as an indicator of habitat or water quality, and to indicate where restoration efforts can help aid in their recovery. The focus of biological assessments will likely be driven by the legal or policy basis for river restoration, such as an Endangered Species Act focusing on single-species restoration or a more holistic act such as the Water Framework Directive focusing on multi-species restoration (Table 3.1). Once a single-species or multi-species approach is chosen, analyses should assess changes in abundance or diversity of species, or ecological and genetic attributes of species that respond to habitat changes or restoration actions. Two basic approaches to understanding the importance of different habitat changes on biota are: (1) field sampling of fish, macro-invertebrates or other biota; or (2) models to assess effects of habitat changes on various species (e.g. Karr 1991; Beechie *et al.* 1994; Wright *et al.* 2000). The purpose of either methodology is to identify which habitat changes have the most significant effects on aquatic biota.

3.5.1 Single-species assessment

Many restoration planning efforts focus on recovery of threatened or endangered species (Table 3.1). Field methods focus on presence/absence or density of fishes or other focal aquatic species (e.g. mussels, crayfish) or, in some cases, attempt to estimate survival in specific habitats or life stages. By contrast, models generally focus on identifying population 'bottlenecks' created by scarcity of specific habitats or high mortality during certain life stages (Beechie *et al.* 1994; Greene & Beechie 2004). Choosing between these two general approaches depends not only on the extent of existing knowledge about the focal species, but also on the purpose of the assessment. Where spatial distributions and abundances of organisms are already well known (e.g. for well-studied or heavily managed species such as salmon or trout), models often provide more new information for restoration planning than additional field surveys. By contrast, where less is known about distribution and abundance of a species (e.g. crayfish), there may be insufficient data on life-stage densities and survivals to parameterize a model, and additional field surveys may be more important for identifying key habitats for protection or restoration.

Field methods to estimate density of aquatic organisms include a variety of enumeration methods, especially for fishes. Methods for counting resident or anadromous fish during rearing life stages include snorkel surveys, electro-fishing, seining, and trapping, whereas adult-counting methods might also include spawner surveys and trapping at weirs. Counting methods for other less-mobile organisms, such as crayfish or mussels, may include surveying transects to estimate densities either by visual counts or trap counts. Estimating density of aquatic organisms by any of these methods requires both an e stimate of the number of individuals and an estimate of the habitat area sampled. Survival estimates are more difficult and costly to acquire, as they require repeat counts or captures to determine decreases in number of individuals through time. Hence, field measurements of survival are rarely a focus of assessments for restoration planning, although they are more commonly used in restoration effectiveness monitoring. Both density and survival estimates are subject to significant errors, but more labor-intensive sampling methods (such as use of mark-recapture methods) can decrease uncertainty in the estimates. However, increased time required at each sample site must be balanced against decreased number of sample sites (see Chapter 8 for discussion of sample size requirements). As with habitat surveys, more sample sites is usually more appropriate for describing watershed-scale patterns of fish abundance (or abundance of other organisms), whereas greater accuracy at a site is more useful for monitoring changes in density at a site (see also Chapter 8).

In contrast to field measurements, models are used to assess the relative importance of various habitat losses to a species and to help identify the restoration actions that are most important to increasing population growth rates or abundance (Leslie 1945; Bolten et al. 2010). Understanding these factors helps managers gain a better understanding of what habitats are most limiting to the species (Kareiva 2002; Bolten et al. 2010) and of the restoration actions that are most important to its recovery, rather than a 'shopping list' of concerns (Lawler et al. 2002). The simplest model for identifying specific

habitat limitations is a limiting factors model, which compares the relative capacity of habitats for each life stage and identifies the possible habitat factors limiting production (Reeves et al. 1989; Beechie et al. 1994). More specifically, this assessment identifies habitat bottlenecks within the life cycle, which are the habitats that limit the size of the population and therefore must be restored in order to increase population size. More complex life-stage models couple spatially and temporally explicit density and survival estimates for each life stage to assess which habitats or life stages have the greatest effect on population size, which is also helpful in identifying potential actions for population recovery (Greene & Beechie 2004; Lawson et al. 2005; Scheuerell et al. 2006). These models have the advantage of identifying not only where restoration of habitat *capacity* might lead to population increases, but also where improvements in *survival* within or between life stages might lead to population increases. In general, habitat capacity in the models is mainly a function of available habitat area and survival is mainly a function of habitat quality, although these two are not completely independent of each other.

It is important to recognize that use of habitat-based models for restoration planning still requires substantial field data to support the modeling effort. Most importantly, a good estimate of habitat area for each important life stage is required, as well as estimates of density and survival of the target organism in each habitat type and life stage (e.g. Reeves et al. 1989; Beechie et al. 1994). Habitat areas can be estimated using methods described in Section 3.4, and estimates of density and survival for various life stages and habitats can be gathered from the literature. If density and survival estimates are not readily available for the target species, modeling efforts should generally be set aside in favor of collecting additional field data on species distribution and abundance by habitat type and land-use impact. Where there are sufficient data to run these models, the important outputs of the model are the identification of key habitat areas or life stages that need restoration in order to improve the status of the target species. For example, the largest habitat losses in the Skagit River basin, USA were removal of side channels in the deltas and floodplains, combined with losses of beaver pond habitats (Table 3.15). However, individual salmonid species use these habitats in different ways depending on their life history, so the importance of restoring each habitat type depends on the target species (e.g. Beechie & Bolton 1999). If the aim is to restore coho salmon for example, then restoration efforts should

Table 3.15 Estimates of coho salmon habitat capacity losses in the Skagit River basin based on a limiting factors analysis (adapted from Beechie *et al.* 1994, 2001).

Habitat type	Historic smolt production	Current smolt production	Change in smolt production	Percent change
Sloughs				
Side channel	688,100	375,800	−312,300	−45%
Distributary	345,000	125,100	−219,900	−64%
Tributaries				
Hydromodified	44,400	37,600	−6,800	−15%
Non-hydromodified	72,700	55,700	−17,000	−23%
Culverts	37,100	0	−37,100	−100%
Dams	3,700	0	−3,700	−100%
Mainstem	379,400	351,400	−28,000	−7%
Lakes	18,300	48,000	+29,700	+162%
Ponds	783,700	177,500	−606,200	−77%
Total	2,372,400	1,171,100	−1,201,300	−51%

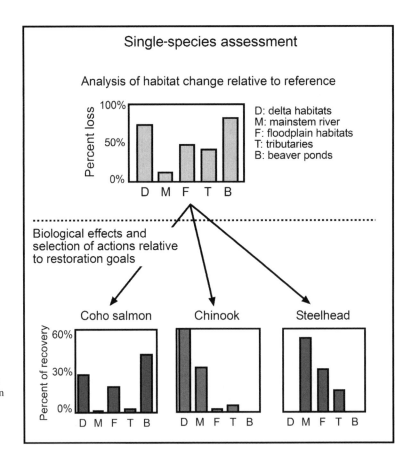

Figure 3.21 Single-species assessments identify important restoration needs based on habitat changes and habitat requirements of individual species. In this example, evaluation of habitat change yields a single result, but choice of species for restoration focus results in differing restoration needs due to differing habitat requirements among species (see text for additional details). Adapted from Beechie *et al.* (2010).

target restoring slow water habitats (e.g. beaver ponds and floodplain or delta habitats) that limit coho salmon winter rearing opportunities (Beechie *et al.* 2010; Figure 3.21). By contrast, restoring Chinook salmon, which do not have an extensive freshwater rearing phase, should focus mostly on mainstem edge habitats and delta habitats that are critical during their early rearing phase. Finally, steelhead make little use of delta and beaver pond habitats, so restoration of steelhead would focus on tributary and mainstem habitats important to both summer and winter rearing of juvenile steelhead.

3.5.2 Multi-species assessment

Ecosystem-focused restoration usually has broad multi-species objectives, which changes the focus of biological analysis from limits on individual populations to degradation of communities or food webs. Fish, invertebrates, amphibians, diatoms and other stream organisms are integral parts of aquatic food webs, and assemblages of these taxa are sensitive to a variety of watershed disturbances expressed over multiple spatial scales (i.e. site, reach, or watershed). Multi-species approaches can therefore include a variety of indicators of degraded habitat quality including species richness, disturbance tolerance, and functional feeding groups, which can detect a wider range of habitat problems than can be detected using a single-species analysis (Davis & Simon 1995). Multi-species assessments can be generally classified as one of two types: multimetric indexes or multivariate models. Multimetric indexes tend to be relatively simple scoring systems that rank the health of stream reaches based on a sum of numerical values assigned to various metrics, whereas multivariate models use statistical techniques to relate biological conditions to environmental variables.

Multimetric indexes have focused on a variety of indicator organisms, but most commonly fishes and benthic invertebrates (Rosenberg & Resh 1993; Merritt & Cummins 1996; Schmutz *et al.* 2000). For example, in western North America a Benthic Index of Biological Integrity (B-IBI) using benthic macro-invertebrates was developed and calibrated with data from both Oregon and Washington (Fore *et al.* 1996). The B-IBI (Karr & Chu 2000) is composed of 10 measures of taxa richness, population structure, disturbance tolerance, and feeding ecology. When scores from these metrics are summed, B-IBI provides a numeric synthesis of site condition that ranges from poor to excellent and indicates a variety of resource conditions or problems (Morley & Karr

2002). Similar metrics have long been in use in Europe as well; maps of these biological conditions within a watershed provide a useful illustration of where habitats have been degraded, and also indicate habitat improvements through time (Figure 3.22). In other areas such as the Midwestern USA and Europe, fish have commonly been used for multimetric indicators of habitat condition (e.g. Schmutz *et al.* 2000). In Europe, Schmutz *et al.* (2000) identified seven fish metrics relative to reference conditions including: number of type-specific species; number of self-sustaining species; whether a shift in fish region occurred; number of guilds; degree of alteration in guild composition; change in biomass and density; and change in population age structure. Ecological integrity scores range from 1 to 5 for each of these criteria, and again the summed scores indicate overall conditions at a site or within reaches (Schmutz *et al.* 2000).

Multivariate models use statistical relationships between measures of instream biological condition and land-use/land-cover patterns to identify specific stressors causing biological impairment (Allan *et al.* 1997; Schmutz *et al.* 2000), which helps in setting realistic recovery goals given current land-use patterns in a river basin. Reference conditions are ideally based on a large number of minimally disturbed sites (c. 200 sites; Reynoldson *et al.* 2001). Using multivariate statistical analyses, reference sites are then matched to a set of reach descriptors (e.g. stream order, elevation, geology, etc.) and classified into physiographic groups. Finally, biological reference conditions can be defined for each group. The level of impairment at new sample sites within a watershed is then determined by comparing its current biological condition with that of the appropriate reference group. This approach has been most widely applied with the development of: RIVPACS (River Invertebrate Prediction and Classification System) in England; AUSRIVAS (Australian River Assessment Scheme) in Australia; and BEAST (Benthic Assessment of Sediment) in Canada (Wright *et al.* 2000). Many regional or provincial approaches have also been developed (e.g. Canale 1999; Reynoldson *et al.* 2001).

The choice of taxa (i.e. fishes versus invertebrates) for either type of assessment depends mainly on regional patterns of species diversity. In some regions, such as the Pacific Northwest USA, fish species diversity is quite low and the most abundant species occupy a relatively wide range of habitat types and qualities, so fishes do not provide a good range of sensitive metrics for multi-species assessments. By contrast, invertebrate communities are very diverse and specialized, making them much better candidate taxa. In other regions such as the eastern

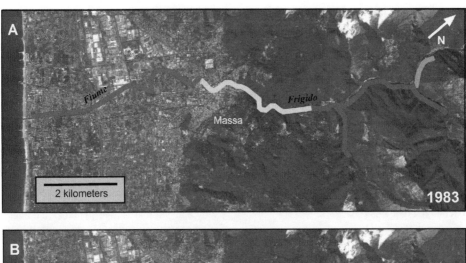

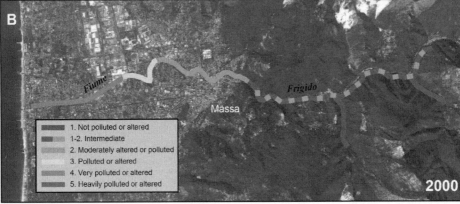

Figure 3.22 Example of using multispecies metrics to display habitat and biological conditions within the watershed of the Fiume Frigido, Italy in (A) 1982 and (B) 2000. In 1982 the river was heavily impacted by sedimentation from marble quarries in the headwaters, as well as agriculture and development pressures in the lower watershed. Efforts to reduce sedimentation and other impacts have significantly improved biological and habitat status in the last two decades. (Adapted from Banchetti *et al.* 2004; CIRF 2006; background image copyright 2012 Google, 2012 TeleAtlas, 2012 Digital Globe.) (See Colour Plate 2)

USA or central Europe, fish taxa are more diverse, and multi-species metrics can be developed for fishes as well as invertebrates (Harris *et al.* 2005).

Outputs of either type of assessment generally include color-coded maps of biological condition in stream reaches, which can be used to identify high-quality areas for protection or degraded reaches needing restoration (Figure 3.22). In many cases, relationships among individual metrics and land-use categories or habitat attributes can also be used to identify general landscape features that lead to degradation, and to help focus assessments of disrupted watershed- and reach-scale processes toward those that are the most likely causes of biological impairment (Beechie & Bolton 1999). For example, in Midwestern USA number of Ephemeroptera, Plecoptera and Trichoptera (EPT) invertebrate taxa and total fish IBI score were both related to percent urban land cover in a catchment, and the IBI metrics were also related to increased concentrations of trace metals in the stream bed (Harris *et al.* 2005; Figure 3.23). These types of results help to focus the spatial extent of subsequent water-quality assessments into areas where water quality data will help identify specific restoration needs. Finally, relationships between multimetric indicators and land-cover attributes can indicate where ecosystem restoration is constrained by existing land uses, suggesting that restoration efforts might best be focused elsewhere until viable restoration actions can be identified.

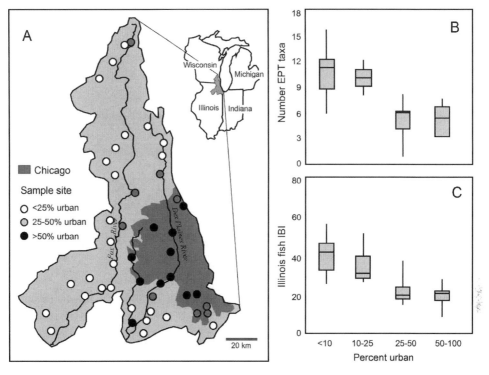

Figure 3.23 Example of multi-species assessments in Midwestern USA, showing (A) distribution of sample sites and percent urbanization, (B) number of Ephemeroptera, Plecoptera and Trichoptera invertebrate taxa relative to percent of the watershed urbanized, and (C) total score for the fish Index of Biological Integrity relative to percent of the watershed urbanized. Based on Harris *et al.* (2005).

3.6 Assessing potential effects of climate change

Ultimately, climate change projections might lead planners to alter restoration plans or actions when there is evidence that future climatic conditions will not support current ecological targets for either habitats or biota (McCarty 2001). There are two important considerations for determining whether restoration plans or actions should be altered in response to climate change (Beechie *et al.* 2012):

1. What is the predicted effect of restoration plans or actions should be altered in response to climate change (usually stream flow or stream temperature)?
2. How will these effects limit habitat restoration effectiveness for targeted biota?

Answering these questions relies on downscaled climate and hydrologic models that simulate changes in future stream flow and temperature, as well as estimates of

future habitat changes that might limit restoration effectiveness for biota (Mantua *et al.* 2010; Beechie *et al.* 2012). If these analyses suggest that future restoration effectiveness will be reduced, then one should re-evaluate planned restoration actions and focus on those actions that either ameliorate a climate change effect or increase habitat diversity and biological resilience (Beechie *et al.* 2012) (see Section 3.7.4). The products of climate change analyses ideally include maps of projected changes in stream flow and temperature, as well as an assessment of how those changes will affect restoration efforts.

Anticipating future constraints on restoration effectiveness first requires scenario modeling of climate change effects on habitats. However, there are considerable uncertainties in prediction of future emissions scenarios, variation in how global climate models translate greenhouse gas emissions into changes in air temperature and precipitation, and uncertainty in models that translate air temperature and precipitation into stream flow and stream temperature. Nevertheless, broad global patterns of climate change effects are consistent across scenarios and models, suggesting that northern latitudes will be substantially warmer and wetter and equatorial latitudes will only be slightly warmer and mostly drier. For river restoration planning, downscaling of global climate models increases the spatial resolution of predictions (e.g. Elsner *et al.* 2010), but additional uncertainties are introduced by downscaling procedures. These uncertainties can only be quantified to a limited degree. First, models can be compared to each other to quantify differences between them, but this does not reveal anything about the accuracy of the predictions. Second, model accuracy can be partially assessed by examining how well the model can predict climate in the recent past, and then quantifying model biases relative to that past climate (e.g. Christensen & Lettenmaier 2007; Hayhoe *et al.* 2007). Both of these steps help understand model uncertainty to some degree, but ultimately it is impossible to know whether any one emissions scenario or climate model is more accurate at future predictions than another.

There are several approaches to translating scenarios of future climate into effects on biota, including bioclimatic envelopes, bioenergetics, and life-cycle modeling. Bioclimatic envelopes define tolerance limits of species for specific environmental variables (e.g. stream temperature). They are developed by correlating current climate and physiographic variables with species ranges, and then modeling how the spatial distribution of those attributes will change in the future (Heikkinen *et al.* 2006). These shifts in attributes are then typically interpreted to mean that the geographic range of a species will shift in response to the shifting environmental variables. In the context of assessments for stream restoration, these bioclimatic envelopes can also be used to evaluate whether habitat conditions within the current species range will likely fall outside a species' tolerance limits in the future, thereby constraining restoration options (McCarty 2001). Heikkinen *et al.* (2006) review a variety of statistical methods and models that can be used for defining bioclimatic envelopes.

Bioenergetics approaches focus on analyzing the potential effects of stream temperature and flow changes on metabolic rates of fishes, their food resources, or both. In general, areas that are already warm may see increased metabolic rates and reduced growth if food resources remain the same or decrease. By contrast, in areas that are near the colder end of thermal tolerances, increased metabolism and food production may result in increased growth rates. Life-cycle models are more complex, but they have a significant advantage over bioclimatic enve-

lopes and bioenergetics analyses in that they integrate multiple climate change effects across life stages to assess likely impacts at the scale of populations. For example, a life-cycle model was used to evaluate the likely future effectiveness of Chinook salmon restoration actions with

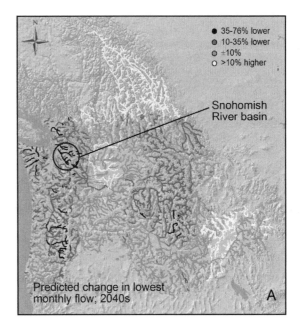

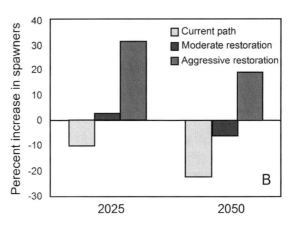

Figure 3.24 Estimates of (A) change in lowest monthly stream flow by the 2040s in Pacific Northwestern USA (data from University of Washington Climate Impacts Group, Seattle, USA, http://www.hydro.washington.edu/2860/report/), and (B) change in Chinook salmon population size by 2025 and 2050 for the Snohomish River basin, Washington State, USA based on varying levels of restoration effort and a moderate CO_2 emissions scenario (adapted from Battin *et al.* 2007). (See Colour Plate 3)

climate change in the Snohomish River basin, USA (Battin *et al.* 2007). In that case, several scenarios of future restoration effort and future climate change were modeled to determine (1) whether climate change might alter future restoration effectiveness for Chinook salmon and (2) which kinds of restoration actions were most robust to climate change. Outcomes of the study indicated that aggressive restoration efforts were more likely to improve population status despite climate change (Figure 3.24).

3.7 Identifying restoration opportunities

The most important (and final) step in watershed assessment is translating the assessment results into a list of necessary restoration actions (Beechie *et al.* 2008a). The list of restoration actions should be spatially explicit, using the full range of assessments to identify restoration actions that are necessary to achieve local restoration goals. Moreover, the watershed assessment summary should clearly identify which processes or functions are most impaired and most responsible for biological degradation, so that limited restoration dollars can be focused on actions and locations that will most improve the status of biota (see also Chapter 6). Important attributes of the summary are that it address each sub-basin or reach separately, and that it ranks the level of impairment for each watershed process to indicate which impairments have the largest habitat effects in each sub-basin or reach. The analysis summary should also identify constraints on restoration options, such as habitats that may not be restorable or that may require prohibitively expensive restoration efforts. Understanding these constraints helps to set realistic expectations for the restored ecosystem, as well as for the potential effectiveness of other restoration actions if key processes cannot be restored (Beechie *et al.* 2010). In this section we briefly discuss each of these summaries, and in the following section we present two brief case examples of watershed assessments to illustrate the use of watershed assessments in restoration planning.

3.7.1 Summarize the watershed assessment results and identify restoration actions

Watershed assessments answer two key questions that link disruptions in causal processes to habitat change, and habitat change to biological responses (Figure 3.2). The summary of the watershed assessment should therefore clearly state the results of: (1) the watershed process assessments (Section 3.3); (2) the habitat change assessments (Section 3.4); and (3) the biological response assessments (Section 3.5). Moreover, key linkages between these assessments should be described as completely as possible so that it becomes clear in developing a restoration plan how each type of restoration action will lead to habitat and biological improvement, and how much biological improvement can be expected from each action type or suite of actions.

Among the most important outcomes of a watershed assessment are maps and summaries of impaired processes or functions causing degradation of habitats and biota, and an indication of the degree to which each process or function is impaired within each area of the watershed (Beechie *et al.* 2008a; Table 3.16). The summary identifies the root causes of river ecosystem degradation (i.e. it answers Question 1 in Figure 3.2), and provides the basis for identifying restoration actions that follow the process-based principles outlined in Chapter 2. This table (or tables) should address all of the driving processes and functions in the assessment, and should be accompanied by maps illustrating key results informing a strategy for restoration. For example, Table 3.16 illustrates that Sub-watershed 4 has relatively few process impairments, whereas Sub-watershed 1 has a broad array of process impairments to address through restoration. Maps of processes in need of restoration help restoration planners visualize the spatial arrangement of restoration needs, and communicate the status of watershed processes to the general public (illustrated in Section 3.8).

Once each of these assessments is summarized, inventories of individual restoration actions or sites should be compiled within each category. Because the watershed-scale assessments are generally based on budgets or models that do not account for site-scale conditions or restoration needs, additional inventories are often needed to identify specific restoration actions. For example, the sediment budget may indicate that forest roads are a significant sediment source in certain sub-basins, but identifying specific road treatments to reduce landslide hazard or surface erosion requires field surveys of road segments to identify areas in need of rehabilitation (e.g. Beechie *et al.* 2003a). By contrast, reach-level process assessments are often based on site-level inventories, and the assessment and inventory are essentially the same (such as a migration barrier assessment or a riparian assessment). In either case, lists of specific restoration actions can be developed based on the inventory results.

Table 3.16 Simplified example summary of process impairments identified in a hypothetical watershed assessment; H: high; M: moderate; and L: low indicate the level of impairment for each process and sub-basin (modified from Beechie *et al.* 2008a).

Process/function	Specific cause of problem	Sub-basin			
		1	2	3	4
Hydrology	Dams reduce channel-forming flows	M	M	L	L
	Levees and tide gates reduce estuary habitat	H	L	L	L
Sediment	Road surface erosion	L	M	M	H
Riparian	Reduced wood delivery	M	M	H	H
	Reduced shade (mainly in agricultural lands)	H	M	L	L
Channel	Bank armoring constrains channel	H	M	L	L
	Channel dredging has reduced rearing habitat	H	L	L	L
Floodplain	Levees disconnect channel from floodplain	H	M	M	L
Connectivity	Fish migration blocked by impassable culverts	L	L	M	M
Water quality	Pesticide input from agriculture and urban zones	M	L	L	L

While the assessment of watershed processes identifies specific causes of habitat degradation and the restoration actions needed to address them, the assessments of habitat change and impacts to biota evaluate the relative importance of the many possible restoration actions. Therefore, the watershed assessment should also summarize habitat change and its effects on biota (i.e. answering Question 2 in Figure 3.2). This summary describes the spatial arrangement of habitat and biological impairments, and highlights which habitat losses or changes have had the greatest effect on targeted biota. These factors are often incorporated into prioritization methods to identify restoration actions that will most improve biological status and therefore be given the highest priority (Chapter 6). Where biological endpoints are driven by single-species needs (e.g. a threatened or endangered species), habitat requirements of the species help to interpret the relative importance of each habitat change and therefore which habitats will have the greatest restoration benefit (from the biological assessment described in Section 3.5.1). As illustrated in Figure 3.21, the same suite of habitat changes can result in a variety of different restoration priorities depending on the species of interest. Where biological endpoints are more broadly defined (e.g. an IBI or other multi-species index), determining which types of restoration actions will result in the greatest biological improvements may be less clear and quantitative. Nevertheless, it should still be possible to make estimates of the likely effects of various restoration actions on biological endpoints.

3.7.2 Develop a restoration strategy

Once the watershed assessment summary is completed, each of the assessment elements can be integrated into a comprehensive plan for watershed restoration. A restoration plan should acknowledge the biological importance of habitat types or areas, and capitalize on an understanding of which restored habitats will achieve the greatest advancement toward a restoration goal. First, an understanding of habitat changes and the biological importance of those changes indicates the spatial arrangement and types of habitats that need to be restored to achieve some level of biological recovery. Second, once the important habitat areas have been identified, the table of process impairments can be translated into a list of restoration needs (Table 3.17). At this stage, restoration needs are identified as general types of restoration actions, their locations, and approximate levels of effort needed to address each of the impaired processes and ultimately to achieve restoration goals. This broad-scale summary lays out important areas for habitat or watershed protection (i.e. areas that are reasonably intact and should be maintained), high-priority areas for restoration, and areas with relatively low potential for improvement. Areas designated as a high priority for either protection or restoration will typically include multiple project types with varying biological benefits.

Within high-priority restoration areas, specific restoration actions for each process or habitat type can then be listed (e.g. which culverts can be removed to allow fish migration, which riparian areas need fencing or silvicul-

Table 3.17 Simplified example summary of restoration needs to address impaired processes and functions summarized for a hypothetical watershed in Table 3.16. At this stage of restoration planning the action descriptions are generalized, and H (high) and M (moderate) indicate the likely level of effort needed to address each process in each sub-basin (modified from Beechie *et al.* 2008a).

Process/function	Restoration action	Sub-basin			
		1	2	3	4
Hydrology	Restore large flood flows	M	M		
	Remove or redesign tide gates for passage	H			
Sediment	Improve road surfacing to reduce erosion		M	M	H
Riparian	Silvicultural treatment to increase wood recruitment	M	M	H	H
	Riparian planting to increase shade	H	M		
Channel	Remove bank armoring where possible	H	M		
	Limit dredging where possible	H			
Floodplain	Remove or set back levees where possible	H	M	M	
Connectivity	Replace blocking culverts with bridges that allow fish migration			M	M
Water quality	Provide riparian buffers to filter pesticides; reduce pesticide use	M			

tural treatment, or which levees might be set back to reconnect a river to its floodplain). There is a wide range of restoration techniques available to address restoration needs, and selecting the most appropriate technique for a given problem will depend on site characteristics and restoration goals (see also Chapters 5 and 7). For a thorough watershed assessment, these project lists can include hundreds or even thousands of potential projects. Therefore, a process for prioritizing these potential restoration actions is essential for cost-effective implementation of a restoration plan (discussed in Chapter 6).

3.7.3 Summarize constraints on restoration opportunities

Constraints include a variety of physical or socioeconomic factors that may limit restoration options. In most cases, constraints are structures or land uses that have economic or social benefits that outweigh ecological values (Baker *et al.* 2004). That is, where local or regional communities place greater value on goods and services provided by human infrastructure than on goods and services provided by a healthy river and its biota, the human infrastructure becomes a constraint on restoration. Such constraints limit the restoration potential of a river ecosystem by precluding full restoration of certain processes or habitats. For example, dams with significant hydropower or irrigation benefits, or levees that protect high-value infrastructure, are often more highly valued than restoration of stream flows and floodplain habitats.

Those dams or levees are therefore not considered for removal or setback in a restoration plan, and their continued existence becomes a constraint on the kinds of restoration actions that are possible. As a consequence, potential habitat and biological improvements are also limited. Understanding these limits is critical for setting realistic expectations for the outcomes of restoration efforts, as constraints may in some cases preclude achievement of biological restoration objectives.

Constraints identified by watershed assessments may also indirectly influence the selection and design of restoration actions, as constraints may alter rates of watershed processes, local restoration potential, or expected attributes of a restored system. For example, a dam that diverts a significant portion of stream flow and eliminates sediment supply to downstream reaches substantially alters expectations for a restored river, and historical conditions do not provide a suitable target condition (Trush *et al.* 2000; Burke *et al.* 2009). Rather, the constraint imposes new sediment and hydrologic regimes on downstream reaches, and therefore the expected restored channel will be smaller and less dynamic than the historical (pre-dam) channel. Recognizing these indirect constraints is important to selecting appropriate types of restoration actions (Chapter 5), and also in designing projects to match the local restoration potential (Chapter 7).

Finally, social and economic constraints often limit the pace and extent of restoration. Most commonly, funding is well short of that needed to implement all (or even most) restoration projects identified in a watershed assessment, and projects are therefore implemented slowly over long periods of time. Moreover, large projects with substantial biological benefits are often the most expensive and require the most time to garner public support. Design and implementation of large high-value projects is therefore often a lengthy process, which further constrains the rate at which restoration can occur. These types of socioeconomic constraints do not limit potential biological outcomes in the same way that infrastructure constraints do, but they do limit the rate at which restoration can occur and ultimately how many restoration actions will be implemented. That is, these constraints do not eliminate the possibility of improvement as an infrastructure constraint does, but they limit the ability to achieve that potential improvement. Therefore, where socioeconomic constraints limit the pace and magnitude of restoration, a transparent process for prioritizing restoration actions becomes more important (Chapter 6).

3.7.4 Climate change considerations

Two key climate change considerations might influence selection of restoration actions: (1) does a proposed action reduce the likely climate change effect? and (2) does the restoration action increase resilience of a population or ecosystems? (Beechie *et al.* 2012). Where climate change might shift either stream flows or temperatures to adversely affect biota, a restoration action that reduces temperature, reduces low-flow effects or decreases the effect of flood flows will likely benefit biota despite a changing climate. If restoration actions will not ameliorate climate change effects, then it is important to select actions that increase ecosystem or population resilience. The ecological concept of resilience refers to the ability of a system to 'absorb disturbance and reorganize while undergoing change so as to still retain essentially the same function, structure, identity, and feedbacks' (Holling 1973). In the case of river restoration, restoring floodplain connectivity, restoring variable flow regimes, or restoring migration pathways between diverse habitats will increase the ability of systems to reorganize both physically and biologically in response to climate change (Waples *et al.* 2009).

Because there are significant uncertainties in predicting future habitat conditions using climate and hydrologic models, a precautionary approach to identifying restoration actions for long-term benefits is to ensure that river ecosystems are physically and biologically able to adjust to a changing environment. Building ecosystem resilience

has been considered critical to ecosystem restoration for the past decade, even without considering climate change (Bottom *et al.* 2011). In a review of restoration actions that either ameliorate climate change effects or increase resilience, Beechie *et al.* (2012) identified a suite of actions that are most likely to maintain their effectiveness in a future climate. Most importantly, actions that increase floodplain connectivity, maintain access to diverse habitats, and restore exchange between surface water and ground water are most likely to improve riverine ecosystem resilience in a changing climate (Waples *et al.* 2009). By contrast, actions that restore watershed processes are moderately likely to ameliorate climate change effects, whereas most instream restoration actions are not likely to sustain their effectiveness in a future climate.

3.8 Case studies

We describe two case studies – one from the USA and one from the UK – to illustrate key products of watershed assessments and their use in identifying restoration actions and developing restoration strategies. The first example uses process-based assessments to identify areas of the river basin in need of restoration actions, whereas the second example uses a risk-based modeling approach to identify restoration needs. In both cases the analyses address root causes of degradation, changes in habitat condition, and the response of biological indicators to guide the development of watershed-scale restoration strategies. Assessment methods for these watersheds were described in Section 3.2 (Table 3.4), and several of the examples of assessment methods and products shown earlier are from these two studies. We refer to those examples as appropriate to indicate linkages between the assessment design, assessment summaries and development of restoration strategies.

3.8.1 Skagit River, Washington State, USA

The first element of assessing causes of habitat change was evaluation of a suite of key processes driving habitat formation in the Skagit River basin, including sediment supply, hydrology, riparian functions, floodplain connectivity, and blockages to salmon migration. Because agricultural land uses significantly overlap salmon habitats – primarily in the floodplains and delta (Beechie *et al.* 1994) – impaired floodplain connectivity, riparian functions, and salmon access to habitats tended to be concentrated in the lower river basin (Figure 3.25; Skagit Watershed Council, unpublished

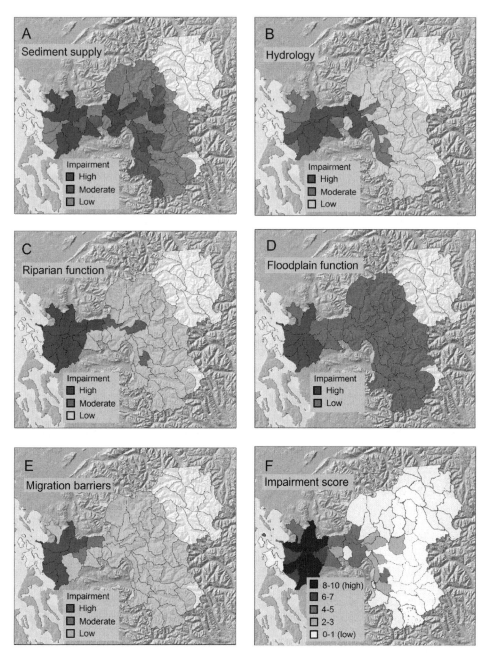

Figure 3.25 Summary maps of analyses of causes of habitat change in the Skagit River basin, Washington State, USA (based on unpublished data from Skagit Watershed Council, Mount Vernon, Washington State, USA). Five panels illustrate altered watershed processes including (A) sediment supply; (B) hydrology; (C) riparian functions; (D) floodplain connectivity; and (E) amounts of habitat blocked to salmon migration by artificial barriers. (F) Impairment rating map illustrates the combined level of impairments across all five factors; each factor receives a score of 2 if impairment was rated high, a score of 1 of impairment was rated moderate, and a score of 0 if impairment was rated low. Score of 0 indicates no or low impairment to any of the five processes and a score of 10 indicates that all five processes are impaired. (See Colour Plate 4)

data). By contrast, increased erosion and runoff were more widely distributed in steeper areas of the basin where forestry activities increased landslide rates and peak flows during fall and winter storms (e.g. Figure 3.5). When considered in combination, the delta and lower basin tributaries were degraded by a larger number of processes than were headwater areas. Notably, a significant proportion of headwater reaches were not rated as impaired for any process.

Analysis of habitat change in the Skagit River basin (the second assessment component) documented large losses of floodplain, delta, and beaver pond habitats (Figure 3.20; Beechie *et al.* 1994, 2001) which are used to varying degrees by the three target salmonid species. Beaver pond habitats were likely reduced significantly by beaver trapping prior to the early 1800s. By contrast, most floodplain and delta habitats were lost in the early 1900s as large areas of floodplain and delta were disconnected from the river by levees built to protect agricultural and urban lands from flooding (Figure 3.20). Mainstem habitat losses were relatively small; only a short section of upriver mainstem channel was inundated by hydroelectric dams, and bank armoring affects a relatively small percentage of river edge habitats (not including the floodplain habitats, assessed separately). Aside from losses of beaver dams, tributary habitats have been primarily affected by loss of in-channel wood and the subsequent loss of pool habitats.

Finally, when each species' habitat requirements were incorporated into the analysis, the importance of habitat losses to restoration of salmon populations varied between species (Figure 3.21). Coho salmon use slow-water habitats in winter, and a limiting factors analysis showed that losses of floodplain, delta, and pond habitats have all significantly reduced habitat capacity for the species (Beechie *et al.* 1994; 2001). The analysis on the whole suggests that riparian and tributary restoration actions may contribute little to increasing population size if the more significant limiting habitats are not addressed. Current habitat constraints on Chinook salmon were evaluated using a combination of analyses that showed that loss of delta habitats is the primary habitat constraint for most Chinook salmon (the ocean-rearing type), but also that losses of mainstem river habitats may be a constraint on the smaller populations of stream-rearing Chinook (Beamer *et al.* 2005a; 2005b). Thus, delta and mainstem habitats are critical restoration areas for Chinook salmon recovery. Finally, a limiting factors analysis for steelhead, a species that makes little use of slow-water habitats at any life stage, showed that the greatest restoration needs are in mainstem, floodplain, and tributary channels.

Using these analyses, the Skagit Watershed Council developed a strategic approach to habitat protection and restoration based on three guiding principles: (1) restore processes that form and sustain salmon habitats; (2) protect functioning habitats from degradation; and (3) focus protection and restoration actions on the most biologically important areas (Skagit Watershed Council 2010). Because funding for the watershed council and its restoration activities is tied to recovery of the Chinook salmon listed under the US Endangered Species Act, the strategic approach considers the most important areas to be those that will most benefit Chinook salmon recovery. Moreover, restoration actions are proposed and carried out by a variety of restoration groups, and most restoration actions are to some extent reliant on finding opportunities with cooperative landowners. Therefore, the strategic approach recognizes that many important projects cannot be enacted in the near term, but also aims to encourage restoration actions in specific target areas. These target areas are organized into three tiers, ranked in order of importance to Chinook salmon recovery (Table 3.18; Figure 3.26). Tier 1 target areas encompass the estuary and delta, as well as floodplain rearing habitats used by multiple salmon populations; Tier 2 target areas include nearshore rearing areas and floodplain rearing habitats used by single populations; and the Tier 3 target area focuses on watersheds with elevated erosion rates or peak flows, both of which affect egg-to-smolt survival in years with high peak flows.

Within these target areas, the Skagit Chinook Recovery Plan (SRSC/WDFW 2005) identifies more than 40 specific project reaches or sub-watersheds, many of which include numerous individual project sites and actions. In addition, the assessments identified more than 600 barriers to salmon migration, 80 km of forest roads needing restoration (in addition to hundreds of kilometers to be treated under current forestry regulations), 98 km of bank armoring that could be restored or modified to increase rearing habitat function, and several hundred hectares of floodplain and delta areas targeted for levee removal or modification. These actions address restoration in each of the target areas, and also address each of the key limiting factors at each life stage. If all these actions are implemented, they are estimated to achieve between 77% and 81% of stated recovery goals (i.e. the population size needed to remove the populations from the Endangered Species list) (Table 3.19). Importantly, the Skagit Chinook Recovery Plan recognizes that land-

Table 3.18 Summary of target areas for the Skagit Watershed Council 2010 Strategic Approach (adapted from Skagit Watershed Council 2010).

	Target area	Importance to Chinook salmon
Tier 1	Skagit estuary	Key habitat features include delta distributaries and blind tidal channels. Critical physiological transition zone for juvenile Chinook and highest growth rates for juvenile Chinook in watershed. Loss of habitat substantially reduces juvenile survival in Puget Sound and ocean.
	Riverine tidal delta	Riverine tidal marshes and wetlands. Historically expansive habitat area for delta-rearing Chinook life history type. Rearing habitats limited due to levee system.
	Large river floodplains (mixed population rearing)	Large river–floodplain areas with highly productive habitats. Historically expansive rearing habitat area for distinct riverine juvenile Chinook life history type; provides rearing habitat for all six independent Chinook populations. Major spawning areas for fall and summer Chinook.
Tier 2	Nearshore pocket estuaries	Isolated and relatively small estuary habitats located along nearshore areas of Skagit Bay. Rearing habitats for fry migrant Chinook salmon emigrating from Skagit River in large numbers.
	River floodplains (single population rearing)	Large tributary floodplains that provided extensive spawning and rearing habitat areas for Chinook salmon. Major spawning areas for single Chinook populations; important to spatial structure and life history diversity of Chinook populations.
Tier 3	Sediment and hydrology impaired watersheds	Increased risk of severe habitat degradation and reduced Chinook survival due to high risk of landslides, road failures, and peak flows caused by historic land management (i.e. logging) and forest road development.

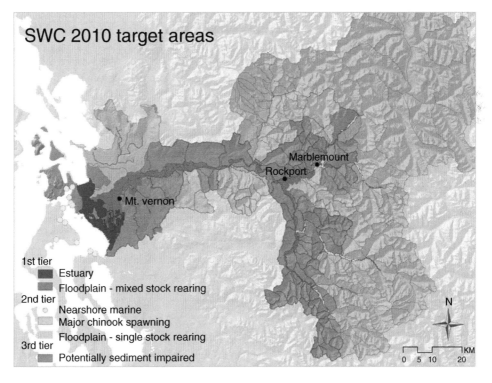

Figure 3.26 Illustration of target restoration areas in the Skagit River basin, USA. Adapted from Skagit Watershed Council (2010). Descriptions of target areas are listed in Table 3.18. (See Colour Plate 5)

Table 3.19 Predicted response of Chinook salmon to planned restoration actions in the Skagit River basin during periods of low or high marine survival (low regime and high regime, respectively). Marine survival rates are 0.01–0.25% for low ocean survival regime and 0.18–3.5% for high ocean survival regime (adapted from SRSC/WDFW 2005).

Marine survival	Recovery goal (adults per year)	Before plan actions (adults per year)	After plan actions (adults per year)	Percent of goal achieved
Low regime	52,430	28,611	40,267	77%
High regime	145,100	83,962	118,168	81%

use constraints will prevent or delay at least some of these actions, and that over the long term societal values may either decrease or increase the importance of salmon restoration relative to land use or economic considerations. Hence, the plan predicts Chinook salmon response to restoration under the optimistic assumption that all planned actions can eventually be completed.

3.8.2 River Eden, England, UK

In the Eden River basin, a risk-based modeling framework focusing on estimating the risk of fine sediment delivery from distributed agricultural sources (SCIMAP or Sensitive Catchment Integrated Modeling Analysis Platform; www.scimap.org.uk) was developed by Durham and Lancaster Universities to assess watershed-scale processes (Reaney *et al.* 2011). Using land-use data (from satellite imagery), rainfall data and 5 m digital elevation data, SCIMAP identified erosion-prone fields that were likely hydrologically connected to the river by overland flow pathways and therefore likely sediment sources. Instream risk at a point in the channel was then calculated by integrating the risk from all upstream sources contributing to that point (Figure 3.6). Based on these analyses the trust was able to identify likely sediment sources (agricultural fields) in need of restoration, as well as reaches that were likely to be most impacted by those land-use effects.

At the reach scale, data on riparian conditions and channel morphology were assessed with high-resolution aerial photography, which could be collected rapidly, was unhindered by land ownership issues, and could be stored and analyzed cost-effectively. Aerial photography for 650 km of main river and tributaries was captured at 20 cm resolution, and then analyzed to identify bank erosion associated with livestock grazing, riparian trees, overhead channel cover, channel width, and substrate. From these data the Eden Rivers Trust was able to identify

important areas for livestock exclusion or riparian restoration (see Figure 3.13). Although focused on sediment delivery, the outputs of the model can be considered a proxy for other forms of diffuse pollution, which follow a similar pathway through the landscape (e.g. phosphorous bound to fine sediment). Habitat assessment at the reach scale made use of a 5 m digital terrain model (DTM) to calculate channel slope and extrapolate associated physical biotopes (e.g. cascade, riffle, run, pool). From these data the Eden Rivers Trust was able to assess the likelihood of suitable age 0+ salmonid habitat throughout the basin (e.g. riffle being present).

Finally, age 0+ Atlantic salmon (*Salmo salar*) and brown trout (*S. trutta*) populations were surveyed as indicators of freshwater habitat condition due to their stringent water-quality and habitat requirements. Timed, semi-quantitative electrofishing counts (Crozier & Kennedy 1994) were used because up to 10 sites a day could be surveyed rather than the typical four sites that could be surveyed using multi-pass removal or mark-recapture methods. Although restricted to age 0+ salmonids, this survey rapidly provided a good overview of salmonid breeding distribution in the Eden, as well as an indication of spawning success. As the least mobile life-stage, abundance of age 0+ salmonids was assumed to be constrained by local habitat pressures and therefore reflective of habitat condition and water quality.

The key output of the assessments was a spatially structured, hierarchical database. This integrated data source allowed relationships between habitat pressures and age 0+ salmon and trout abundance to be examined and prioritized using multivariate statistics at multiple scales (Figure 3.27). At the watershed scale, age 0+ Atlantic salmon occupied wide low-gradient streams with more spawning gravel, fewer migration barriers, and less siltation, whereas age 0+ brown trout occupied narrower, steeper streams with overhead cover from riparian trees. The most significant variable explaining age 0+ salmon

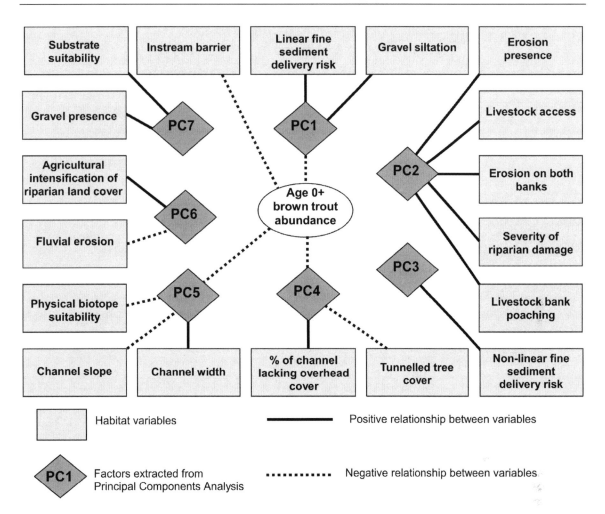

Figure 3.27 Schematic diagram of the multivariate statistical analysis used to evaluate habitat condition and landscape influences on brown trout in the Eden River basin, England, UK.

abundance was fine sediment delivery risk, although the relationship was non-linear in that both extreme low- and high-risk correlated with reduced abundance. Age 0+ brown trout abundance was also related to fine sediment delivery risk and population fragmentation due to instream barriers, and overall there were fewer high-abundance trout sites than salmon sites (i.e. Atlantic salmon seem to be performing better overall than brown trout). At the sub-basin scale, bank erosion and gravel presence became important predictors of age 0+ salmon and trout abundance, although the importance of each process varied by sub-basin.

Integrating these assessments in a multivariate statistical analysis showed the overall significance of fine

sediment delivery risk for both salmonid species. Further analysis of this relationship found that arable land cover types were relatively unimportant as drivers of age 0+ salmonid abundance, whereas improved pasture was very important. Improved pasture is associated with a number of stressors including nutrient applications, fecal material, and fine sediment (Reaney et al. 2011). Addressing these diffuse large-scale pressures had previously been considered infeasible because high-risk land covers occupy the vast majority of the landscape and it was believed that the required changes in land management would be too economically burdensome. However, the watershed analysis considered both land-cover impacts and hydrological connectivity, which allows development

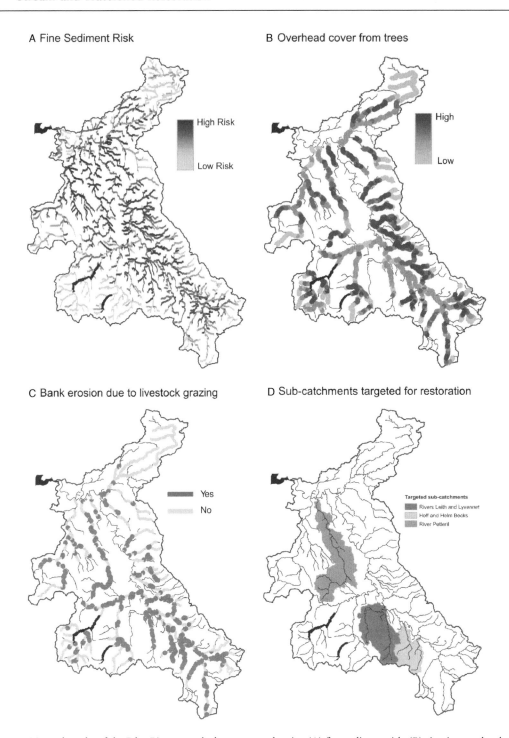

A Fine Sediment Risk

B Overhead cover from trees

High Risk

Low Risk

High

Low

C Bank erosion due to livestock grazing

D Sub-catchments targeted for restoration

Yes

No

Targeted sub-catchments

Rivers Leith and Lyvennet
Hoff and Helm Becks
River Petteril

Figure 3.28 Mapped results of the Eden River watershed assessment showing (A) fine sediment risk; (B) riparian overhead cover; (C) bank erosion due to livestock grazing; and (D) the three sub-catchments targeted for restoration efforts. See also Table 20 for description of targeted sub-basins shown in (D). (See Colour Plate 6)

Table 3.20 Rationale for selection of restoration focus areas in the Eden River basin, England, UK.

Sub-basin	Area (km²)	Rationale for selection
River Petteril	163	– Predicted to have a high to very high risk of fine sediment delivery from agricultural sources – High levels of riparian damage due to intensive livestock grazing; 19.5 km of surveyed river bank was eroding due to livestock and 12.8 km of surveyed channel had less than 25% cover from overhanging trees and riparian vegetation – Poor age 0+ Atlantic salmon and brown trout populations
Hoff Beck and Helm Beck	70	– Highest concentration of riparian damage within the Eden watershed due to intensive livestock grazing; 30–40% of the surveyed riparian habitat was classified as severely degraded and 15 km of channel had less than 25% cover – Predicted to have a moderate to high risk of fine sediment delivery from agricultural sources – Poor age 0+ Atlantic salmon and brown trout populations
River Leith and River Lyvennet	129	– Predicted to have a high risk of fine sediment delivery – High levels of riparian damage due to intensive livestock grazing; 17 km of the surveyed river bank was eroding due to livestock and 12 km of surveyed channel had less than 25% cover – Highly variable age 0+ Atlantic salmon and brown trout populations

of more targeted remedies so that large-scale changes to land management may not be required (Lane *et al.* 2006). For example, restoration strategies such as buffer zones to reduce hydrological connectivity or small-scale land-cover changes at highly sensitive locations can be used to reduce diffuse pollution without affecting the majority of agricultural lands in the basin (Burt 2001).

An advantage of the approach was that habitat data could be extrapolated across the basin and converted into management strategies targeting locations where they were likely to be most beneficial (Table 3.20, Figure 3.28). Moreover, habitat data could be readily visualized in map and photographic form, creating a powerful tool for illustrating and justifying restoration strategies to the public, as well as for engaging local communities, landowners, and funders in restoration efforts. Based on these data, the Eden Rivers Trust developed a suite of five management plans focused around three key strategies:

1. a species-specific plan targeting juvenile brown trout habitat restoration in small streams concentrated on increasing cover, reducing sediment delivery, and reconnecting fragmented populations;

2. three sub-basin restoration plans targeting highly degraded areas of the basin where hydrological connectivity, livestock grazing, and associated fine sediment delivery were significantly impaired for both salmonid species at a range of scales; and

3. a plan aimed at protecting a particularly productive sub-basin for juvenile salmonids in response to a specific funding opportunity.

Each of these plans helped the Eden Rivers Trust identify and target the most important restoration needs within each sub-basin, which led the Trust to develop capabilities for providing soil and nutrient management advice and reducing diffuse pollution from farms. While the assessment techniques adopted for the River Eden did not quantify the status of salmonid stocks and habitat quality across the basin, they illustrated relationships between habitat and salmonid abundance at a range of scales, which indicated where and at what scale restoration actions were needed.

3.9 Summary

Assessments needed to identify restoration actions and develop a restoration strategy must answer two key questions: (1) how have habitats changed and altered biota? and (2) what are the causes of those habitat changes? Setting restoration goals is a critical first step in developing an effective restoration strategy, as restoration goals guide the selection of methods for assessing watershed conditions and identifying potential restoration actions. Restoration goals are usually linked to specific legislative mandates, which tend to fall into two broad categories: legislation to protect individual species, or legislation to improve riverine ecosystems. When restoration goals target protection or recovery of endangered species, assessments of habitat conditions (the *effects* of degradation) focus on habitat indicators relevant to species at risk. By contrast, when restoration goals target holistic restoration of water quality or biological communities, assessments of effects of degradation focus on a broad suite of habitat and biological indicators. In both cases, assessments to identify *causes* of degradation remain essentially the same regardless of the biological restoration goals.

Assessments that identify causes of degradation include evaluations of the watershed template (geologic and topographic controls), watershed-scale processes (erosion, runoff, and delivery of nutrients or pollutants), and reach-scale processes (floodplain interactions, riparian functions, and fluvial processes). Understanding the watershed template helps to identify the range of habitat and biological conditions that can be expressed in each reach of the river basins. Watershed-scale assessments identify non-point causes of habitat degradation and actions that require restoration at multiple sites across a watershed to restore habitats and biota. Reach-scale assessments identify restoration actions for riparian areas or floodplain connections, and those reach-scale actions address localized problems. Together, these assessments identify key restoration actions needed to achieve recovery goals, and also indicate the scale at which those actions must be taken. The habitat and biological assessments help to ascertain which actions are most likely to help achieve recovery goals. Finally, the watershed assessment results and restoration plan should not only identify key habitat losses that constrain biological recovery, but also the causes of those losses and which restoration actions will most contribute to biological recovery.

The summary of the watershed assessment should include maps and data indicating where each process is impaired and what kinds of restoration actions are needed. Maps of impairments portray the spatial arrangement of restoration needs, and a tabular summary of the data helps communicate the relationship between process impairments and the key restoration strategies or action types that are needed to address those impairments. Field inventories can then be used to identify and plan specific restoration actions within targeted areas. Finally, land-use constraints and climate change considerations may help identify actions that are either most feasible, or most likely to remain effective in a future climate. These considerations may function as filters that reduce the number of potential projects to be considered in subsequent application of prioritization schemes (Chapter 6).

3.10 References

Allan, D.J., Erickson, D.L. & Fay, J. (1997) The Influence of catchment land use on stream integrity across multiple spatial scales. *Freshwater Biology* **37**, 149–161.

Amy, J. & Robertson, A.I. (2001) Relationships between livestock management and the ecological condition of riparian habitats along an Australian floodplain river. *Journal of Applied Ecology* **38**, 63–75.

Arheimer, B. & Brandt, M. (1998) Modeling nitrogen transport and retention in the catchments of southern Sweden. *Ambio* **27**(6), 471–480.

Arheimer, B. & Olsson, J. (2003) Integration and Coupling of Hydrological Models with Water Quality Models: Applications in Europe. World Meteorological Organization (WMO) RA VI – Europe working group on hydrology report (HOME Component K55.1.02 (Mar 03). Available at: www.wmo.ch.

Baker, J.P., Hulse, D.W., Gregory, S.V. *et al.* (2004) Alternative futures for the Willamette River basin, Oregon. *Ecological Applications* **14**, 313–324.

Banchetti, R., Ceccopieri, N., & Bombardieri, G. (2004) Valutazione della qualita delle acque del Fiume Frigido (Toscana) mediante l'indice I.B.E. (Indice biotic esteso). *Atti Soc. Tosc. Sci. Nat.* **111**, 55–64.

Barber, W.E. & Taylor, J.N. (1990) The importance of goals, objectives, and values in the fisheries management process and organization: A review. *North American Journal of Fisheries Management* **10**, 365–373.

Bartz, K.L., Lagueux, K., Scheuerell, M.D., Beechie, T.J., Haas, A. & Ruckelshaus, M.H. (2006) Translating restoration scenarios into habitat conditions: An initial step in evaluating recovery strategies for Chinook salmon (*Oncorhynchus tshawytscha*). *Canadian Journal of Fisheries and Aquatic Sciences* **63**, 1578–1595.

Battin, J., Wiley, M.W., Ruckelshaus, M.H. *et al.* (2007) Projected impacts of climate change on salmon habitat restoration. *Proceedings of the National Academy of Sciences* **104**, 6720–6725.

Beamer, E., McBride, A., Greene, C. *et al.* (2005a) *Delta and nearshore restoration for the recovery of wild Skagit River Chinook salmon: Linking estuary restoration to wild Chinook salmon populations.* Appendix D of the Skagit Chinook Recovery Plan. Skagit River System Cooperative, La Conner, Washington.

Beamer, E., Hayman, B. & Smith, D. (2005b) *Linking freshwater rearing habitat to Skagit Chinook salmon recovery.* Appendix C of the Skagit Chinook Recovery

Plan. Skagit River System Cooperative, La Conner, Washington.

Beechie, T.J. & Bolton, S. (1999) An approach to restoring salmonid habitat-forming processes in Pacific Northwest watersheds. *Fisheries* **24**, 6–15.

Beechie, T., Beamer, E. & Wasserman, L. (1994) Estimating coho salmon rearing habitat and smolt production losses in a large river basin, and implications for restoration. *North American Journal of Fisheries Management* **14** 797–811.

Beechie, T.J., Collins, B.D. & Pess, G.R. (2001) Holocene and recent geomorphic processes, land use and salmonid habitat in two north Puget Sound river basins. In: Dorava, J.B., Montgomery, D.R., Fitzpatrick, F. & Palcsak, B. (eds) *Geomorphic Processes and Riverine Habitat*, Water Science and Application Volume 4, American Geophysical Union, Washington DC, pp. 37–54.

Beechie, T.J., Pess, G., Beamer, E., Lucchetti, G., & Bilby, R.E. (2003a) Roles of watershed assessments in recovery planning for threatened or endangered salmon. In: Montgomery, D., Bolton, S., Booth, D. & Wall, L. (eds) *Restoring Puget Sound Rivers*, University of Washington Press, Seattle, pp. 194–225.

Beechie, T.J., Steel, E.A., Roni, P.R. & Quimby, E. (eds) (2003b) *Ecosystem Recovery Planning for Listed Salmon: An Integrated Assessment Approach for Salmon Habitat*. NOAA Technical Memorandum NMFS-NWFSC-58, National Marine Fisheries Service, Seattle, Washington.

Beechie, T.J., Liermann, M., Beamer, E.M. & Henderson, R. (2005) A classification of habitat types in a large river and their use by juvenile salmonids. *Transactions of the American Fisheries Society* **134**, 717–729.

Beechie, T.J., Liermann, M., Pollock, M.M., Baker, S. & Davies, J. (2006) Channel pattern and river-floodplain dynamics in forested mountain river systems. *Geomorphology* **78**(1–2), 124–141.

Beechie, T., Pess, G., Roni, P. & Giannico, G. (2008a) Setting river restoration priorities: A review of approaches and a general protocol for identifying and prioritizing actions. *North American Journal of Fisheries Management* **28**, 891–905.

Beechie, T.J., Pollock, M.M. & Baker, S. (2008b) Channel incision, evolution and potential recovery in the Walla Walla and Tucannon River basins, northwestern USA. *Earth Surface Processes and Landforms* **33**, 784–800.

Beechie, T.J., Sear, D., Olden, J. *et al.* (2010) Process-based principles for restoring river ecosystems. *BioScience* **60**, 209–222.

Beechie, T., Imaki, H., Greene, J. *et al.* (2012) Restoring salmon habitat for a changing climate. *River Research and Applications*. doi: 10.1002/rra.2590 (in press).

Benda, L., Miller, D., Andras, K., Bigelow, P., Reeves, G. & Michael, D. (2007) NetMap: A new tool in support of watershed science and resource management. *Forest Science* **53**, 206–219.

Benejam, L., Angermeier, P.L., Munne, A. & Garcia-Berthou, E. (2010) Assessing effects of water abstraction on fish assemblages in Mediterranean streams. *Freshwater Biology* **55**, 628–642.

Bennett, E.M., Reed-Andersen, T., Houser, J.N., Gabriel, J.R. & Carpenter, S.R. (1999) A phosphorous budget for the Lake Mendota watershed. *Ecosystems* **2**, 69–75.

Berenbrock, C. & Tranmer, A. (2008) Simulation of Flow, Sediment Transport, and Sediment Mobility of the Lower Coeur d'Alene River, Idaho. USGS Scientific Investigations Report 2008–5093. US Geological Survey, Reston, Virginia.

Beschta, R.L. (1987) Sediment transport in gravel-bed rivers. In: Thorne, C.R., Bathurst, J.C. & Hey, R.D. (eds) *Sediment Transport in Gravel-Bed Rivers*. John Wiley and Sons, London, pp. 387–408.

Beschta, R.L., Pyles, M.R., Skaugset, A.E. & Surfleet, C.G. (2000) Peakflow response to forest practices in the western Cascades of Oregon, U.S.A. *Journal of Hydrology* **233**, 102–120.

Bisson, P.A., Montgomery, D.R. & Buffington, J.M. (2006) Valley segments, stream reaches, and channel units. In: Hauer, F.R. & Lamberti, G.A. (eds) *Methods in Stream Ecology*. Elsevier, New York, pp. 23–50.

Bolten, A.B., Crowder, L.B., Dodd, M.G. *et al.* (2010) Quantifying multiple threats to endangered species: An example from loggerhead sea turtles. *Frontiers in Ecology and the Environment* **9**, 295–301.

Booth, D.B. & Jackson, C.R. (1997) Urbanization of aquatic systems: Degradation thresholds, stormwater detention, and the limits of mitigation. *Journal of the American Water Resources Association* **33**(5), 1077–1090.

Booth, D.B., Hartley, D. & Jackson, R. (2002) Forest cover, impervious-surface area, and the mitigation of stormwater impacts. *Journal of the American Water Resources Association* **38**(3), 835–947.

Booth, D.T., Cox, S.E. & Simonds, G. (2007) Riparian monitoring using 2-cm GSD aerial photography. *Ecological Indicators* **7**, 636–648.

Bormann, H. & Elfert, S. (2010) Application of WaSiM-ETH model to Northern German lowland catchments: Model performance in relation to catchment characteristics and sensitivity to land use change. *Advances in Geosciences* **27**, 1–10.

Borsuk, M.E., Stow, C.A. & Reckhow, K.H. (2002) Predicting the frequency of water quality standard violations: A probabilistic approach for TMDL development. *Environmental Science and Technology* **36**, 2109–2115.

Bottom, D.L., Jones, K.K., Simenstad, C.A., Smith, C.I. & Cooper, R. (2011) *Pathways to Resilience: Sustaining Salmon Ecosystems in a Changing World*. Oregon Sea Grant, ORESU-B-11-001, Corvallis, Oregon.

Bowling, L.C., Storck, P. & Lettenmaier, D.P. (2000) Hydrologic effects of logging in Western Washington, United States. *Water Resources Research* **36**, 3223–3240.

Brierley, G.J. & Fryirs, K.A. (2005) *Geomorphology and River Management: Applications of the River Styles Framework*. Blackwell, Oxford.

Brierley, G.J. & Fryirs, K.A. (2008) *River Futures: An Integrative Approach to River Repair*. Island Press, Washington.

Brierley, G.J., Fryirs, K., Outhet, D. & Massey, C. (2002) Application of the River Styles framework as a basis for river management in New South Wales, Australia. *Journal of Applied Geography* **22**, 91–122.

Brierley, G.J., Fryirs, K.A., Boulton, A., & Cullum, C. (2008) Working with change: the importance of evolutionary perspectives in framing the trajectories of river adjustment. In: Brierley, G.J. & Fryirs, K.A. (eds) *River Futures: An Integrative Approach to River Repair*. Island Press, Washington, pp. 66–84.

Brooks, A., Lymburner, L., Dowe, J. *et al.* (2008) *Development of a Riparian Condition Assessment Approach for Northern Gulf Rivers Using Remote Sensing and Ground Survey*. Final Report, Project number GRU38. Land and Water Australia, Canberra.

Buijse, A.D., Coops, H., Staras, M. *et al.* (2002) Restoration strategies for river floodplains along large lowland rivers in Europe. *Freshwater Biology* **47**, 889–907.

Burke, M., Jorde, K. & Buffington, J.M. (2009) Application of a hierarchical framework for assessing environmental impacts of dam operation: Changes in hydrology, channel hydraulics, bed mobility and recruitment of riparian trees in a western North American river. *Journal of Environmental Management* **90**, S224–S236.

Burt, T.P. (2001) Integrated management of sensitive catchment systems. *Catena* **42**, 275–290.

Canale, G.A. (1999) *BORIS: Benthic Evaluation of Oregon Rivers*. Technical Report BIO98-003. Oregon Department of Environmental Quality, Laboratory Division, Biomonitoring Section, Portland.

CH2M HILL. (2010) *Enhanced Conceptual Site Model for the Lower Basin of the Coeur d'Alene River*. Report to U.S. Environmental Protection Agency Seattle, Washington. CH2M HILL, Boise, Idaho.

Christensen, N. & Lettenmaier, D.P. (2007) A multimodel ensemble approach to assessment of climate change impacts on the hydrology and water resources of the Colorado River basin. *Hydrology and Earth System Sciences* **11**, 1417–1434.

Church, M. (2002) Geomorphic thresholds in riverine landscapes. *Freshwater Biology* **47**, 541–557.

CIRF (Centro Italiano per la Riqualificazione Fluviale) (2006) *La riqualificazione fluviale in Italia. Linee guida, strumenti ed esperienze per gestire i corsi d'acqua e il territorio*. In: Nardini, A. & Sansoni, G. (eds) Mazzanti Editori, Venezia.

Clark, G.M. & Woods, P.F. (2001) *Transport of Suspended Solids and Bedload Sediment at Eight Stations in Coeur d'Alene River Basin, Idaho*. USGS Open-File Report 00-472. U.S. Geological Survey, Boise, Idaho.

Clary, W.P. & Leninger, W.C. (2000) Stubble height as a tool for management of riparian areas. *Journal of Range Management* **53**, 562–573.

Collins, B.D. & Montgomery, D.R. (2001) Importance of archival and process studies to characterizing pre-settlement riverine geomorphic processes and habitat in the Puget Lowland. In: Dorava, J.M., Montgomery, D.R., Palcsak, B.B. & Fitzpatrick, F.A. (eds) *Geomorphic Processes and Riverine Habitat*, Water and Science Application 4, American Geophysical Union, Washington, DC, pp. 227–243.

Collins, B.D., Montgomery, D.R. & Haas, A.D. (2002) Historical changes in the distribution and functions of large wood in Puget Lowland rivers. *Canadian Journal of Fisheries and Aquatic Sciences* **59**, 66–76.

Crozier, W.W. & Kennedy, G.J.A. (1994) Application of semi-quantitative electrofishing to juvenile salmonid stock surveys. *Journal of Fish Biology* **45**, 159–164.

Cullum, C., Brierley, G.J. & Thoms, M. (2008) The spatial organization of river systems. In: Brierley, G.J. & Fryirs, K.A. (eds) *River Futures: An Integrative Approach to River Repair*, Island Press, Washington, pp. 41–64.

Cuo, L., Lettenmaier, D.P., Alberti, M. & Richey, J.E. (2009) Effects of a century of land cover and climate change on the hydrology of Puget Sound basin. *Hydrological Processes* **23**, 907–933.

Cupp, C.E. (1989) *Identifying Spatial Variability of Stream Characteristics Through Classification*. MS Thesis, University of Washington, Seattle.

Davis, W.S. & Simon, T.P. (1995) *Biological Assessment and Criteria: Tools for Water Resource Planning and Decision Making.* Lewis Publishers, Boca Raton, Florida.

Dietrich, W.E., Kirchner, J.W., Ikeda, H. & Iseya, F. (1989) Sediment supply and the development of the coarse surface layer in gravel-bedded rivers. *Nature* **340**, 215–217.

Dixon, I., Douglas, M., Dowe, J.L. & Burrows, D. (2006) *Tropical Rapid Appraisal of Riparian Condition, Version 1 (for use in tropical savannas).* River and Riparian Land Management, Technical Guideline No. 7, Land & Water Australia, Canberra.

Donato, M.M. (1998) *Surface-Water/Ground-Water Relations in the Lemhi River Basin, East-Central Idaho.* USGS Water-resources Investigations Rep. 98-4185. United States Geological Survey, Washington, D.C.

Downs, P.W., & Thorne, C.R. (1996) The utility and justification of river reconnaissance surveys. *Transactions of the Institute of British Geographers* **21**, 455–468.

Downs, P.W., Dusterhoff, S.R. & Sears, W.A. (2006) Outside the bankfull realm: challenges in developing a flood-based management and restoration strategy for a large, semi-arid river system: Santa Clara River, Ventura County, California, In: *Eos Transactions AGU* **87**(52), Fall Meeting Supplement, Abstract H44C-03.

Draut, A.E., Logan, J.B. & Mastin, M.C. (2010) Channel evolution on the dammed Elwha River, Washington, USA. *Geomorphology* **127**, 71–87.

Dunham, J., Chandler, G., Rieman, B. & Martin, D. (2005) Measuring stream temperature with digital data loggers: A user's guide. United States Department of Agriculture General Technical Report RMRS-GTR-150WWW, Fort Collins, Colorado.

Dunne, T. & Leopold, L.B. (1978) *Water in Environmental Planning.* Freeman, San Francisco, California.

Eaton, B.C., Church, M. & Millar, R.G. (2004) Rational regime model of alluvial channel morphology and response. *Earth Surface Processes and Landforms* **29**, 511–529.

Ebersole, J.L. & Liss, W.J. (1997) Restoration of stream habitats in the western United States: Restoration as reexpression of habitat capacity. *Environmental Management* **21**, 1–14.

Ehrenfeld, J.G. (2000) Defining the limits of restoration: The need for realistic goals. *Restoration Ecology* **8**, 2–9.

Elmore, W. (1992) Riparian responses to grazing practices. In: Naiman, R.J. (ed.) *Watershed Management: Balancing Sustainability and Environmental Change.* Springer-Verlag, New York, New York, pp. 442–457.

Elsner, M.M., Cuo, L., Voisin, N. *et al.* (2010) Implications of 21st century climate change for the hydrology of Washington State. *Climatic Change* **102**, 225–260.

Environment Agency (2003) *River Habitat Survey in Britain and Ireland: Field Survey Guidance Manual.* Environment Agency, Bristol, United Kingdom.

Ewen, J., Parkin, G. & O'Connell, P.E. (2000) SHETRAN: Distributed river basin flow and transport modeling system. *ASCE J. Hydrologic Eng.* **5**, 250–258.

Federal Register Notice of Puget Sound Chinook ESU listing (FR50 CFR Parts 223 and 224), Volume 64, No. 56, March 24, 1999.

Federal Register Notice of Puget Sound steelhead ESU listing (FR50 CFR Parts 223), Vol. 72, No. 91, May 11, 2007.

Feller, M.C. & Kimmins, J.P. (1984) Effects of clearcutting and slash burning on streamwater chemistry and watershed nutrient budgets in southwestern British Columbia. *Water Resources Research* **20**(1), 29–40.

Fitzpatrick, F.A., Waite, I.R., D'Arconte, P.J., Meador, M.R., Maupin, M.A. & Gurtz, M.E. (1998) Revised methods for characterizing stream habitat in the National Water Quality Assessment Program. United States Geological Survey Water-Resources Investigations Report 98-4052, Raleigh, North Carolina.

Fitzpatrick, F.A., Knox, J.C. & Schubauer-Berigan, J.P. (2009) Channel, floodplain, and wetland responses to floods and overbank sedimentation, 1846–2006, Halfway Creek Marsh, Upper Mississippi Valley, Wisconsin. *Geological Society of America Special Papers* **451**, 23–42.

Fore, L.S., Karr, J.R. & Wisseman, R.W. (1996) Assessing invertebrate responses to human activities: Evaluating alternative approaches. *Journal of the North American Benthological Society* **15**, 212–231.

Frissell, C.A., Liss, W.J., Warren, C.E. & Hurley, M.D. (1986) A hierarchical framework for stream habitat classification: Viewing streams in a watershed context. *Environmental Management* **10**, 199–214.

Frissell, C.A., Liss, W.J., Gresswell, R.E., Nawa, R.K. & Ebersole, L. (1997) A resource in crisis: Changing the measure of salmon management. In: Stouder, D.J., Bisson, P.A. & Naiman, R.J. (eds) *Pacific Salmon and Their Ecosystems: Status and Future Options.* Chapman and Hall, New York, pp. 411–446.

Fullerton, A.H., Beechie, T.J., Baker, S.E., Hall, J.E. & Barnas, K.A. (2006) Regional Patterns of Riparian Characteristics in the Interior Columbia River Basin, Northwestern USA: Applications for Restoration Planning. *Landscape Ecology* **21**(8), 1347–1360.

Geist, C. & Galatowitsch, S.M. (1999) A reciprocal model for meeting ecological and human needs in restoration projects. *Conservation Biology* **13**, 970–979.

Gergel, S.E., Stange, Y., Coops, N.C., Johansen, K. & Kirby, K.R. (2007) What is the value of a good map? An example using high spatial resolution imagery to aid riparian restoration. *Ecosystems* **10**(5), 688–702.

Goetz, S.J., Wright, R.K., Smith, A.J., Zinecker, E. & Schaub, E. (2003) IKONOS imagery for resource management: Tree cover, impervious surfaces, and riparian buffer analysis in the mid-Atlantic region. *Remote Sensing of the Environment* **88**, 195–208.

Graf, W.L. (2000) Locational probability for a dammed, urbanizing stream: Salt River, Arizona, USA. *Environmental Management*, **25**, 321–335.

Greene, C.M. & Beechie, T.J. (2004) Consequences of potential density-dependent mechanisms on recovery of ocean-type Chinook salmon (*Oncorhynchus tshawytscha*). *Canadian Journal of Fisheries and Aquatic Sciences* **61**, 590–602.

Grimm, N.B., Foster, D., Groffman, P. *et al.* (2008) The changing landscape: Ecosystem responses to urbanization and pollution across climatic and societal gradients. *Frontiers in Ecology and the Environment* **6**, 264–272.

Groffman, P.M., Law, N.L., Belt, K.T., Band, L.E. & Fisher, G.T. (2004) Nitrogen fluxes and retention in urban watershed ecosystems. *Ecosystems* **7**(4), 393–403.

Gurnell, A.M., Downward, S.R. & Jones, R. (1994) Channel planform change on the River Dee meanders, 1876–1992. *Regulated Rivers: Research and Management* **9** 187–204.

Hamada, Y., Stow, D.A., Coulter, L.L., Jafolla, J.C. & Hendricks, L.W. (2007) Detecting Tamarisk species (Tamarix spp.) in riparian habitats of Southern California using high spatial resolution hyperspectral imagery. *Remote Sensing of Environment* **109**, 237–248.

Hankin, D.G. (1984) Multistage sampling designs in fisheries research: Applications in small streams. *Canadian Journal of Fisheries and Aquatic Sciences* **41**, 1575–1591.

Hankin, D.G. & Reeves, G.H. (1988) Estimating total fish abundance and total habitat area in small streams based on visual estimation methods. *Canadian Journal of Fisheries and Aquatic Sciences* **45**, 834–844.

Harding, J., Clapcott, J., Quinn, J. *et al.* (2009) *Stream Habitat Assessment Protocols for Wadeable Rivers and Streams of New Zealand*. University of Canterbury, Christchurch, New Zealand.

Harris, M.A., Scudder, B.C., Fitzpatrick, F.A. & Arnold, T.L. (2005) *Physical, Chemical, and Biological Responses to Urbanization in the Fox and Des Plaines River basins of Northeastern Illinois and Southeastern Wisconsin*. United States Geological Survey Scientific Investigations Report 2005-5218, 72 pp.

Harris, R. (1999) Defining reference conditions for restoration of riparian plant communities: Examples from California, USA. *Environmental Management* **24**(1), 55–63.

Harris, R. & Olson, C. (1997) Two-stage system for prioritizing riparian restoration at the stream reach and community scales. *Restoration Ecology* **5**, 34–42.

Harwell, M.A., Meyers, V., Young, T. *et al.* (1999) A framework for an ecosystem integrity report card. *BioScience* **49**, 543–556.

Hauer, F.R. & Lamberti, G.A. (2006) *Methods in Stream Ecology*. Elsevier, New York.

Hayhoe, K., Wake, C., Huntington, T.G. *et al.* (2007) Past and future changes in climate and hydrological indicators in the US Northeast. *Climate Dynamics* **28**, 381–407.

Heikkinen, R.K., Luoto, M., Araujo, M.B., Virkkala, R., Thuiller, W. & Sykes, M.T. (2006) Methods and uncertainties in bioclimatic envelope modelling under climate change. *Progress in Physical Geography* **30**, 751–777.

Hendry, K. & Cragg-Hine, D. (1996) Restoration of Riverine Salmonid Habitats – A Guidance Manual. Environment Agency Fisheries Technical Manual 4, HO-11/97-B-BAHB, Environment Agency 1996, Bristol, UK.

Hohensinner S., Habersack, H., Jungwirth, M. & Zauner, G. (2003) Reconstruction of the characteristics of a natural alluvial river-floodplain system and hydromorphological changes following human modifications: The Danube River (1812–1991). *River Research and Applications* **20**, 25–41.

Hohensinner, S., Jungwirth, M., Muhar, S. & Habersack, H. (2005) Historical analyses: A foundation for developing and evaluating river-type specific restoration programs. *International Journal of River Basin Management* **3**, 1–10.

Holling, C.S. (1973) Resilience and stability of ecological systems. *Annual Review of Ecology and Systematics* **4**, 1–23.

Hughes, R.M., Larsen, D.P. & Omernik, J.M. (1986) Regional reference sites: A method of assessing stream potentials. *Environmental Management* **10**, 629–635.

Hyatt, T.L., Waldo, T.Z. & Beechie, T.J. (2004) A watershed-scale assessment of riparian forests, with implications for restoration. *Restoration Ecology* **12**, 175–183.

Kareiva, P.M. (2002) Applying ecological science to recovery planning. *Ecological Applications* **12**, 629.

Karr, J.R. (1991) Biological integrity: A long-neglected aspect of water resource management. *Ecological Applications* **1**(1), 66–84.

Karr, J.R. (2006) Seven foundations of biological monitoring and assessment. *Biologia Ambientale* **20**(2), 7–18.

Karr, J.R. & Chu, E.W. (2000) Sustaining living rivers. *Hydrobiologia* **442**, 1–14.

Kaufmann, P.R., Levine, P., Robison, E.G., Seeliger, C. & Peck, D.V. (1999) *Quantifying Physical Habitat in Wadeable Streams*. EPA/620/R-99/003. United States Environmental Protection Agency, Research Triangle Park, North Carolina.

Kellerhals, R., Church, M. & Bray, D.I. (1976) Classification and analysis of river processes. *Journal of the Hydraulics Division* **102**(7), 813–829.

Ketcheson, G.L., Megahan, W.F. & King, J.G. (1999) "R1-R4" and "BOISED" sediment prediction model tests using forest roads in granitics. *Journal of the American Water Resources Association* **35**, 83–98.

Kilgour, B.W. & Stanfield, L.W. (2006) Hindcasting reference conditions in streams. In: Hughes, R.M., Wang, L. & Seelback, P.W. (eds) *Landscape Influences on Stream Habitats and Biological Assemblages*, American Fisheries Society Symposium 48, Bethesda, Maryland, pp. 623–639.

Kloehn, K.K., Beechie, T.J., Morley, S.A., Coe, H.J. & Duda, J.J. (2008) Influence of dams on river-floodplain dynamics in the Elwha River, Washington. Special Issue on Dam Removal and Ecosystem Restoration in the Elwha River Watershed, Washington State. *Northwest Science* **82**, 224–235.

Knighton, A.D. & Nanson, G.C. (1993) Anastomosis and the continuum of channel pattern. *Earth Surface Processes and Landforms* **18**, 613–625.

Kondolf, G.M. (1997) Hungry water: Effects of dams and gravel mining on river channels. *Environmental Management* **21**, 533–551.

Kondolf, G.M. (2006) River restoration and meanders. *Ecology and Society* **11**, article 42. Available at: http://www.ecologyandsociety.org/vol11/iss2/art42.

Kondolf, G.M. & Piegay, H. (2003) *Tools in Fluvial Geomorphology*. John Wiley and Sons, Ltd., Chichester, UK.

Kondolf, G.M., Montgomery, D.R., Piegay, H. & Schmitt, L. (2003) Geomorphic classification of rivers and streams. In: Kondolf, G.M. & Piegay, H. (eds) *Tools in Fluvial Geomorphology*. John Wiley and Sons, Ltd., Chichester, UK, pp. 171–204.

Krause, S., Jacobs, J. & Bronstert, A. (2007) Modelling the impacts of land-use and drainage density on the water balance of a lowland–floodplain landscape in northeast Germany. *Ecological Modelling* **200**(3–4), 475–492.

Lane, S.L., Flanagan, S. & Wilde, F.D. (2003) *Selection of Equipment for Water Sampling* (ver. 2.0): United States Geological Survey Techniques of Water-Resources Investigations, Book 9, Chapter A2. Available at: http://pubs.water.usgs.gov/twri9A2.

Lane, S.N., Brookes, C.J., Heathwaite, A.L. & Reaney, S. (2006) Surveillant science: Challenges for the management of rural environments emerging from the new generation of diffuse pollution models. *Journal of Agricultural Economics* **57**, 239–257.

Lawler, J.J., Campbell, S.P., Guerry, A.D. *et al.* (2002) The scope and treatment of threats in endangered species recovery plans. *Ecological Applications* **12**, 663–67.

Lawson, P.W., Logerwell, E.A., Mantua, N.J., Francis, R.C. & Agostoni, V.N. (2005) Environmental factors influencing freshwater survival and smolt production in Pacific Northwest coho salmon (*Oncorhynchus kisutch*). *Canadian Journal of Fisheries and Aquatic Sciences* **61**, 360–373.

Leopold, L.B. & Wolman, M.G. (1957) River channel patterns: braided, meandering and straight. In: *Physiographic and Hydraulic Studies of Rivers*. United States Geological Survey Professional Paper 282-B, United States Government Printing Office, Washington, DC, pp. 39–85.

Leslie, P.H. (1945) On the use of matrices in certain population mathematics. *Biometrika* **33**, 183–212.

Liang, X., Lettenmaier, D.P., Wood, E.F. & Burges, S.J. (1994) A simple hydrologically based model of land surface water and energy fluxes for GSMs. *Journal of Geophysical Research* **99** (D7), 14415–14428.

Likens, G.E., Bormann, F.H., Johnson, N.M., Fisher, D.W. & Pierce, R.S. (1970) Effects of forest cutting and herbicide treatment on nutrient budgets in the Hubbard Brook watershed-ecosystem. *Ecological Monographs* **40**, 23–47.

Lisle, T.E. & Hilton, S. (1992) The volume of fine sediment in pools: n index of sediment supply in gravel-bed streams. *Water Resources Bulletin* **28**, 371–383.

Lohmann, D., Raschke, E., Nijssen, B. & Lettenmaier, D.P. (1998) Regional scale hydrology: II. Application of the VIC-2 L model to the Weser river, Germany. *Hydrological Sciences Journal* **43**(1), 143–158.

Lower Columbia Fish Recovery Board (2010) *Washington Lower Columbia Salmon Recovery and Fish and Wildlife*

Sub-basin Plan. Lower Columbia Fish Recovery Board, Longview, Washington.

Lowrance, R.R., Leonard, R.A., Asmussen, L.E. & Todd, R.L. (1985) Nutrient budgets for agricultural watersheds in the southeastern coastal plain. *Ecology* **66**, 287–296.

Lunetta, R.S., Cosentino, B.L., Montgomery, D.R., Beamer, E.M. & Beechie, T.J. (1997) GIS-Based evaluation of salmon habitat in the Pacific Northwest. *Photogrammetric Engineering and Remote Sensing* **63**, 1219–1229.

Madrid, Y. & Zayas, Z.P. (2007) Water sampling: traditional methods and new approaches in water sampling strategy. *Trends in Analytical Chemistry* **26**, 293–299.

Mantua, N., Tolver, I. & Hamlet, A. (2010) Climate change impacts on streamflow extremes and summertime stream temperature and their possible consequences for freshwater salmon habitat in Washington State. *Climatic Change* **102**, 187–223, doi: 10.1007/s10584-010-9845-2.

Mattheussen, B., Kirschbaum, R.L., Goodman, I.A., O'Donnell, G.M. & Lettenmaier, D.P. (2000) Effects of land cover change on streamflow in the interior Columbia river basin (USA and Canada). *Hydrological Processes* **14**, 867–885.

Maupin, M.A. (1999) *Methods to Determine Pumped Irrigation-Water Withdrawals from the Snake River Between Upper Salmon Falls and Swan Falls Dams, Idaho, Using Electrical Power Data.* USGS Water-Resources Investigations Report 99-4175. United States Geological Survey, Boise, Idaho.

McBain & Trush, Inc. (2007) *Coarse Sediment Management Plan: Lewiston Dam to Douglas City, Trinity River, California.* Trinity River Restoration Program, Weaverville, California.

McCarty, J.P. (2001) Ecological consequences of recent climate change. *Conservation Biology* **15**, 320–331.

McElhany, P., Ruckelshaus, M.H., Ford, M.J., Wainwright, T.C. & Bjorkstedt, E.P. (2000) *Viable Salmonid Populations and the Recovery of Evolutionarily Significant Units.* NOAA Technical Memorandum NMFS-NWFSC-42. United States Department of Commerce, Seattle, Washington, 156 p.

Merritt, R.W. & Cummins, K.W. (eds) (1996) *An Introduction to the Aquatic Insects of North America.* Kendall Hunt Publishing Company, Dubuque, Iowa.

Millar, R.G. (2000) Effect of bank vegetation on alluvial channel patterns. *Water Resources Research* **36**, 1109–1118.

Milner, N., Pinder, A., Fraser, D., & Hendry, K. (2008) The Development of Salmonid Catchment Management Plans for Northern Ireland: Best Practice, Protocols and Background Information. APEM Scientific Report 410600, Belfast, UK.

Montgomery, D.R. (1999) Process domains and the river continuum. *Journal of the American Water Resources Association* **35**, 397–410.

Montgomery, D.R. & Buffington, J.M. (1997) Channel-reach morphology in mountain drainage basins. *Geological Society of America Bulletin* **109**(5), 596–611.

Montgomery, D.R. & MacDonald, L.H. (2002) Diagnostic approach to stream channel assessment and monitoring. *Journal of the American Water Resources Association* **38**, 1–16.

Montgomery, D.R. & Bolton, S.M. (2003) Hydrogeomorphic variability and river restoration. In Montgomery, D., Bolton, S., Booth, D. & Wall, L. (eds) *Restoring Puget Sound Rivers*, University of Washington Press, Seattle.

Morgan, R.P.C., Quinton, J.N., Smith, R.E. *et al.* (1998) European Soil Erosion Model (EUROSEM): A dynamic approach for predicting sediment transport from fields and small catchments. *Earth Surface Processes and Landforms* **23**, 427–544.

Morley, S.A. & Karr, J.R. (2002) Assessing and restoring the health of urban streams in the Puget Sound basin. *Conservation Biology* **16**, 1498–1509.

Muhlfeld, C.C., Glutting, S., Hunt, R., Daniels, D. & Marotz, B.L. (2003) Winter diel habitat use and movement by subadult bull trout in the upper Flathead River, Montana. *North American Journal of Fisheries Management* **23**, 163–171.

Muhlfeld, C.C., Jones, L.A., Kotter, D. *et al.* (2011) Assessing the impacts of river regulation on native bull trout (*Salvelinus confluentus*) and westslope cutthroat trout (*Oncorhynchus clarkii lewisi*) habitats in the upper Flathead River, Montana, USA. *River Research and Applications*, doi: 10.1002/rra.1494.

Munné, A., Prat, N., Solà, C., Bonada, N. & Rieradevall, M. (2003) A simple field method for assessing the ecological quality of riparian habitat in rivers and streams: QBR index. *Aquatic Conservation: Marine and Freshwater Ecosystems* **13**, 147–163.

Naiman, R.J., Lonzarich, D.G., Beechie, T.J. & Ralph, S.C. (1992) General principles of classification and the assessment of conservation potential in rivers. In: Boon, P.J., Calow, P. & Petts, G.E. (eds) *River Conservation and Management.* John Wiley and Sons. New York, pp. 93–124.

Naiman, R.J., Decamps, H. & McClain, M.E. (2005) *Riparia: Ecology, Conservation, and Management of Streamside Communities.* Elsevier, Burlington, Massachusetts.

Nakamura, K. & Tockner, K. (2004) River and wetland restoration in Japan. *Proceedings of the 3rd European Conference on River Restoration*, 17–21 May, 2004, Zagreb, Croatia.

Nanson, G.C. & Hickin, E.J. (1986) A statistical analysis of bank erosion and channel migration in western Canada. *Geological Society of American Bulletin* **97**, 497–504.

Nearing, M.A., Foster, G.R., Lane, L.J. & Finkner, S.C. (1989) A process-based soil erosion model for USDA water erosion prediction project technology. *Transactions of the American Society of Agricultural Engineers* **32**, 1587–1593.

Negishi, J.N., Sagawa, S., Kayaba, Y. *et al.* (2010) Using airborne scanning laser altimetry (LiDAR) to estimate surface connectivity of floodplain water bodies. *River Research and Applications* doi: 10.1002/rra.1442.

Newson M.D., Harper, D.M., Padmore, C.L., Kemp, J.L. & Voge, B. (1998) A cost-effective approach for linking habitats, flow types, and species requirements. *Aquatic Conservation: Marine and Freshwater Ecosystems* **8**, 431–446.

Nickelson, T.E. & Lawson, P.W. (1998) Population viability of coho salmon, Oncorhynchus kisutch, in Oregon coastal basins: application of a habitat-based life cycle model. *Canadian Journal of Fisheries and Aquatic Sciences* **55**, 2383–2392.

Omernik, J.M. (1986) Ecoregions of the conterminous United States. *Annals of the Association of American Geographers* **77**, 118–125.

Parker, V.T. (1997) The scale of successional models and restoration objectives. *Restoration Ecology* **5**, 301–306.

Paulson, K. (1997) *Estimating Changes in Sediment Supply Due to Forest Practices: A Sediment Budget Approach Applied to the Skagit River Basin, Washington.* MS Thesis, College of Forest Resources, University of Washington, Seattle.

Pess, G.R., Beechie, T.J., Williams, J.E., Whitall, D.R., Lange, J.I. & Klochak, J.R. (2003a) Chapter 8: Watershed assessment techniques and the success of aquatic restoration activities. In: Wissmar, R.C. & Bisson, P.A. (eds) *Strategies for Restoring River Ecosystems: Sources of Variability and Uncertainty in Natural and Managed Systems.* American Fisheries Society, Bethesda, Maryland, pp. 185–201.

Pess, G.R., Montgomery, D.R., Beechie, T.J. & Holsinger, L. (2003b) Anthropogenic alterations to the biogeography of salmon in Puget Sound. In: Montgomery,

D.R., Bolton, S., Booth, D.B. & Wall, L. (eds) *Restoration of Puget Sound Rivers.* University of Washington Press, Seattle, pp. 129–154.

Peterson, N.P. & Reid, L.M. (1984) Wall-base channels: their evolution, distribution, and use by juvenile coho salmon in the Clearwater River, Washington. In: Walton, J.M. & Houston, D.B. (eds) *Proceedings of the Olympic Wild Fish Conference*, March 23–25, 1983. Fisheries Technology Program, Peninsula College, Port Angeles, Washington, pp. 215–225.

Platts, W.S. (1991) Livestock grazing. In: Meehan, W.R. (ed.) *Influences of Forest and Rangeland Management on Salmonid Fishes and Their Habitats*, American Fisheries Society, Special Publication 19, Bethesda, Maryland, pp. 389–423.

Poff, N.L. & Ward, J.V. (1990) Physical habitat template of lotic systems: recovery in the context of historical pattern of spatiotemporal heterogeneity. *Environmental Management* **14**(5), 629–645.

Poff, N.L., Richter, B.D., Arthington, A.H. *et al.* (2010) The ecological limits of hydrologic alteration (ELOHA): a new framework for developing regional environmental flow standards. *Freshwater Biology* **55**, 147–170.

Pollock, M.M., Pess, G.R., Beechie, T.J. & Montgomery, D.R. (2004) The importance of beaver ponds to coho salmon production in the Stillaguamish River basin, Washington, USA. *North American Journal of Fisheries Management* **24**, 749–760.

Pollock, M.M., Beechie, T.J. & Jordan, C.E. (2007) Geomorphic changes upstream of beaver dams in Bridge Creek, in incised stream channel in the interior Columbia River basin, eastern Oregon. *Earth Surface Processes and Landforms* **32**, 1174–1185.

Quigley, T.M., & Arbelbide, S.J. (eds) (1997) *An Assessment of Ecosystems Components in the Interior Colombia Basin and Portions of the Klamath and Great Basins: Volume III.* General Technical Report PNW-GTR-405. United States Department of Agriculture, Forest Service, Portland, Oregon.

Raven, P., Holmes, N., Dawson, F. & Everard, M. (1998) Quality assessment using River Habitat Survey data. *Aquatic Conservation: Marine and Freshwater Ecosystems* **8**, 477–499.

Reaney, S.M., Lane, S.N., Heathwaite, A.L. & Dugdale, L.J. (2011) Risk-based modelling of diffuse land use impacts from rural landscapes upon salmonid fry abundance. *Ecological Modelling* **222**, 1016–1029.

Reeves, G.H., Everest, F.H. & Nickelson, T.E. (1989) *Identification of Physical Habitats Limiting the Production of Coho Salmon in Western Oregon and Washington.*

General Technical Report PNW-245. United States Forest Service, Pacific Northwest Research Station, Corvallis, Oregon.

Reid, L.M. (1998) Calculation of average landslide frequency using climatic records. *Water Resources Research* **34**, 869–877.

Reid, L.M. & Dunne, T. (1995) *Rapid Evaluation of Sediment Budgets*. Catena Verlag, Reiskirchen, Germany.

Renard, K.G., Foster, G.R., Weesies, G.A. & Porter, J.P. (1991) RUSLE: Revised universal soil loss equation. *Journal of Soil and Water Conservation* **46**(1), 30–33.

Renard, K.G., Foster, G.R., Weesies, G.A., McCool, D.K. & Yoder, D.C. (1997) *Predicting Soil Erosion by Water: A Guide to Conservation Planning with the Revised Universal Soil Loss Equation (RUSLE)*. United States Department of Agriculture, Washington, DC, 384 p.

Rennison, W. (1998) Risky business. *Engineering Field Notes* **30**, 7–20. US Forest Service.

Reynoldson, T.B., Rosenberg, D.M. & Resh, V.H. (2001) A comparison of models predicting invertebrate assemblages for biomonitoring in the Fraser River catchment, British Columbia. *Canadian Journal of Fisheries and Aquatic Sciences* **58**, 1395–1410.

Richter, B.D., Baumgartner, J.V., Powell, J. & Braun, D.P. (1996) A method for assessing hydrologic alteration within ecosystems. *Conservation Biology* **10**, 1163–1174.

Richter, B.D., Baumgartner, J.V., Braun, D.P. & Powell, J. (1998) A spatial assessment of hydrologic alteration within a river network. *Regulated Rivers: Research and Management* **14**, 329–340.

Rojas, R., Velleux, M., Julien, P.Y. & Johnson, B.E. (2008) Grid scale effects on watershed soil erosion models. *Journal of Hydrologic Engineering* **13**, 793–802.

Rosenberg, D.M. & Resh, V.H. (eds) (1993) *Freshwater Biomonitoring and Benthic Macroinvertebrates*. Chapman and Hall, New York.

Rosgen, D.L. (1994) A classification of natural rivers. *Catena* **22**, 169–199.

Ryder, D., Brierley, G.J., Hobbs, R., Kyle, G. & Leishman, M. (2008) Vision generation: What do we seek to achieve in river rehabilitation? In: Brierley, G.J. & Fryirs, K.A. (eds) *River Futures: An Integrative Approach to River Repair*, Island Press, Washington, pp. 16–27.

Scheuerell, M.D., Hilborn, R., Ruckelshaus, M.H. *et al.* (2006) The Shiraz model: A tool for incorporating anthropogenic effects and fish-habitat relationships in conservation planning. *Canadian Journal of Fisheries and Aquatic Sciences* **63**, 1596–1607.

Schmutz, S., Kaufmann, M., Vogel, B., Jungwirth, M. & Muhar, S. (2000) A multi-level concept for fish-based, river-type-specific assessment of ecological integrity. *Hydrobiologia* **422/423**, 279–289.

Schumm, S.A. (1985) Patterns of alluvial rivers. *Annual Review of Earth and Planetary Sciences* **13**, 5–27.

Sear, D.A. & Arnell, N.W. (2006) The application of paleohydrology in river management. *Catena* **66**, 169–183.

Sear, D.A., Millington, C.E., Kitts, D.R. & Jeffries, R. (2009), Logjam controls on channel: floodplain interactions in wooded catchments and their role in the formation of multi-channel patterns. *Geomorphology* **116**, 305–319.

Sedell, J.R. & Froggatt, J.L. (1984) Importance of streamside forests to large rivers: the isolation of the Willamette River, Oregon, U.S.A., from its floodplain by snagging and streamside forest removal. *Verhandlung Internationale Vereinigung Limnologie* **22**, 1828–1834.

Shafroth, P.B. (1999) *Downstream Effects of Dams on Riparian Vegetation: Bill Williams River, Arizona*. PhD Dissertation, Arizona State University, Tempe, Arizona.

Sheer, M.B. & Steel, E.A. (2006) Lost watersheds: Barriers, aquatic habitat connectivity, and species persistence in the Willamette and Lower Columbia basins. *Transactions of the American Fisheries Society* **135**, 1654–1669.

Simon, A. (1989) A model of channel response in disturbed alluvial channels. *Earth Surface Processes and Landforms* **14**, 11–26.

Skagit Watershed Council (2010) *Skagit Watershed Council Year 2010 Strategic Approach*. Skagit Watershed Council, Mount Vernon, Washington.

Slocombe, D.S. (1998) Defining goals and criteria for ecosystem-based management. *Environmental Management* **22**, 483–493.

Spence, B.C., Lomnicky, G.A., Hughes, R.M. & Novitzki, R.P. (1996) *An Ecosystem Approach to Salmonid Conservation*. Report number TR-4501-96-6057. Funded jointly by the United States Environmental Protection Agency, the United States Fish and Wildlife Service and National Marine Fisheries Service. Man Tech Environmental Research Services Corp., Corvallis, Oregon.

SRSC/WDFW (Skagit River System Cooperative and Washington Department of Fish and Wildlife) (2005) *Skagit Chinook Recovery Plan*. Skagit River System Cooperative, La Conner, Washington.

Stanford, J.A., Ward, J.V., Liss, W.J. *et al.* (1996) A general protocol for restoration of regulated rivers. *Regulated Rivers: Research and Management* **12**, 391–413.

Steel, E.A. & Lange, I.A. (2007) Using wavelet analysis to detect changes in water temperature regimes at multiple scales: Effects of multi-purpose dams in the Willamette River basin. *River Research and Applications* **23**, 351–359.

Stoddard, J.L., Larsen, D.P., Hawkins, C.P., Johnson, R.K. & Norris, R.H. (2006) Setting expectations for the ecological condition of streams: The concept of reference condition. *Ecological Applications* **16**, 1267–1276.

Tear, T.H., Kareiva, P., Angermeier, P.L. *et al.* (2005) How much is enough? The recurring problem of setting measurable objectives in conservation. *BioScience* **55**, 835–849.

Thorne, C.R. (1998) *Stream Reconnaissance Handbook: Geomorphological Investigation and Analysis of River Channels.* John Wiley & Sons, Chichester.

Thorne, C.R., Allen, R.G. & Simon, A. (1996) Geomorphological river channel reconnaissance for river analysis, engineering and management. *Transactions of the Institute of British Geographers* **21**, 469–483.

Tiegs, S.D. & Pohl, M. (2005) Planform channel dynamics of the lower Colorado River: 1976–2000. *Geomorphology* **69**, 14–27.

Torgerson, C.E., Faux, R.N., McIntosh, B.A., Poage, N.J. & Norton, D.J. (2001) Airborne thermal remote sensing for water temperature assessment in rivers and streams. *Remote Sensing of Environment* **76**, 386–398.

Toth, L.A. (1995) Principles and Guidelines for Restoration of River/Floodplain Ecosystems – Kissimmee River Florida. In: Carins, Jr., J. (ed.) *Rehabilitation of Damaged Ecosystems*, CRC Press, Boca Raton, Florida, pp. 49–74.

Toy, T.J., Foster, G.R. & Renard, K.G. (2002) *Soil Erosion: Processes, Prediction, Measurement, and Control.* John Wiley and Sons, Inc., New York.

Trush, W.J., McBain, S.M. & Leopold, L.B. (2000) Attributes of an alluvial river and their relation to water policy and management. *Proceedings of the National Academy of Sciences* **97**(22), 11858–11863.

Turner, D.L. & Clary, W.P. (2001) Sequential sampling protocol for monitoring pasture utilization using stubble height criteria. *Journal of Range Management* **54**, 132–137.

US Fish and Wildlife Service & Hoopa Valley Tribe. (1999) *Trinity River Flow Evaluation.* Unite States Department of the Interior, Washington, D.C.

Verbunt, M., Zwaaftink, M.G. & Gurtz, J. (2005) The hydrologic impact of land cover changes and hydropower stations in the Alpine Rhine basin. *Ecological Modelling* **187**(1), 71–84.

Walter, R.C. & Merritts, D.J. (2008) Natural streams and the legacy of water-powered mills. *Science* **319**, 299–304.

Waples, R.S., Beechie, T.J. & Pess, G.R. (2009) Evolutionary history, habitat disturbance regimes, and anthropogenic changes: What do these mean for resilience of Pacific salmon populations? *Ecology and Society* **14**(1), article 3. Available at: http://www.ecologyandsociety.org/vol14/iss1/art3/

Ward, J.V., Tockner, K., Uehlinger, U. & Malard, F. (2001) Understanding natural patterns and processes in river corridors as the basis for effective river restoration. *Regulated Rivers: Research and Management* **17**, 311–323.

Ward, J.V., Tockner, K., Arscott, D.B. & Claret, C. (2002) Riverine landscape diversity. *Freshwater Biology* **47**, 517–539.

Whitehead, P.G., Butterfield, D. & Wade, A.J. (2009) Simulating metals and mine discharges in river basins using a new integrated catchment model for metals: Pollution impacts and restoration strategies in the Aries-Mures river system in Transylvania, Romania *Hydrology Research* **40**, 323–345.

Wigmosta, M.S., Vail, L. & Lettenmaier, D.P. (1994) A distributed hydrology-vegetation model for complex terrain. *Water Resources Research* **30**, 1665–1679.

Wilcox, D.B. (1993) *An Aquatic Habitat Classification System for the Upper Mississippi River System.* US Fish and Wildlife Service, Onalaska, Wisconsin.

Wischmeier, W.H. & Smith, D.D. (1965) *Predicting Rainfall-Erosion Losses from Cropland East of the Rocky Mountains.* Agriculture Handbook No. 282, United States Department of Agriculture.

Wischmeier, W.H. & Smith, D.D. (1978) *Predicting Rainfall Erosion Losses – A Guide to Conservation Planning.* Agriculture Handbook No. 357, United States Department of Agriculture. United States Government Printing Office, Washington, DC.

Wolman, M.G. 1954. A method of sampling coarse riverbed material. *Transactions of the American Geophysical Union* **35**, 951–956.

Wright, J.F., Sutcliffe, D.W. & Furse, M.T. (eds) (2000) *Assessing the Biological Quality of Fresh Waters: RIVPACS and Other Techniques.* Freshwater Biological Association, Ambleside, United Kingdom.

4

The Human Dimensions of Stream Restoration: Working with Diverse Partners to Develop and Implement Restoration

Jon Souder

Coos Watershed Association, Charleston, Oregon, USA

4.1 Introduction

Even with the best plans and assessments, knowledge and work to restore streams is of little value if you cannot get people to implement projects. This chapter will bring in the human element in order to broaden our perspective beyond just the physical and biological aspects of planning and implementing stream restoration. The importance of the role that socio-political concerns play in such efforts was raised in Chapter 1. Humans have clearly played a major role in creating existing conditions in streams, and successfully involving a broad range of people into restoration efforts will likely be key to whether rigorous programs can be implemented and effective.

The traditional New World view is that humans are the source of most problems for streams and that, if human effects could be removed, then everything would return to 'normal.' More contemporary perspectives recognize that humans are, and have always been, an integral part of the ecosystem, and that their actions need to be considered in restoration strategies (Cronin 1996; Cowx & Welcomme 1998). Not only are humans a part of the ecosystem, but if people are properly engaged their support will facilitate restoration efforts often with sub-

stantial results (Higgs 2003). Alternately, if stakeholders are not properly engaged, failure is much more likely.

The science of human behavior and how to influence it is not novel, although it has not been consistently applied in stream restoration programs (Higgs 2003). This is not surprising because stream restoration is usually promoted by scientists and agency managers who have physical or biological science rather than social science backgrounds. The tools and techniques in this chapter will enable stream restoration practitioners to better understand the motivations of, and construct effective approaches for, the various types of human-dominated systems within which their programs will either succeed or fail.

Stream restoration occurs at many scales, ranging from working with individual property owners on projects such as streambank protection and riparian revegetation at the smallest scales to whole-basin restoration involving multiple agencies, landowners, and funders over the course of multiple decades. There are tools and approaches outlined in this chapter that will be useful at any of these scales, but some will be more pertinent at the individual project level while others more useful for larger restoration programs. However, individuals are involved at all scales; the regulatory approvals and funding generally require the involvement of some sort of government agency for even the smallest of projects, and larger restoration programs will involve significant negotiation and cooperation among individuals and agencies.

The models introduced in this chapter have been used in other contexts for over 80 years; they have been tested over time and space, and have demonstrated their continuing utility (but see the caution in Box 4.1). Most important and relevant is the agricultural extension model where new ideas and techniques are transferred from university research settings to farmers and communities. The extension model, which was developed in Europe and the United States, has proliferated worldwide.

Box 4.1 A caution about stereotyping

In this chapter there will be numerous cases where we will categorize people's behavior with descriptions that might be interpreted negatively. It is important to avoid stereotyping individuals because it will limit your ability to work with them. Research also shows that people who fit one category in one situation may act differently in others. The categorizations are meant to help and not hinder insight in how to best approach people to increase the likelihood of cooperation.

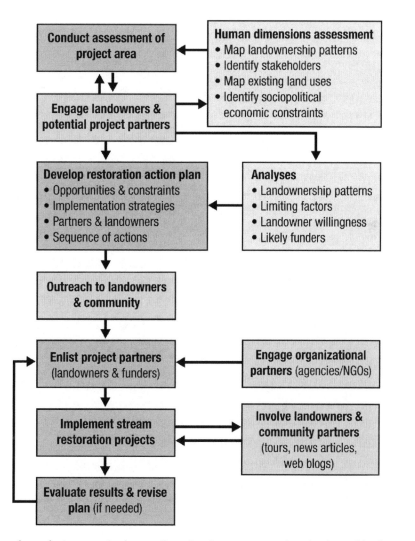

Figure 4.1 Flow chart of steps for incorporating human dimensions in stream restoration planning and implementation.

The original agriculture focus has expanded to include the diffusion of all types of innovations, including the spread of various technologies (e.g. cell phones and personal computers) as well as advances in medicine (Rogers 1995). A second model appropriate to stream restoration is 'social forestry' that arose from the Chipko movement in northern India in the early 1970s (Shiva 1991) and has continued through today (Wiersum 1995). Many of the early social forestry schemes explicitly recognized that the community played a vital role in protecting the integrity of the watershed for its benefits (principally water for direct human use and irrigation). Fitting stream restoration into a larger societal framework in the communities where the work will occur is essential if it is to become accepted and sustained over time. The value of involved communities is that once practices are adopted and become accepted by some individuals, they will be easily replicated and likely require reduced effort over time.

How do we incorporate the human element into stream restoration? Figure 4.1 provides a flowchart of one common approach towards doing so. The value of doing an assessment prior to developing restoration projects has been emphasized in the Chapter 3, but here we assess the human dimension that identifies land ownership patterns, stakeholders, land uses, and other socio-political constraints not discussed in detail previously. This information is then used in developing a restoration action plan to highlight opportunities, develop strategies to engage needed landowners and funding partners and, ideally, identify a sequence of proposed actions needed to implement the restoration plan. Subsequent to the action plan are outreach activities needed to engage these partners and, of course, periodic evaluation of the restoration results to ensure that needed corrective responses are made.

This chapter provides an outline of how to incorporate human aspects into stream restoration. We will discuss this in the context of the interrelationships among individuals, organizations, and geography. The following section covers the socio-political geography of stream restoration, emphasizing the important influence that land ownership patterns have on potential stream restoration strategies. Section 4.3 introduces the idea that stream restoration can be characterized as an 'innovation' in land management, and that effective strategies can build upon existing knowledge of how people accept new ideas. Section 4.4 moves from ideas to organizations and how people perform in them. Section 4.5 presents the implementation of restoration programs as a series of negotiations, where there may be only limited coop-

eration on the part of the various parties involved. Section 4.6 provides information about further readings that expand beyond what can be covered here, and the chapter concludes with a summary in Section 4.7.

4.2 Setting the stage: Socio-political geography of stream restoration

Understanding property and the rights to property and resources is fundamental in the design and implementation of stream restoration projects. This is especially true as stream restoration moves from simple single-site projects to more complex and area-wide projects in a watershed. All projects are going to take place on some type, or types, of property whose ownership comes with a variety of rights; paying attention to this as projects are developed is likely to increase the potential for success while decreasing the risk that a party with an interest (or a 'stake') in a piece of property will suddenly appear and object to what is being proposed.

4.2.1 Nature of the challenge

There are numerous critiques of stream restoration: many projects are grouped in the upper, headwater reaches without regard for habitat connectivity with downstream reaches (Lake *et al.* 2007); many are focused solely on single species or life stage rather than being ecosystem-process oriented (Beechie *et al.* 2010); and decisions tend to be made on an opportunistic rather than systematic basis (Bernhardt *et al.* 2005). While there are multiple causes leading to this situation, patterns of landownership often influence where projects are placed. There will likely be a mix of ownerships in most watersheds where restoration will occur. Understanding ownership patterns, property rights, and the objectives of various types of owners provides a foundation on which to build an effective approach to enlist cooperation, increase the likelihood that a stream restoration project can be implemented, and reduce the transactions costs (time, treasure, and temper) associated with project design and implementation.

4.2.2 Understanding property and property rights

Although land ownership may seem like a clear and simple concept at first glance, it becomes quite complex when dealing with the types of properties commonly needed to implement stream restoration projects. Because property rights – and their strength and enforceability – vary by country, making generalizations is hazardous. It is important to recognize that property rights need to be considered in designing stream restoration, and that they can potentially be quite complex. Depending upon the situation, property rights may be different for the surface versus any minerals located beneath the surface. There may be leases, easements or other tenure relationships that need to be considered in planning and project execution. Water flowing through a given property may have different rights associated with its ownership and use; the same situation may exist for wildlife and fish on the property as well. There may be legal and/or traditional aboriginal use rights associated with various resources located on a given piece of property.

Most land ownership systems are based on a sovereign (either king or government) asserting control over territory and subsequently divesting this property through grants or sales. However, there may be residual claims that have the potential to affect your restoration project: for example, the Mabo v. Queensland (No. 2) decision in 1992 in Australia returned lands to aboriginal groups and similar settlement agreements have provided lands to native groups in Canada and New Zealand. Security of tenure is even less certain in many areas of Asia, South America and Africa.

There may be additional legal or traditional – called 'usufruct' – rights to access and use of particular properties or resources on them. The Boldt decision in the United States Pacific Northwest affirmed a claim by the Native American tribes to half of the fish (principally salmon) produced, and gave the tribes a stake in increasing this production through both natural and artificial means (Wilkinson 2000). There may be strongly held views, and in some cases rights, on the part of the dominant cultures to access public lands for recreational use, including vehicular access on roads and off-road vehicle use in streams and riparian areas. In the United States, while the beds and banks of navigable streams and tidelands up to the high tide line are publicly owned, adjacent landowners can control access through their properties.

We generally assume that these various rights are clearly defined, but this is not always the case. Being sensitive to the property rights in your project area is a foundation for sound restoration programs; it is very easy to antagonize landowners and other stakeholders if they feel their rights or interests are at risk from your project. Our experience has been that, even if you have a legal right of entry, taking the 'neighborly' step of asking permission usually sets the stage for a productive interaction as compared to an argument.

4.2.3 Landscapes of restoration

The challenges related to landownership patterns are analogous to those found in conservation and landscape ecology. It is well known that patch size, connectivity, and patch turnover rates influence ecosystem performance (cf. Bennett & Saunders 2010 for a review). The same situation exists when the landownership pattern is placed over these habitats: the size of an ownership has important implications, as does its location in the watershed. Connectivity is important both within and between different ownerships. Turnover rates in landownership have an effect on the long-term prospects and success of stream restoration projects. Stream restoration projects require an understanding of ownership patterns and a strategy for approaching landowners that are most likely to be successful and stable over time.

As early as the 1870s in the United States, John Wesley Powell lamented that the arrangement of governmental subdivisions of watersheds with complex patterns made ecosystem-based management difficult (Stegner 1953). It is common throughout the world that land tends to be owned in larger parcels in the upper reaches of basins, and that this land is more likely to be in some form of public ownership. Historically, land was often subdivided until it reached the smallest economically viable size and so land on steeper slopes is more likely to be in larger parcels compared to gentler-sloped lands. Lowland areas (including valley floors) are often privately owned, of smaller parcel size, and have higher-intensity existing uses such as residential or smallholder[1] agriculture. Even if publicly owned, the ownership in lowland areas may be in narrow strips along stream banks and beds that inhibit a number of desirable types of restoration projects (i.e. remeandering, dike removal, floodplain connectivity).

The costs associated with land acquisition are much greater where higher numbers of smaller parcels are needed for a project. Compulsory acquisition of land

---

[1] 'Smallholders' is a term used to represent landowners with only small parcels of land (from 0.5 ha to c. 20 ha). These are sometimes called 'hobby farms' or 'rural residential', but they may also be subsistence farmers (Netting 1993).

Box 4.2 The Varde River restoration

The Varde River in the southwest Jutland peninsula of Denmark demonstrates the evolution of stream restoration to meet multiple, and changing, objectives over time. Societal desires to have more environmental and aesthetic land uses, as well as legal mandates from the European Union, were catalysts for stream restoration in Danish Rivers (Pedersen 2009). Restoration projects in the Varde River occurred in three distinct phases which had different objectives but in aggregate resulted in restoring virtually the complete stream. The Varde River projects began prior to better-known Danish efforts such as the Skjern River project (Pedersen *et al.* 2007), but continued afterwards and were affected by experiences there and in other areas of Denmark (Pedersen 2009, 2010).

Background. The Varde River has a catchment area of 1092 km^2 with a mean annual flow 16 m^3/s. It is one of nine streams in Denmark with historic Atlantic salmon (*Salmo salar*) populations and contains one of seven relict houting (*Corgonus oxyrhunchus*) populations in the Wadden Sea, which are protected under the EU Habitat Directive (Jensen *et al.* 2003). The Varde is one of the last streams in Denmark where freshwater pearl mussels (*Margaritifera margaritifera*) reside (Kann 2006). These mussels require adequate numbers of salmonids for one of their life stages (Skinner *et al.* 2003). Wetlands and fields in the lower Varde River where it enters the Ho Bugt estuary provide valuable habitat for the corncrake (*Crex crex*), an avian species that was listed in the EU Birds Directive, Annex I (Jensen undated). The significance of the tidal wetlands has resulted in their inclusion as a Ramsar site of international significance as well as being a EU Bird Protection Area and an EU Habitat Site (Frikke 1999).

Pre-restoration conditions in the Varde were similar to those seen throughout much of Europe and North America. Floodplains had been cleared, wetlands ditched and drained, and the river channelized so that agriculture could be practiced (Frikke 1999). In 1921 The Karlsgårde hydropower plant was constructed on a tributary to the Varde (Manøe 2011), and in 1929 thirteen meanders in the Varde below the power plant were cut off to increase the head for the turbines with the spoils used as embankments (Kann 2006). Over 90% of Varde flows were diverted into the Ansager Canal built in 1945 to feed the hydropower facility. Another 38 meanders above Karlsgarde Lake were straightened in the late 1940s and 1950s to increase farmland. Aquaculture ponds were built in the floodplain to take advantage of the return water downstream from the power plant. These practices resulted in environmental effects such as land subsidence, water quality degradation, the loss of both aquatic and terrestrial habitat, barriers and impediments to fish passage, and entrainment of juvenile fish in the canal and fish farm.

Restoration. Restoration projects in the Varde River Valley occurred in three distinct periods (see Figure 4.2). The first, begun in 1995 and completed by 2002, focused on restoring tidal wetlands and floodplain fields in the lower Varde River for the corncrake (Frikke 1999; Jensen undated). Activities began in 1995 with cooperation among the Varde Farmer's Union, Ribe County, the Danish Agriculture Ministry, and the Environment and Energy Ministry to raise groundwater levels and practice 'environmentally friendly' farming in 2700 ha of saltmarshes and meadows governed by the Danish Nature Protection Act (Frikke 1999). The project ultimately resulted in over 250 landowners enrolling over 92% of the land in the project area (Jensen undated). Redistributing farm lands to aggregate cooperating landowners into better habitat lands and entering into 20 year agreements to subsidize farming practices consistent with the corncrake's needs were key project features (Frikke 1999).

A second 2-year restoration phase started in 1999 to create freshwater pearl mussels and Atlantic salmon spawning habitat in 1.4 km of channelized stream just above the town of Varde, but below the power plant. Four meanders were recreated based on historic scar patterns, doubling the length of stream to 2.8 km, and 1500 m^3 of gravel were placed to improve spawning substrate (Kann 2006). Half the funding for the project (c. €270,000) was provided from a conservation group,

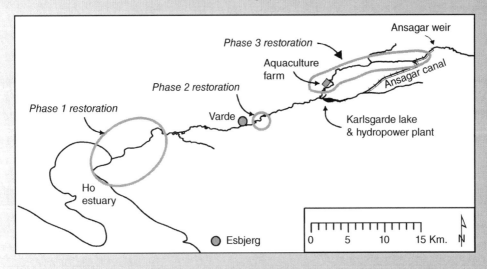

Figure 4.2 Map of the Varde River, Denmark and three phases of restoration.

with the remainder principally from the local county and municipality. Ribe County provided the designs, and Varde Municipality worked with landowners (Kann 2006).

The third phase in the Varde River restoration, began in 2005 and completed in 2010, was the most extensive of the three because it focused on restoring passage for houting to spawn in the upper river while also improving juvenile nursery habitat (Houting Project 2011). This project was identified and prioritized in the National Management Plan for the Houting (Jensen *et al.* 2003) and funded primarily by the EU LIFE program (€4.5 million for the Varde projects) (Strategic Restoration and Management or STREAM 2009). Four years of negotiations with landowners were concluded in 2009 (only two landowners out of 80 declined to participate). The project included remeandering 13 km of stream into 18 km of length, decommissioning of the Ansager Canal and the Karlsgårde power station to restore full stream flows to the Varde for the first time since 1945, and installing screening devices to prevent fish from entering the aquaculture ponds. As a result of this third phase, access to 75% of potential spawning areas was restored, stream flows were restored over 18 km of stream, and the channel embankments that previously limited nursery use of the floodplain were removed.

Lessons learned

1. Phase 1 in the lower, tidal reaches was conceived with the Varde Farmers' Union and explicitly recognized their interests in its third project objective, 'to ensure compensation of the owners or users of the land for any loss of income and to give them a high level of influence' (Frikke 1999). This set the stage for further cooperation.

2. Joint benefits between farmers and restoration practitioners were incorporated in Phase 1 by classifying lands most suitable for farming versus better restored to wetlands, identifying those farmers who were willing to adjust their practices to meet conservation goals, and exchanging lands outside the project area or buying out those who didn't want to participate. It still took almost four years to acquire lands for Phase 3, but in the end the two significant landowners who declined to participate (including the aquaculture farm) did not stop the project. This is in contrast to the Skjern River restoration where land expropriation was ultimately required (Pedersen 2009). The willingness to provide lands for exchange was identified as a key factor in the project success (Elbourne 2009).

3. There was a strong local presence during all three phases of the projects. The initial cooperation of the Varde Farmers' Union and on-the-ground personnel slowly, but steadily, enrolled landowners over six years. Local meetings, newsletters, and other outreach efforts provided assistance (Frikke 1999). Phase 2 brought together a Steering Committee that ixncluded the local municipality, the County, and the Society for Nature Conservation, each of which utilized their comparative strengths to implement the project. Ultimately, the trust between the farmers and local restoration practitioners, who had a good track record, helped achieve the larger Phase 3 project objectives (STREAM 2009).

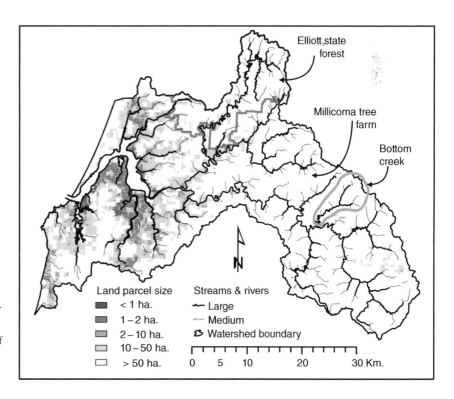

Figure 4.3 Pattern of land ownership in the Coos Watershed in southcentral coastal Oregon, USA showing blocked parcels in the upper basin with smaller parcels along streams and the estuary. As the parcel sizes become quite small, especially along the estuary and in the cities of Coos Bay and North Bend, the area becomes completely black.

for restoration projects (variously known as expropriation or eminent domain) is difficult, always time-consuming, and usually controversial. All these factors may lead to the development of restoration programs with larger landowners, which are thus likely to occur higher in watersheds. The comparative ease in getting projects started is especially apparent during the early phases of stream restoration.

There are significant differences in land ownership patterns in all three of the case studies discussed in this chapter. The Coos watershed on the south-central Oregon coast in the United States has diversity of ownerships, parcel sizes (see Figure 4.3), and land uses including public and industrial timber management, agriculture, smallholder residences, and urban areas surrounding the estuary. That diversity and parcel size decreases lower in the watershed adjacent to the estuary. In this case, the Coos Watershed Association was able to work initially with two large owners in the upper basin (the Elliott State Forest and Weyerhaeuser Timber Company) to build capacity and demonstrate success that could be transferred to other areas of the watershed. Ownership in the Varde Rivers watershed on the Jutland Peninsula in Denmark (Figure 4.2) was primarily private lands in agricultural use (pasture, croplands, and aquaculture farms), although there was additional infrastructure for hydropower generation. It was relatively straightforward to value these properties and identify replacement locations or compensate the private owners whose lands were needed for the restoration projects (see Box 4.2 for the Varde River case study). Finally, the Rio Puerco watershed in the western part of New Mexico in the United States has a highly complex mix of federal, state, tribal, and private ownerships, often with different surface and mineral estates. Its large scale (over $17,000\,km^2$), overlapping and sometimes competing authorities, and lack of adequate funding have impeded restoration efforts in this watershed over the last 30 years.

4.2.4 Understanding landowner/manager and agency objectives

In addition to understanding ownership geography, an awareness of some basic characteristics of each type of ownership will assist in developing an effective strategy to build support for collaborative stream restoration. Table 4.1 shows some of these characteristics for four of the major ownership groups: State/federal, aboriginal, large corporate, and smallholder. Note that while these traits are simplified, they are still valuable to consider when assessing potential approaches. Each of these four

ownership types presents advantages and disadvantages when creating a stream restoration program or project.

In many cases governments have the capacity, time, knowledge, and resources to implement the projects although they may have difficulty building a consensus within their own organization on the most effective restoration approach on lands they own or control. They are also likely to own much of the infrastructure (roads and water control) affected by a project. In many cases, governments also play a regulatory role both in projects on their lands as well as other ownerships, and there may be multiple layers of government that have the potential to affect a stream restoration project.

Depending on the country and tribe, aboriginal ownerships may be managed by an outside federal agency or one within the tribal organization. In either case, most decisions are made by a council whose membership and leadership is either elected or appointed through traditional means. If more than one tribe's property is involved, each one needs to be included in the partnership; do not expect that one tribe will speak for another. Social structures in tribal communities and governance are complex, and it is likely that it will take concomitantly greater time to get projects approved. While stream restoration is often consistent with tribal values, you will need to be very diplomatic in approach and will typically need approval from the traditional elders to be successful. However, as you will see in the Rio Puerco Case Study, working with tribes in their project area has been one of the most successful and rewarding experiences they have had.

The benefit of working with large corporate ownerships, whether in timber, agriculture, or mining, is that when they decide to do something they usually want it done quickly, safely, and with a minimum of disturbance to their core operations. As we will see below (and in the Coos case study), when your interest and theirs coincide it is possible to build a highly successful restoration program. However, large corporate owners will expect a comparatively high standard of sophistication, expertise, and professionalism on the part of restoration proponents.

Relationships and trust are crucial for success in approaching small holders of farms, timberlands, and home sites. These smallholdings may have some of the most intensively used land in any category, and their owners may have limited ability and knowledge to manage them in ways consistent with good stream qualities. Although smallholders present challenges, their location in a watershed often requires their participation

Table 4.1 Land management characteristics among various types of ownerships.

Characteristic	State/federal agency	Aboriginal	Large corporate: timber, agriculture, mining	Small holder: non-industrial timber/agricultural/rural residential
Organizational culture	May be agencies with overlapping authorities or purposes	Strong traditional, cultural focus; often tension between revenue-generating interests and other uses	Attuned to corporate culture; relationships are important; limited time to assist with projects; need to be self-sufficient: 'no hassle-no drama'	Reputation as good land steward is important; limited available area means that meeting their land management objectives is critical
Decision-making process	Multiple levels of review and approval; set procedures and regulatory requirements	Often complex, lengthy and incomprehensible (at least to outsiders) decision-making processes	May be made at the local level, or may need to go up their hierarchy. As more trust is gained, the decision is likely to be made at closer to the local level	Trust and credibility necessary to start building relationship; may need to involve family members.
Timescale	Long lead times	Deliberate, not fast	Fast to act: 'get-it-done' mentality	Generally relatively quick
Budgeting	Needs to fit into their budget cycles	Funds will probably be provided by third party	Identified benefits – bottom-line driven	May have limited ability to cost-share.
Project implementation	Often eager to provide assistance – 'multiple cooks' dilemma	Desire, but not necessarily the capacity, to manage their own resources	Will want to have control, or at least oversight, over project implementation, whether directly or through their approved contractors	If they own equipment, may want to provide services and may want to be compensated for this; otherwise, will probably want you to manage
Stakeholder involvement	May require public notices and other stakeholders to be involved	There may not be a single, consistent approach among groups; may have rivalries	Want to be seen as good managers and neighbors; otherwise not likely to be great except if your project conflicts with other existing uses such as fishing or OHV use	Concerned with how neighbors view collaboration

in stream restoration activities. These landowners will likely need different strategies to engage them in stream restoration.

4.2.5 Why understanding socio-political geography is important

The challenge of constructing a restoration project increases for every additional landowner whose permis-

sion is required. Objections from landowners have stalled projects, reduced opportunities or stymied restoration programs in a number of places, including Denmark (Pedersen 2010) and the United States (Alexander & Allan 2007). If you cannot obtain voluntary access to needed property, you may be reduced to more coercive measures such as expropriation or condemnation, which both take more time and are politically more difficult. Ultimately, coercive measures may actually make the proposed project – or subsequent projects – impossible, as we saw in the Varde River case study from Denmark.

We will go into more depth in the remainder of this chapter on ways to approach these different constituencies for stream restoration; however, lessons to keep in mind include the following.

1. It is likely that the stream restoration project that you are considering will occur on someone else's property; even if it is primarily on your property, others may have some sort of interest (i.e. hunting, fishing, or recreation access) in its outcome. It is also important to recognize that your project may have some effects on adjacent properties or in the larger region (e.g. there may be a 'tipping point' where an activity such as agriculture or forestry collapses if insufficient landowners participate to maintain supplier and purchaser networks). Recognizing these ahead of time may allow you to adequately address them in your plans.

2. Because landownership patterns overlay geomorphological and biological systems, it will be difficult to focus solely on one side of the equation (bio-physical versus socio-political) to create a viable stream restoration program. For the highest success, strategies must be developed that incorporate social, biological, and physical considerations into stream restoration programs.

3. People are important both as participants as well as supporters of stream restoration. People already have considerable interests (property and otherwise) in streams and in the lands that surround them. Only by adequately understanding their interests can we work out how to reconcile our programs. Examples of such interests range from protecting their stream banks from erosion, to maintaining a 'park'-like appearance and being considered a good land steward in the community. Interests such as these, and others, can be used to find common ground when establishing relationships with land owners.

4. Social justice and equity concerns need to be identified and addressed as stream restoration programs expand across the landscape (Hillman 2004, 2005). There may be losers as well as winners in large-scale stream restoration programs. Insuring that both the benefits and the costs of these activities are spread equitably has important long-term implications for the ongoing acceptability of stream restoration in the larger society.

5. While building a large public constituency to support stream restoration is important for long-term sustainability (Spink *et al.* 2010), our immediate interest is to focus on tools and techniques to work with the primary partners who will determine the success or failure of individual projects. Long-term sustainability is built on a record of individual successes. In the concluding section of this chapter we will provide suggestions for further reading to gain an understanding of these broader issues.

4.3 How stream restoration becomes accepted

Stream restoration is still in its infancy in the larger forum of natural resource management (Higgs 2003). In the United States, dust storms and loss of prairies at the beginning of the Depression provided the impetus for early upland restoration efforts (Chapter 1; Higgs 2003). Many restoration actions are as yet unproven, although promising (see Chapter 5). In this respect these restoration techniques can be considered innovations or experiments. Stream restoration practitioners have the challenge of convincing others of both the need for the actions they are proposing and the efficacy of those actions. We are assuming in this discussion that the decision to adopt a stream restoration project is an individual action, and not a collective or authoritative action. These decisions will be made by individuals who are acting within organizations – agencies or corporations – as well as by private landowners. While these individuals have their own characteristic traits (but recall the cautions in Box 4.1), their actions will also reflect their organization's culture, needs and interests, and decision-making processes.

Effectively targeting your investments of time and resources where they are likely to be most successful will accelerate the acceptance of stream restoration programs, while reducing frustration and ameliorating conflict. There are specific situations where you will need to interact with, and try to persuade, a wide range of different types of individuals to cooperate[2] with your plans. Understanding individuals' traits should allow you to evaluate whether it is worthwhile to make an investment to persuade them to cooperate and, if so, what to expect.

---

[2] 'Cooperate' means that the other party does not resist what is being proposed, and may assist in your efforts. 'Collaborate' goes beyond this to reflect jointly working together to achieve a common goal.

We will start by examining how innovations gain acceptance and how to understand the likelihood that different individuals will be receptive, and in the next section focus on organizations themselves. Section 4.3 is based largely on the pioneering work of Everett Rogers (1995) as found in four editions (since 1962) of his *Diffusion of Innovations*.

4.3.1 Restoration as innovation

The process of asking someone (or some agency) to implement a stream restoration project is not unlike marketing a new product or innovation. In this context, the 'product' is a new alternative, a new means of solving problems, or simply a new idea (Rogers 1995). The entity that you are requesting to cooperate may or may not know about either the need for the restoration project or the benefits of the specific action you are proposing. The idea may be completely new to them. What is not new is the understanding of how people accept new ideas and products.

To be accepted, a restoration action that is considered innovative must be perceived to:
1. provide a relative advantage to the person or entity over existing techniques;
2. be compatible with existing land uses and practices;
3. not be too complex to understand or use;
4. be testable on an experimental or limited basis; and
5. be observable so that the results can be clearly articulated.

4.3.2 Innovation diffusion through networks

Knowledge about innovations is diffused through networks of interconnected people. New ideas are typically brought into a local network through connections that one member has with other outside networks or sources of information. These ideas are then exchanged among others within the local network through personal connections as well as through more formal mechanisms such as meetings, newsletters, and plans. A way to visualize these networks is that the external connections radiate outward (similar to a star), while the local internal connections form an internal web of interrelationships.

Networks provide a structure that allows communication of ideas to lead to decisions to adopt innovations as well as to validate the adoption of innovations among peers. Within any given network there will be various levels of value given to a member's ideas. 'Opinion leaders' in the network provide a credible source of new ideas, which are

typically adopted by their followers over time. However, to maintain credibility, opinion leaders must not stray too far: their ideas must be based on criteria already understood and approved by their followers and peers and assumptions must not be too broad or complicated, as complication implies risk.

4.3.3 Process of innovation adoption

As innovation becomes accepted, its adoption spreads throughout a community in a progression of five stages with sixteen steps (Box 4.3; Rogers 1995). The first stage revolves around an increasing awareness of the innovation or technique, initially with its description followed by a broader understanding of its potential impacts. Once there is general knowledge about a technique, a potential adopter often needs some persuasion prior to making a decision. Your leverage in this second stage is that the potential adopter is open or receptive to consider an innovation's potential. This incremental adoption of restoration practices can be seen in the Coos Bay case study described later in this chapter. By the time a decision is made about whether to adopt a practice, there are really only two choices: try it out, or defer until further

---

Box 4.3 Innovation diffusion stages

Knowledge stage
  1. Recall of information
  2. Comprehension of message
  3. Knowledge or skill for effective adoption of the innovation
Persuasion stage
  4. Liking the innovation
  5. Discussion with others
  6. Acceptance of message
  7. Formation of a positive image
  8. Support for the innovation
Decision stage
  9. Intention to seek additional information
  10. Intention to try innovation
Implementation stage
  11. Acquisition of additional information
  12. Use on a regular basis
  13. Continued use of innovation
Confirmation stage
  14. Recognition of the benefits
  15. Integration into ongoing routine
  16. Promotion to others

(Based on Rogers, 1995)

inducements (information, incentives, etc.) tip the scale. In truth, the initial implementation of an innovation is a trial on the part of the adopter, who is evaluating whether the technique acceptably meets their needs and objectives. Ultimately, the best outcome is that the project is highly successful and that the participating landowner (or agency) gives it a ringing endorsement, expands its use in their own operations, and encourages others in their network to adopt the practice.

4.3.4 Innovation acceptance

The stages of innovation diffusion represent an idealized, linear pattern from first notice to final acceptance in the larger community. However, not all potential recipients of information and persuasion are equally likely to adopt a stream restoration practice when it is newly introduced. Identifying 'opinion leaders' is an important early task because they form the key target audience and are likely considered credible in their own networks. Understanding their characteristics – as well as those of other potential candidates to implement restoration projects – will enable you to be effective in targeting who to enlist for support when introducing innovative stream restoration practices.

The cumulative rate of adoption of innovative practices is widely recognized to take the form of an S-shaped curve where the initial acceptance is slow, then rises rapidly as the practice's visibility and effectiveness become recognized, and then slows again as the majority of people who will adopt the practice already have (see Figure 4.4). The rate of adoption by numbers of individuals at any given period of time resembles a bell-shaped curve that represents a normal distribution with typical statistics such as mean (the amount of time by which half of the people who will adopt a practice have done so) and standard deviation (Rogers 1995). The intervals between the means and standard deviations in the adoption curve have been given names that are now in general usage: innovators, early adopters, early majority, late majority and laggards. It is important to recognize that these categories identify solely the intervals in the normal frequency distribution, and that a particular individual may switch his or her category depending upon the issue at hand. Nonetheless, the categories and descriptions of their characteristics provide a useful insight in targeting potential stream restoration partners by understanding who is – and who is not – a good candidate for investment of time and resources and who is or is not likely to adopt a practice and provide a successful example that will be copied by others.

Innovators are venturesome, willing to take risks and accept that failure is possible. They are often well connected to the outside world (i.e. 'cosmopolitan,' but may

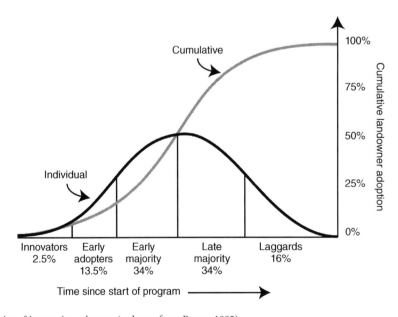

Figure 4.4 Categories of innovation adopters (redrawn from Rogers 1995).

lack integration into the local social system. They tend to have a high level of resources, and are generally able to understand and apply complex technical knowledge. *Early adopters* are often respected local opinion leaders – the 'locals' as opposed to the 'cosmopolitans.' They tend to be upwardly mobile, well-educated and prosperous. Generally, early adopters are empathetic, not dogmatic, and maintain strong contacts with local change agents. Through these contacts with change agents they bring new ideas into their local network. *Early majority* are 'deliberate,' watching to see how an innovation plays out before making a decision, but are not considered leaders. *Late majority* can be thought of as 'skeptical' of new ideas and practices. They generally need to be pushed into accepting the innovation or forced into it for regulatory or financial reasons. *Laggards* are 'traditionalists'; they are almost completely outside the local social networks, and tend to focus on the past. They tend to be suspicious of new practices and of the people who encourage them.

The first three categories of adopters cumulatively represent half of the people who are potentially ever going to adopt a stream restoration project. Of these, the early adopters provide the best investment of time and effort. This is because the innovators, while likely to accept something new, are not considered by the community to be representative, i.e. they are 'outliers' in local social networks. While the early majority will adopt a restoration practice, they will wait until they see others who they respect do it before making a commitment themselves. Adding incentives to encourage participation is one way to accelerate their acceptance.

The final two categories represent people who tend to lag behind others in their willingness to adopt an innovative practice. These types generally give less return on time invested, especially in the early stages of a restoration program, because persuading them will take significant time with little likelihood of success. Recognizing these limitations, however, there are also situations where an effort is justified if their property or involvement is deemed critical to the restoration program. It is important to be aware that different strategies (such as direct payments, regulatory requirements, or ultimately land purchase) may be needed to increase the probability that they will cooperate.

4.3.5 Why understanding innovation diffusion is important

There are specific situations where it is necessary to interact with, and try to persuade, individuals in each of these five categories to cooperate. The descriptions in the previous section characterize the individuals' traits that should allow one to evaluate whether it is worthwhile to make a particular persuasive investment and, if so, what to expect.

While the research shows that the best investment of effort will be to work with early adopters, there are situations where working with innovators may be the best strategy. For example, if one is trying to test out different techniques and knows that some are likely to fail (perhaps even significantly), then an innovator's willingness to take risks and suffer failures makes them a preferable project partner. One would not want to try out something risky with an early adopter because, if it failed for them, the technique and the project proponent will gain a bad reputation, souring later efforts. An early majority person would be unwilling to accept the practice at all because it had not been demonstrated as previously successful.

It is important to recognize that the best potential project sites are not necessarily going to be owned by those most likely to accept and adopt a restoration practice. One must strategize accordingly: if the preferred site is owned by an early majority landowner, then it may make sense to identify that person's social network to find an early adopter to take on a project recognizing that, over time, you will get to your preferred location. If it is absolutely necessary to access a site that is owned by a landowner classified as a late majority or a laggard, then it is important to recognize that voluntary measures and persuasion are unlikely to be quickly successful. In those cases, more forceful approaches such as regulatory mechanisms or purchasing the property may be needed to obtain access.

4.4 Organizations and the behaviors and motivations of those who work for them

Most, if not all, stream restoration programs and projects occur within organizations of one type or another. Often multiple organizations – different levels of governmental agencies, non-profits with various objectives, and sometimes even professional associations – are involved. Understanding who you are dealing with on a personal basis, and especially their motivations, will lead to more productive (and less frustrating) interactions. Understanding different stages of organizations will also assist in setting realistic expectations in your dealings with them. We will present a matrix of behavioral types and organizations at their different life stages, along with some concepts on how to evaluate their effectiveness. One significant value in understanding these two major concepts (behaviors and organizations) is that you – as a stream restoration practitioner – will be dealing with individuals with specific behavioral types but who are also associated with an organization at its given life stage. Understanding how a particular behavioral type fits into an organization provides insight into whether that person is an outlier in their organization and whether that person can potentially bring support to your project.

4.4.1 Organizational behaviors and motivations

The process of implementing stream restoration projects typically requires supportive interaction with a multitude of players who bring various strengths and limitations to the process. Many, if not all, of these individuals will be working for various institutions, some public, others private. While the previous section characterized individuals' traits based on their acceptance of innovation, this second set of characterizations describes how people function within organizations. Understanding these types with whom you are going to be interacting should help determine the best process to enlist their support or avoid their disapproval.

Government officials'[3] motivations were described as early as the 15th century by the Italian Niccolò Machiavelli (1469–1527). Modern organizational behavior characterizations – particularly for government officials – were made by Max Weber (1864–1920) in the early 20th century and form the basis for the study of administrative organizations (Weber 1962). Anthony Downs classified motivations of officials working for agencies in *Inside Bureaucracy*, published in 1967. It is worth noting that while Downs used the term 'bureaucracy,' he meant any large organization (with some caveats). While there are certainly alternative approaches and much subsequent research, Downs' categorizations continue to be used because they can be so readily observed. We will draw liberally from his descriptions in this section.

4.4.1.1 Motivations of officials

People work for organizations for a number of reasons: because they believe in the work they do; for the pay and benefits; or even because of the situational characteristics of the job such as physical location, co-workers, or ease of work. Downs divides these motivations into two general classes: those people who are purely self-interested (*Zealots* or *Climbers*), and those who have mixed motives (*Advocates*, *Conservers*, or *Statesmen*). Those who are purely self-interested seek benefits only for themselves, while people with mixed motives have a concern for others (or society, or the agency) as well as for themselves. People have varying characteristics that need to be considered in how you interact with them, but generally fall within one of the five following categories.

1. *Zealots* seek power so that they can affect their own view of the world. Often their perspectives of what should be done are relatively narrow, and many times they do not see how what they want fits into the larger picture.

2. *Climbers* are people who are focused on how they can get ahead. In some cases this means that they are very attuned to policy and performance so that they can stand out. When advancement is not forthcoming, these officials may seek to aggrandize their current position to make themselves appear more powerful.

3. *Advocates* are loyal to organizations, policies, and causes and will protect these interests against outsiders who threaten them. They are strong team players.

4. *Conservers* are people who are primarily interested in protecting their security and convenience. They are generally resistant to change if it might affect their current situation. At worst, these are the people who have 'retired in place.'

5. *Statesmen* seek power and authority in furtherance of broad governmental or society or corporate goals; they enjoy being recognized for the influence they wield. Statesmen are likely to mediate well with others of their type, thus sometime confusing personality with job title.

4.4.1.2 Leveraging organizational behaviors

It is important to know who you are working with and figure out what it is that motivates them. What are their goals and objectives? How can you play to that objective

---

[3] We will use the term 'official' rather than 'bureaucrat' for the same reason as Downs (1967): to avoid the pejorative connotation that bureaucrat has among the general public.

in order to meet your needs? This approach is smart and relevant no matter what field you are in or whose team you are on.

New organizations and initiatives are frequently established or quickly sought out by *zealots*. Zealots have a charismatic vision that enables them to enlist others, motivating their adherents to expend effort and resources to further the cause. Zealots are also useful if a single-minded persistent focus is needed to get a program off the ground. However, zealots are also capable of antagonizing those who may be indifferent or opposed to their ideas.

Climbers can advance the mission of a restoration project if their approach is to attract attention from higher levels in the organization or use their success as a launching point. In this respect, they can be entrepreneurial innovators if it benefits them to create high-visibility programs. Climbers also commonly devote significant efforts to building relationships with those above them in organizational or professional hierarchies. This can be beneficial to advertising one's restoration program, but do not expect climbers to defend the program if they meet resistance from people they want to impress. Climbers are unlikely to exhibit the persistence and staying power needed to sustain a restoration program in the long term. However, they may use their affiliation with a successful program to bolster their reputation as they move up in power.

Once a restoration program is established, *advocates* are the people who are most valuable to sustain it. Advocates are likely to be open to the program within their own work group, especially once they see its virtues. These are the people who will negotiate, design, and build your projects. Advocates will help write briefing papers, prepare project budgets, and sell the program internally. They will be strongly supportive of the program with external entities, and are more likely to be effective in collaborating with outsiders than zealots, especially when seeking joint benefits or compromises are necessary.

While *conservers* are focused on preserving their own status quo, they can be appealed to if a case is made that, to do otherwise, would result in changes to their work environment. In the sense, 'do this or else . . .' may gain grudging acquiescence, a subtler and gentler approach can also be used. For example, if their agency might be threatened by a lawsuit, an implicit suggestion that implementing a specific type of restoration project or program – such as road sediment reduction or fish passage – might be a sufficient catalyst to get them

involved. Conservers are also valuable allies when more routine tasks within their job scopes are needed, such as procurement and financial services.

Finally, *statesmen* can help deal with powerful interests who may be needed to support a restoration program. They generally have the tenure, stature, and skills to interact diplomatically with their peers in other agencies. Conflicts, if not resolved, will rise to the statesman level; however, intermediary levels of authority (the conservers and climbers particularly) may resist passing conflicts upwards because it might reflect badly on their performance. Be cautioned, however, that statesmen are likely to trade away things that you think might be valuable (e.g. specific regulatory requirements, enforcement actions, or project components) if, in their opinion, it protects the agency, the government, or high officials.

It is important to build relationships and be aware of the people with whom you are dealing. The caution in Box 4.1 about stereotyping equally applies to identification of bureaucratic behaviors: people may respond differently in different situations, and behavioral characteristics are known to change over time for a number of reasons (age, family responsibilities, wealth, etc.).

4.4.2 Understanding your own and other organizations

How organizations form, sustain themselves and finally renew or deteriorate is important because the stage at which an organization is in can profoundly affect how it approaches projects and conducts itself. As with individual behavior, understanding and accommodating these stages will improve the success of your projects.

There is extensive literature on organization theory and organizational life cycles. In the interest of keeping our discussion to a reasonable length, we will rely on two perspectives. The first is from Anthony Downs and is included in his *Inside Bureaucracy* (1967) that we used in the previous section. The second perspective is from Quinn & Cameron (1983) in their synthesis of nine different life-cycle models (including Downs') with a particular eye towards organizational effectiveness.

Common among most theories is that any organization goes through a cycle of creation, performance and renewal or demise. Quinn & Cameron (1983) put this cycle into four different phases (depicted in Figure 4.5) called entrepreneurial, collectivity, formalization and control, and elaboration of structure. At the end of the cycle there are two alternatives: the organization can renew itself

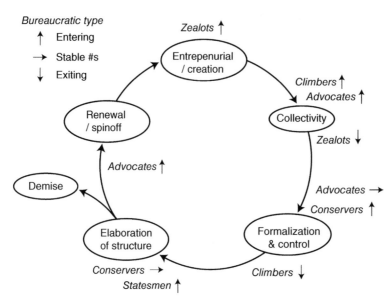

Figure 4.5 Organizational life stages and the various types of behavior of officials typically associated with them.

(possibly spinning off to a new organization), or it can face its demise. Because there is a direct correlation between behaviors of officials in organizations and organizational life stages, the dominant types at any one given stage of an organization are identified in Figure 4.5. At any given stage of an organization, different officials' behaviors are likely to be rewarded while others may be in conflict, thus giving rise to tensions, ineffectual work habits, and difficulty on your part figuring out how to get things done. Understanding these patterns will make your efforts more effective and satisfying.

The *entrepreneurial* phase, also called 'creation' by other authors, begins when an organization is created or breaks off from a parent entity, needing to quickly find or create a 'niche' for itself. Generally, the organization has more ideas than resources at this stage, so gaining needed resources and organizational autonomy are prime activities. Zealots tend to be attracted to organizations in this phase.

Organizations that survive their first stage move into a phase called 'collectivity' which refers to their rapid growth with a related high level of staff commitment. Innovation and expansion are the focus, with the expectation that products (in our case restoration projects) will be emphasized. The organization has an informal structure, with long hours being willingly expended by

staff that place high value in the mission and sense of camaraderie among themselves. Zealots tend to withdraw at this stage, being replaced by climbers and advocates.

As an organization grows during the collectivity phase, it becomes necessary to increase structure and functions. During the *formalization and control* phase an organization emphasizes efficiency and stability. Coordination and communication becomes more difficult as the organization grows larger, thus necessitating formal rules and procedures. Planning and performance measures rise in importance, with more work being done by specialists. Personnel policies – such as regular evaluations and performance rewards – become standardized as it becomes increasingly difficult for everyone to know each other and their relative contributions. Advocates still have important roles, but conservers provide a significant proportion of the workforce.

The formalization and control phase can last a long time, although organizational effectiveness tends to decline as personnel and policies become more entrenched and less responsive to changing conditions. At some point an organization will proceed into the *elaboration of structure* phase, which can be thought of as renewal. This stage emphasizes adaptability to changed (and changing) circumstances, a search for new markets or products and a reflection on the organization's purpose. The outcome of these actions may result in decentralization or spinoff of organizational units (thus creating new organizations) and possibly, but rarely, a decision to terminate the organization. Statesmen tend to lead organizations at this stage of their life cycle because they are less tied to the minutia of its day-to-day affairs and are more likely to take a broader perspective on how the organization best fits into overall societal needs. However, having inadequate leadership at this stage may lead to the timely (or untimely) demise of the organization.

4.4.3 Why understanding organizational patterns is important

The success of restoration practitioners is often determined by how well they navigate interactions with individuals and organizations over both time and space. Understanding how a particular behavioral type fits into an organization at its life stage provides insight into whether that person is an outlier in their organization, or is likely to bring the organization's support to a restoration project or program. For example, if you are considering partnering up with a *zealot* who is employed by an agency at the *formalization and control* life stage, you should be aware that this person may be viewed as

a 'loose cannon' by the employer. While you may enjoy that perspective on a particular aspect of stream restoration, the zealot may have antagonized some other people in their organization from whom you need support. This 'mismatch' between employee motivations and organizational life stages is not uncommon, particularly in governmental organizations where jobs are more secure.

Your organization will be dealing with other organizations in order to get stream restoration implemented; the capabilities and effectiveness of all participating organizations will be tested during this process. For example, if your organization is in the *entrepreneurial* or *collectivity* stage, and your funder organization is similarly situated, it is quite possible that neither organization will be able to effectively manage the financial responsibilities required for large projects. Your organization may not be able to keep the financial records required by the grantor for reimbursement, and the grantor organization may be unable to process payment requests in a timely manner because its systems (human and machine) are not sufficiently sophisticated. Being aware of this in advance could help avoid potential problems. The Rio Puerco Management Committee case study (Box 4.4 and 4.5) provides an illustration of organizational cycles in a very complex restoration environment.

Box 4.4 Rio puerco management committee members

Federal agencies	Department of Transportation
Army Corps of Engineers	Environment Department
Bureau of Indian Affairs	Mid-region Council of Governments
Bureau of Land Management	NMSU Cooperative Extension Service
Bureau of Reclamation	State Engineer
Environmental Protection Agency	State Land Office
Fish and Wildlife Service	Cuba Soil & Water Conservation District
Forest Service	Ciudad Soil & Water Conservation District
Geological Survey	Lava Soil & Water Conservation District
Natural Resources Conservation Service	Valencia Soil & Water Conservation District
Tribal	*NGOs and private*
Jicarrilla Apache Tribe	Albuquerque Wildlife Federation
Pueblo of Acoma	Cabezon Water Pipeline Association
Pueblo of Isleta	Quivira Coalition
Pueblo of Jemez	Sierra Club, Rio Grande Chapter
Pueblo of Laguna	Rio Puerco Alliance
Navajo Nation	Rio Puerco Watershed Committee
State of New Mexico	WildEarth Guardians
Bureau of Mines	Private Landowners
Department of Game and Fish	Public-at-Large

Box 4.5 Restoring the Rio Puerco: Challenges of coordinating across broad scales and complex ownerships

This case study illustrates the challenges faced both by scale and complex land ownership patterns in restoring a severely degraded, semi-arid watershed. Such a scale and ownership pattern mean that large numbers of parties need to be involved in restoration efforts and, given the slow pace of recovery in arid environments, it is very challenging to sustain a restoration program over time, space, and culture. The Rio Puerco Management Committee's 15 year history aptly illustrates the effects of organization life cycles and the people they attract.

Background. The Rio Puerco was once called the 'breadbasket of New Mexico,' lined with villages and agricultural fields fed by acequias (traditional community-operated irrigation systems) (Scurlock 1998). The watershed has a long history of land-use and property disputes dating to original Spanish colonization in the early 1600s. Land-use practices – principally grazing and road building – have reduced vegetative cover, altered drainage, and reduced riparian vegetation. Climate changes may have originally triggered channel incisions, but land uses have exacerbated these effects leading to significant gullying and high sediment yields. The Rio Puerco is estimated to produce about 6.5 m^3/ha of sediment annually, and its highest recorded suspended solid concentration of greater than 600,000 parts per million places it in the top four sediment producers worldwide and the highest in the United States (Gellis et al. 2004).

The 19,000 km^2 Rio Puerco watershed, located in the Southeastern Colorado Plateau Province of the American Southwest Region, has 359 km of streams in four basins and nine sub-basins (Figure 4.6). Wholly within the state of New Mexico, the Rio Puerco drains portions of seven counties,

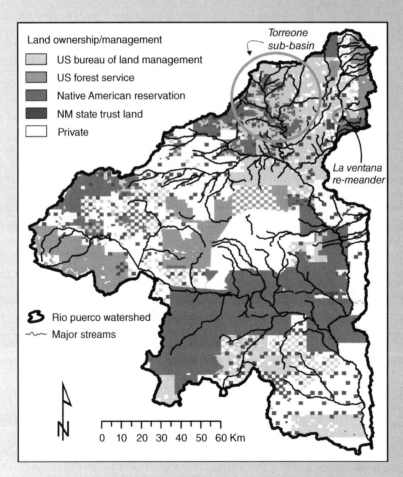

Figure 4.6 Map of the Rio Puerco basin in New Mexico (USA) demonstrating diverse ownership.

contains parts (or all) of six Indian Reservations, federal and state lands managed by four different agencies, and has at least five regional economic development and soil and water conservation districts operating within its boundaries (RPMC 2001; NMED 2010). The RPMC includes these entities, along with many other organizations (see Box 4.4).

The extensive history of restoration planning (see Box 4.6) has been characterized as disjointed, disorganized, largely non-collaborative, and not holistic (RPMC 2001). To overcome this, the original Rio Puerco Committee headquartered in Cuba, New Mexico convinced the United States Congress to pass the Rio Puerco Watershed Act (Act) in 1996 which designated the US Bureau of Land Management (BLM) as the lead agency. The Act also: (1) created the inter-agency and stakeholder Rio Puerco Management Committee (RPMC) to develop and implement a watershed management program; (2) established a clearinghouse for research and an inventory of best management practices and monitoring efforts; and (3) required plans to be based on best management practices. The Act authorized $7.5 million over 10 years (although this amount was never entirely appropriated), and was reauthorized for another 10 years in 2007.

Restoration. The Rio Puerco Management Committee has had numerous successes over its past 15 years, of which four are significant. First, the RPMC has brought parties in conflict over livestock grazing in the Rio Puerco to the same table, and their continuing relationships have consistently been identified as the most important outcome. Honest disagreement among the parties was acceptable, but not personal attacks. Meetings were held in the evening and on Saturdays in local communities throughout the watershed to make it easier for non-government people to attend, and the facilitator was the only person who was paid to be there.

Second, restoration projects were an early priority for the RPMC. Early projects were small: riparian fencing and reseeding. The RPMC's top priority was to remeander a 3.5 km stretch of the Rio Puerco that had been channelized in the mid-1960s to avoid having to build two bridges on State Highway 44 (now US 550). The 1.75 km channelized section doubled the stream gradient, resulting in headcutting and lateral channel widening that is estimated to have contributed almost 850,000 metric tons of sediment since its creation. This US$4.5 million project was completed as a partnership between the RPMC and the New Mexico State Highway Department, with approximately $1 million in funding provided by the USEPA for the channel reconstruction.

Third, the bourgeoning involvement of Native American (aboriginal) tribes in watershed restoration is significant due to both their cultural reticence and the extent and ownership in the Rio Puerco watershed. The RPMC worked with tribal members to reduce the number of horses and burros on overgrazed lands, helped the Navajo community of Torreon to create a Youth Conservation Crew to work on watershed projects, and cooperated with a number of tribes

on planning and implementing restoration projects. One RPMC member remarked, 'tribes were here before any of us, and will be the ones who keep this effort going in the long term.'

Finally, the RPMC successfully accessed scientific capacity for assistance in its strategic planning and to evaluate restoration effectiveness. The United State Geological Survey (USGS) conducts studies in the watershed (Gellis et al. 2004), hosts a website devoted to Rio Puerco data, and provides monitoring design and analysis for the Torreon restoration. The RPMC supported a graduate student who evaluated Rio Puerco sediment sources that identified road-related sedimentation as the greatest single problem (Pippen & Wohl 2003).

Lessons learned. In Section 4.2 we introduced the idea that organizations (including government agencies) have life cycles that go through various stages and identified various types of personal behavioral characteristics who seem attracted to these different stages. The RPMC illustrates this pattern, and the Act recognized that restoration efforts required at least 10 years. The founder of the RPMC, a BLM District Manager from 1993 to 1998, supported the organization in its entrepreneurial/creation stage. While not entirely a zealot – and with some of the characteristics of a climber – it was his vision and drive that brought the parties together originally in 1993, and his political acumen that moved the original Rio Puerco Committee to seek the legislation that created the RPMC. During the RPMC's 'collectivity' stage (c. 1998–2001), he recruited additional BLM staff members and the RPMC developed its Action Strategy (RPMC 2001), received its first Congressionally designated funding ($300,000), and was recognized for its collaboration with two national awards (RPMC 2001).

The 'formalization and control' stage occurred during 2002–2005 when the RPMC received significant funding from the USEPA, and a partnership with the New Mexico Department of Transportation allowed it to accomplish is highest priority restoration project. All major participants during this period could be characterized as advocates and many are still involved in either the RPMC or the Rio Puerco Alliance (RPA).

The RPMC attained the 'elaboration of structure' life stage when, in 2006, it founded the RPA in order to receive funds from sources that would not fund government agencies. The RPA provided continuity through its board, which was previously lacking as agency representatives on the RPMC cycled through or retired. The RPMC is presently at the fork in the organizational life cycle where it can either renew itself or is likely to cease its operations. The spinoff of the RPA is a sign of renewal, as is the re-convening of the partners that occurred in 2011.

Based on the RPMC experiences, it is possible to overcome both geographic challenges (size and ownership pattern) and cultural barriers to stream restoration. It is however apparent that it takes time, dedication, funding, and the acknowledgment that organizations need to be resilient and entrepreneurial to be successful.

Box 4.6 Timeline of actions in the Rio Puerco

1250	Chaco (Anasasi) culture dies out from overuse of resources and climate change
1599	Spanish Conquistadores arrive in Rio Puerco
1680	Pueblo Revolt drives Hispano settlers out of NM until 1692 reconquest
1693	Hispano settlement begins in the Rio Puerco
1740	Cattle and sheep grazing begins
1760s	Rio Puerco channel incision begins
1846	First Anglo-Americans arrive in New Mexico
1850	New Mexico becomes a US territory
1880s	Rio Puerco channel incision accelerates
1898	Irrigated fields cover 7236 ha and are served by 62 ditches
1912	New Mexico becomes a state
1927	USGS report identifies the Rio Puerco as having significant sediment problems
1978–1984	US Army Corps of Engineers studies watershed treatment as a 'non-structural' alternative to flood and sediment control dams on the Rio Puerco
1985	US Forest Service completes Cibola and Santa Fe National Forest Plans
1986	Bureau of Land Management completes Resource Management Plan for its lands in the Rio Puerco basin
1993	US Bureau of Reclamation begins review and study of Rio Puerco sedimentation impacts on Rio Grande and Elephant Butte Reservoir
1993	Rio Puerco Watershed Committee (RPWC) formed by local stakeholders as a sub-committee of the Cuba (NM) Region Economic Development Board
1996	Rio Puerco Watershed Act of 1996 (PL 104-333) passed; requires management committee and provides $7,500,000 over 10 years to establish restoration program
1997	Rio Puerco Management Committee (RPMC) formed pursuant to PL 104-333.
1998	Rio Puerco identified as a Category 1 'In need of restoration' in New Mexico Unified Watershed Assessment prepared under the federal Clean Water Act (CWA)
1999	Sub-basin prioritization initiated by RPMC
1999	Decision to construct two bridges on US 550 to remeander the Rio Puerco at La Ventana
2001	Watershed Restoration Action Strategy (WRAS) completed and submitted to USEPA to qualify for CWA §319(h) project funding
2006	Rio Puerco Watershed Alliance (RPA) formed
2007	NM Environment Department completes TMDL for Rio Puerco
2007	Rio Puerco Management Committee reauthorized under PL 111-11 §2501 for another 10 years

---

4.5 Approaches to elicit cooperation

Eliciting and building cooperation among landowners and other restoration project partners is the key to success in the long term. Cooperation can take many forms: grudging acceptance of a proposal; agreeing to help with, but failing to take action; eagerly encouraging and participating in projects; or even to funding stream restoration projects. In many cases, cooperation will ultimately transcend into collaboration, where your partners are actively working to support your goals. While the importance of individual relationships has been emphasized, it is also useful to understand that there are organizational and institutional structures that can aid in the acceptance of stream restoration. In this section we will first begin by highlighting the rise of these institutions, then afterwards evaluate a number outreach,

information and communication or environmental education tools that can be used to engage landowners and project partners. We will also discuss more deeply the concept of reciprocity and the lessons that can be learned from theoretical outcomes to build a successful stream restoration program.

4.5.1 Institutions to support stream restoration

The importance of involving the public and stakeholders in decisions, environmental policy, and resources management gained stature after the Second World War (Parkins & Mitchell 2005). In the United States, the requirement for public participation began with passage of the Federal Administrative Procedures Act in 1947, which evolved into the framework for stakeholder 'standing' in legal review of agency actions that was subsequently incorporated in the National Environmental Policy Act (NEPA) of 1970 and the Clean Water Act of 1973 (Fairfax 1978). The involvement of the public in government decisions in Europe goes back further in certain countries, and for specific issues such as water management and drainage (Enserink *et al.* 2007). Broad policies requiring public involvement within the European Community took force through the Economic Commission for Europe Aarhus Convention on Access to Information, Public Participation in Decision-making and Access to Justice in Environmental Matters (http://www.unece.org/env/pp/documents/cep43e.pdf) (Enserink *et al.* 2007). The Water Framework Directive incorporates this by requiring public participation in its mandatory integrated river basin management plans (European Commission 2000, 2003).

As stakeholders and the general public began to receive legal rights to participate in governmental environmental policies, there became an interest in determining institutional mechanisms that would best achieve these goals. Beginning in the 1980s and accelerating through the 1990s and early 2000s, collaborative groups involving stakeholders, landowners, and agencies developed as a way to bring together interested parties to both meet regulatory requirements to improve watershed conditions, and to build longer-term relationships among the partners (Leach & Pelkey 2001). In the United States, these collaborative groups are commonly called 'watershed councils' (Leach & Pelkey 2001); in Australia they are 'catchment management committees' (Hillman *et al.* 2003); in England and Scotland they may be 'river trusts' (www.associationofriverstrusts.org.uk); 'Wupperverband' in Germany (Moellenkamp *et al.* 2010), and in South Africa 'catchment management forums' (Pollard & du Toit 2011).

While structures and authorities vary widely, almost all groups consist of a broad range of stakeholders (including government agencies) who meet periodically to discuss and negotiate specific watershed conditions, practice consensus-based decision making and, in many cases, may actually implement restoration projects (Leach & Pelkey 2001). There are process outcomes that result from the exchange of views and ability to be heard and influence discussions, and there are content outcomes where improved decisions are achieved through joint problem solving, additional information and making decisions on priorities and projects (European Commission 2003). One important long-term outcome from this process is *social learning* whereby stakeholders, through sharing perspectives and identifying interdependencies among them, improve their ability to identify mutually beneficial outcomes (Bouwen and Taillieu 2004; Pahl-Wostl *et al.* 2008).

4.5.2 Techniques to engage landowners

Given the long and world-wide deployment of the previously discussed agricultural extension model, it is not surprising that numerous techniques to engage landowners in adopting new techniques have been developed and evaluated over the last 50 years. Examples of commonly used techniques, often generally known as 'outreach,' are:

1. demonstration projects;
2. field days and tours;
3. newsletters;
4. displays in public places and at events;
5. talks to community groups;
6. newspaper, radio and television stories;
7. web sites, social media and e-mail contacts; and
8. work with schools and youth groups.

All these techniques can be effective in broadening knowledge about your plans, ideas, and potential restoration projects. There are a number of excellent guides for adapting and applying outreach techniques that can be found on the internet. Because many of these techniques need to be adapted for a specific country, culture, and objective, it makes sense to first look for local practices and experiences to identify preferable methods.

A potential project partner becoming aware of the technique that is being proposed is a pre-condition for them considering its adoption. The general outreach techniques listed above can work within local networks to increase the visibility of a project, while more targeted techniques such as demonstration projects and tours are particularly useful to enlist specific potential project partners. Rosenberg & Margerum (2008), who evaluated how landowners made decisions about whether to implement stream restoration projects, found that family, friends, and neighbors were the most trusted sources of information (particularly for the less well-educated) but that there were also high levels of trust in the agricultural extension network and local watershed groups. Newsletters were considered to be the best communication tool (Rosenberg & Margerum 2008).

Even more targeted outreach can be employed by more direct engagement through individual meetings in a person's home or small gatherings of neighbors, sometimes called 'kitchen table' meetings and 'coffee klatches,' respectively. Directly visiting, or being hosted by a local in the area where you want to work, allows for more personal interactions with a targeted audience, allows back-and-forth discussions, and gives your project partners a chance to get to know you as a person rather than just someone trying to get a stream restoration

Box 4.7 Working with Weyerhaeuser

Weyerhaeuser Timber Company (WeyCo) provides an example of a 'push-pull' dynamic that resulted in considerable stream restoration on their 85,000 ha Millicoma Tree Farm in southwestern Oregon, USA the 'push' being their legal compliance requirements and the 'pull' their self-image as a progressive land steward with high standards for its self-directed forest management operations. This push-pull dynamic ultimately provided support for the formation of the Coos Watershed Association (CoosWA) and a joint partnership that has resulted in over US$3,000,000 in stream restoration projects.

Background. The regulatory push on Weyerhaeuser began in the early 1970s as the first significant forest practices laws came into being. Tensions and rancor accelerated during the 1980s as these rules were strengthened and federal Endangered Species Act (ESA) requirements further constrained operations. The pull emerged in the 1990s with their preparation of watershed analyses, creation of the first industrial timberland habitat conservation plan (HCP) to protect the northern spotted owl (*Strix occidentalis caurina*) and marbled murrelet (*Brachyramphus marmoratus*), and ISO14001 environmental management certification. Also reflecting the push-pull dynamic was the 1994 formation of CoosWA in an attempt to avoid coho salmon (*Oncorhynchus kisutch*) becoming listed under the Endangered Species Act. The push-pull tension continues into the 21st century with ever-greater demands for financial and environmental performance in the management of the Millicoma Tree Farm.

CoosWA began putting their plans in place, with a growing appreciation that implementing restoration projects brought a wide range of stakeholders together. The culture that developed in CoosWA reflected its origins: the desire to work cooperatively with landowners and managers to create 'win-win' situations, avoid increased regulatory burdens through its assessments and monitoring, and being very cognizant and respectful of private property owners.

Restoration program. During the formative period for CoosWA, Weyerhaeuser began exploring the potential to place stream restoration projects on the Millicoma Tree Farm. Early efforts reflected common practices at the time: channel-spanning boulder weirs to capture gravel and create spawning habitat; placement of individual or small groups of logs over short stream reaches; and construction of jump weirs to help fish pass perched culverts. Early projects were focused solely on benefits to salmon (see Figure 4.7). Severe floods in 1996 highlighted the need to improve culverts to prevent their failing from being undersized. This recognition of joint benefits to Weyerhaeuser's facilities and fish habitat led to a change in the types of projects, increased the amount of overall expenditures on restoration projects, and altered the cost-sharing proportions to better reflect the comparative benefits to each party (Figure 4.8).

In 2003 another 'push' occurred when Weyerhaeuser was issued a civil citation for inadequate road maintenance and log hauling during wet weather that caused sediment to flow into an adjacent stream. Weyerhaeuser agreed to institute best management practices (BMPs) to prevent sediment from entering streams, which led to a significant expansion in cooperative restoration projects and a change to upgrading whole road systems rather than spot treatments (see Figure 4.7). The Dellwood mainline upgrade, 97 km of riparian-adjacent heavy-use hauling route was the first of these whole-road projects. When Weyerhaeuser began planning in 2006 to move their harvesting operations into the Bottom Creek sub-basin (4622 ha with 23.2 km of fish-bearing streams), CooWA proposed that the project be used as a

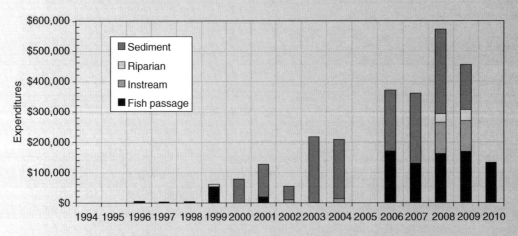

Figure 4.7 Changes in stream restoration project types on the Millicoma Tree Farm, 1994–2010.

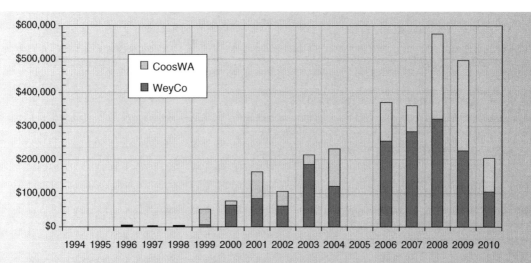

Figure 4.8 Relative contributions of Weyerhaeuser (WeyCo) and Coos Watershed Association (CoosWA) to stream restoration projects on the Millicoma Tree Farm.

demonstration of 'whole watershed restoration.' Projects constructed during 2008–2010 included 47 logjams totaling 462 trees placed over 12 km of streams, decommissioning 1.5 km of unneeded logging roads, five fish passage barriers removed that opened up 5.3 km of streams to anadromous fish, and installing 141 cross drains and 30 culverts on 19.3 km of roads to reduce road-related sediment.

Lessons learned. Weyerhaeuser's strong corporate standards pushed their staff to improve watershed conditions, while CoosWA brought technical assistance (surveys and designs), funding, and a non-regulatory approach. Through this cooperation we have learned a number of lessons about why this partnership is so successful.

1. *Share goals.* Early projects that focused solely on CoosWA objectives were limited in scope and extent. Once Weyerhaeuser's needs in terms of road upgrades were included, cooperative projects expanded rapidly and Weyerhaeuser increased their funding.
2. *Share work.* Coordinating work – in this case logging, road construction, and large wood placement – is highly efficient and leverages each partner's involvement. Trees that are being placed into streams now come from new road and landing construction.
3. *Keep your needs clear.* There may be times in the relationship when one party feels taken advantage of. We learned to better identify the relative fish benefits compared to timber harvest benefits and use that ratio to determine financial contributions.
4. *Not everything will be transparent.* When working with private businesses as partners you have to realize that their internal communications and policies will not necessarily be shared with you.

5. *Look out for each other.* There have been incidents where the strength of the collaborative relationship was demonstrated. For example, during a recent review of a large consolidated restoration grant, the funders initially removed part of the grant that would have built a bridge in order to reduce the cost of the project. However, WeyCo was providing a matching funds for other parts that were not a priority for them. CoosWA successfully appealed the decision since the project was designed as a package.
6. *Keep a clean house.* From the Weyerhaeuser perspective, the relationship has worked well because the watershed council has been very organized in its planning. This allows Weyerhaeuser the lead time that it needs to budget and allocate staff time.
7. *Work to nurture future collaboration.* The collaborative relationships between CoosWA and Weyerhaeuser staff have evolved, strengthened over time, and become part of our mutual working culture, strengthening our stream restoration program as new staff are brought on by both CoosWA and WeyCo.

This case study demonstrates that it is possible to work with a large, multi-national corporation on stream restoration projects when each party's needs and objectives are met. The key to meeting those needs and objectives is to understand the conditions under which your partner operates, and what its expectations for you are. As trust is built through experiences working with each other, there should be a point when both parties are working to support each other's interest without the need for constant reciprocation. Ultimately the working arrangement becomes part of each partners' organizational culture and is transferred through generations of employees as they rotate through jobs related to the restoration projects.

project implemented. This approach allows you to intersect with local networks, providing an avenue for local participants in these networks to become acquainted with stream restoration.

4.5.3 Achieving agreement with project partners

There will come a point in a proposed stream restoration project where you will need to obtain the agreement from project partners, particularly landowners, for project implementation. This section will stress the importance of reciprocity, that is, making sure that the objectives and needs of project partners are understood to ensure they are met as the stream restoration program is developed. In many cases, cooperation can be achieved even if a partner does not necessarily wholeheartedly believe in the cause, as long as that partner stands to gain benefit by working with you. While formal negotiations can achieve this outcome, often on a case-by-case basis, it is also possible to build a cooperative relationship over time by choosing a set of behaviors on your part that will lead to a mutually successful collaboration. The Coos Watershed Associations experiences in the 'Working with Weyerhaeuser' case study (Box 4.7) describes this process with a large multi-national timber company in the Pacific northwest of the United States.

The practice of implementing stream restoration projects is in many ways similar to diplomacy: you want the other party to cooperate so that you gain something you want (or avoid something you do not want). Game theory – which was developed during World War II and used widely during the Cold War – is designed to examine potential decision-making strategies when the actions of the other party cannot be controlled or known in advance (von Neumann & Morgenstern 1947). In application to stream restoration, the results from 60 years of game theory simulations provide insight into how best to work with potential project partners, whether they be agency collaborators or landowners on whose property you want to install a project. This approach may seem inapplicable for stream restoration (see Box 4.8), but we have found it a useful way to evaluate potential choices and strategies.

4.5.3.1 The Prisoner's Dilemma

The principal game theory tool, called the Prisoner's Dilemma, was developed in the early 1950s to illustrate the difficulties in overcoming rational self-interest to obtain cooperation for mutually beneficial outcomes. It

> **Box 4.8 Restoration is not a game!**
>
> Understandably, the use of the term 'game' to describe strategies to restore streams may seem sacrilegious to those who are devoting their lives and careers to this cause. We can assure you that you are not alone in this. However, we have found the rules and guidelines discussed here to be very useful during the years of implementing restoration projects.

is premised on police interrogating two suspects about a crime, each of whom is focused in their own self-interest to receive the least sentence. If the suspects cooperate with each other they can be convicted only of a minor crime due to limited evidence. However, if one or the other confesses and implicates the other one, the confessor receives the little or no penalty while the other serves a long sentence. If both confess, however, they are each convicted and both receive long sentences (Poundstone 1992). In the context of restoration, we frequently work with landowners who usually have their own gains in mind when they cooperate on a project, while at the same time we also have our own goals to achieve. Table 4.2 shows a matrix for the Prisoner's Dilemma in this situation where the 'payoffs' are in net gains (i.e. absent any benefits given to the other party in the project).

Axelrod (1984) described which strategies in the Prisoner's Dilemma are likely to induce stable cooperation. The Prisoner's Dilemma game rules can be largely transferred to the situation facing stream restoration proponents, or at least to those who are working solely on a voluntary basis with landowners and managers: (1) there is no mechanism to make enforceable threats or commitments; (2) there is no way to know what the other player will do on a given move; (3) there is no way to eliminate the other player or leave the interaction; and (4) there is no way to change the other player's payoffs (Axelrod 1984). While these rules can in fact be overcome by various mechanisms (such as legal contracts and easements), other commonly used strategies such as incentive payments are internal and considered in the payoffs shown in Table 4.2.

4.5.3.2 Guidelines to build and maintain cooperation

From his simulations, Axelrod (1984) derived four rules that he found consistently lead to desired outcomes:
1. *Don't be envious.* Success is when both partners gain from cooperating. As long as you are getting what you need, the fact that a partner may be gaining even more

Table 4.2 Net gains between parties in the Prisoner's Dilemma of stream restoration.

		Landowner (L)	
		Cooperate	*Defect*
Restoration practitioner (R)	*Cooperate*	L = 2; R = 2 Each party gets the majority of what they want, but not everything.	L = 3, R = 0 Landowner solely benefits but restoration ineffective because landowner commitments not fulfilled.
	Defect	L = 0; R = 3 Restoration implemented, but landowner feels taken advantage of and restoration gets bad reputation.	L = 1; R = 1; Status quo where nobody gains but there is a minimal potential for loss. In some cases, this could even be zero or negative for both parties.

from the relationship does not necessarily mean that you are being taken advantage of. A partner's success is what allows the cooperative relationship to continue and build over time. The proper standard for comparison is what is achievable in the absence of this cooperation.

2. *Be nice!* Do not be the first to withdraw (defect) from cooperating. In the initial phases of the relationship it is likely that a partner will be looking for indications that you will not fulfill commitments. In this respect, reputation is important, as is the likelihood that you specifically – and your organization generally – will stay involved over the long term. If the other party does not think that they will see you in the future, or if you might be replaced by someone else in your organization who might not be agreeable to the existing arrangement, then obtaining cooperation will be more difficult if not impossible. Most landowners desire an ongoing relationship so that if something does not go as planned they are confident that you will be around and willing to fix whatever is wrong.

3. *Reciprocate both cooperation and defection.* If someone cooperates with you it is clearly in your best interest to reciprocate. In the initial stages of a relationship there is an implied *quid pro quo*: 'I help you, you help me.' Over time, as trust is built, there may be less need for a strict accounting of each party's contribution, though some documentation may be helpful in the future. The relationship between the Coos Watershed Association and the Weyerhaeuser Timber Company in the case study provides a good example of how reciprocating cooperation builds over time.

Knowing how to reciprocate if a partner reneges on an agreement or quits the partnership (Table 4.2) is more difficult. The challenge here is to avoid having your nice-ness misinterpreted as being a 'pushover' or being willing to do anything to establish or maintain the relationship. Remember, the object here is to build cooperation, not create a condition where your potential project partner takes advantage of you or refuses to work with you. If the relationship is on a formal contractual basis it may be possible to enforce compliance. This strategy will probably be more successful with larger organizations and agencies than with individual landowners with whom the personal relationship is more important. In this latter situation, a more passive reciprocation of defection may bring your partner back to the partnership. Examples of such reciprocity would be to decline to participate in future projects with them, allow your partner to be exposed to regulatory burdens (although it is not a good strategy to report them to authorities!), or explicitly and visibly work with others on the same type of projects that the partner desires. The goal in all this is to avoid burning bridges so that you can keep future options for cooperation open.

Your relationship with a new partner may begin by being put into a 'good cop, bad cop' situation, where your partner cooperates with you (the good cop) because you are preferable to the alternative (regulatory enforcement actions). The leverage that you gain in this situation is offset against an implicit reciprocity that you will help protect your partner's interests. Also, be aware that if you are stereotyped with organizations similar to your own, it is in your best interest to make sure that these parties play nice with your partner. All too frequently, a partner will withdraw from cooperation due to the actions of a third party that you have little control over. Part of the 'reciprocation of cooperation' is your willingness to assist your partner to resolve these difficulties. This approach, however, can frequently take regulatory agencies by surprise and leave them wondering where your loyalties lie. It is important to remember that it is not about taking sides, but rather about getting a restoration program implemented and seeking 'win-win' solutions (Fisher et al. 1991).

4. *Don't be too clever!* Other players will be watching for signs of whether you will reciprocate cooperation or not, and therefore your own behavior is likely to be mirrored back to you (Axelrod 1984). Simple is usually preferable to complex to avoid a cycle of increasingly nuanced demands. There are times when elaborate procedural mechanisms are necessary to obtain cooperation in the absence of mutual trust, but the effort in preparing and negotiating such agreements can be large and may not be needed. Some form of agreement and limitation of liability is usually necessary when implementing restoration projects, but once a relationship of trust is achieved this documentation becomes secondary to the mutual interest in moving the project forward and can actually impede progress by siphoning resources and trust.

Game theory and practical experience indicate that it is not beneficial to 'permanently retaliate' when your partner fails to cooperate at some point. One reason that this is not a successful strategy is that it is likely to be reciprocated, i.e. you will never again be able to work with this person. In addition, one's reputation will also suffer when your partner tells others, 'Look what they did to me!' Your interest is in maintaining a reputation for fairness over the long term, which is especially important when other potential partners are watching. Sometimes silence is the best alternative.

4.5.4 Why understanding cooperation is important

The work of Robert Axelrod and other game theorists who have used the Prisoner's Dilemma to examine behavior (ranging from bacteria to humans) provide useful lessons for restoration practitioners. First, stream restoration should not be thought of as a single event (or project), but instead a recurring set of interactions where partners and cooperators meet one another over significant periods of time (usually years, if not decades). The actions taken today can benefit or haunt you and your successors in the future because they may transfer to other natural resource efforts, multiplying the effects of any single project. A recurring cooperative relationship is not necessarily a zero-sum outcome (one person's win is another's loss), but rather should have positive-sum out-comes, i.e. the classic win-win situation for both parties (Fisher et al. 1991).

Trust and credibility take time to build, and can easily be lost by unwise actions on your or others' part. However, both parties do not necessarily have to agree on all points for the response to be mutually beneficial. The limited options available to us mean that trust and goodwill are assets worth creating and protecting. It is always a good strategy to 'take the high road.' Make sure that your organization has policies and training to reduce the potential for damage to your reputation. It may be advantageous to allow your partner to get the first benefits from cooperation. Implicit in this approach is the likelihood that the other party will cooperate with you the next time you need something out of a sense of obligation. Just as in the 'be nice' strategy, reciprocity may not occur in all cases but you will still receive more benefits over the longer term by using this approach.

The durability and frequency of interactions leads to the potential for cooperation. The people you are trying to convince to undertake a restoration project need to see you in the community: at the post office, church, or grocery store. There may be an added benefit if you grew up in the community (assuming you did not cause too much trouble in your youth!) because you and your family will have had long-term relationships with the same families with whom you are now trying to work. For example, at the Coos Watershed Association we had been trying to install monitoring equipment in a stream on land owned by a very skeptical dairy farmer. He would only agree to do something when a staff person who had grown up in the neighborhood asked, but would continue to allow the monitoring only at the same site with subsequent staff. When we wanted to change locations and types (i.e. replace a juvenile fish traps with PIT tag antennas), he refused until a new staff member who had gone to high school with his kids asked.

When entering into new cooperative relationships, break larger projects into smaller pieces while trust and reciprocity are built. This allows each party to observe the behavior of the other, to test whether commitments will be fulfilled and to judge if cooperating is in their best interest. The Weyerhaeuser case study (Box 4.7) is an example of how, over the last 10–15 years, restoration projects have become larger and more complex as the cooperative relationship has strengthened by building on a history of smaller successes. It is unlikely that a whole sub-basin restoration program would have been palatable to Weyerhaeuser if it had been the first project proposed.

People will cooperate if it is in their long-term interest to do so, even if it is not necessarily in their short-term interest. The Coos Watershed Association (and the Oregon Plan for Salmon and Watersheds) brought together timber and environmental interests who would not normally cooperate; the timber representatives did not want to see additional limits on their operations if coho salmon were listed as threatened under the United States Endangered Species Act (ESA). They believed that by cooperating in the short-term they would have greater management flexibility in the long-term. By the time coho salmon were listed under the ESA, the timber industry had sufficient commitment to the Oregon Plan that cooperating still appeared a preferable option when compared to defecting and going it alone. Often restoration practitioners think that our project partners – particularly landowners – should allow stream restoration projects on their property because it is the right thing to do. Expecting such altruism is an unrealistic foundation for a restoration program; when it occurs it is wonderful, but it is unlikely to be widespread enough to meet the restoration needs. However, not all benefits that a partner may receive necessarily have to be in monetary terms. Examples of non-monetary values can include your partners' reputation in the community (wanting to be seen as a good land steward by his or her peers) and the aesthetic benefits of vegetated riparian zones and stable stream banks.

It often feels easier to deal with people who are like us because we believe we know how they think and believe we understand their values. Even if this attitude does not stigmatize those who think or act differently, labeling – or stereotyping – unnecessarily limits our opportunities. Modeling results show that we are far more likely to improve our situation by cooperating with those of different perspectives than if we limit ourselves to those who believe as we do. This does not mean compromising our values, but rather placing ourselves in other people's positions to understand their perspectives.

4.6 Moving forward: Further reading in human dimensions of stream restoration

This chapter has provided some fundamental tools and approaches to incorporate the human dimension into stream restoration. However, we have really just scratched the surface of a rich and diverse field of study and practice. For those interested in delving deeper, we offer the following suggestions on further reading and investigation.

4.6.1 Collective action

Over the last 20 years there has been a large and swift rise in the creation of organizations whose purpose is to restore streams and watersheds (catchments). A good, albeit fairly advanced, introduction to understanding organizations with a focus on natural resources is Elinor Ostrom's *Understanding Institutional Diversity* (2005). Ostrom shared the 2009 Nobel Prize in Economics for her work on institutions to manage common property resources, thus there are strong correlations with organizations to advance stream restoration. The *International Journal of the Commons* (http://www.thecommonsjournal.org) is an open-access source for current theory and case studies.

There have been a number of studies of the effectiveness of watershed councils, beginning with Julia Wondolleck and Steve Yaffee's *Making Collaboration Work: Lessons From Innovation in Natural Resources Management* (2000), which provides a good overview. They include a short discussion of the Prisoner's Dilemma and cite Axelrod's (1984) work, but emphasize group formation and group dynamics using a suite of case studies to bolster their points. Paul Sabatier and his colleagues (2005a) in *Swimming Upstream: Collaborative Approaches to Watershed Management* provide a political science-focused evaluation focusing on the critical role that trust plays in successful stakeholder collaborations (recall, however, that we also think reciprocity plays a key role). Sabatier *et al.* (2005b) test various theories based on a quantitative analysis of surveys of watershed councils in California, Oregon and Washington in the United States.

4.6.2 Social capital and the triple bottom line

In its most elemental form, stream restoration builds 'natural capital' by improving ecological function. Both Wondollek & Yaffee (2000) and Sabatier *et al.* (2005b) note the benefits of cooperative actions for stream restoration to build 'social capital,' that network of relationships and trust that allows groups and communities to achieve common goals, and whose loss was described by Robert Putnam in his *Bowling Alone: The Collapse and Revival of American Community* (2000). In Section 4.5.2 we highlighted social learning as a desired outcome from collaborative efforts to implement stream restoration, and the process through which this social learning occurs and builds social capital. We also discussed the importance of reciprocity and understanding landowners' management objectives, one of which is commonly meeting their economic objectives, i.e. helping them build and maintain *economic capital*, including good infrastructure.

Restoring ecosystems, building social capital and working with landowners to meet their economic (and other) objectives lends itself to placing stream restoration work within the 'triple bottom line' that came out of the business world to account for environmental, economic, and social benefits and impacts from your restoration projects and programs (Elkington 1998). Aronson *et al.* (2007) provide a good introduction and case studies on restoring natural capital. Placing stream restoration in the triple bottom line framework has the added benefit of signaling to your project partners that you care about their concerns rather than being singularly focused on meeting your goals and objectives, which may result in them being more open to hearing your ideas.

4.6.3 Environmental justice

While everyone targets 'win-win' solutions as their ideal outcome, in any large program of stream restoration there are likely to be winners and losers. This is often the case when property is condemned or expropriated to implement projects, and is even more problematic where there is questionable title to the property (cf. Section 4.2.2). Distributive outcomes about who wins and who loses are frequently called 'environmental justice'; Mick Hillman (2004) and his colleagues in Australia and New Zealand have been at the forefront of explicitly considering these effects in stream restoration. Their *River Futures: An Integrative Scientific Approach to River Repair* (Brierley and Fryirs 2008) provides good coverage as well as case studies. Boyce *et al.* in *Reclaiming Nature: Environmental Justice and Ecological Restoration* (2007) provide a broad survey.

4.6.4 Resilience

Restoring 'resilience' or the ability of a system to withstand shocks is an emerging focus for all types of ecological restoration projects, particularly given expected climate changes. Ensuring that economies and communities are also resilient to system shocks is a desirable goal as part of a triple bottom line approach, however. An excellent and broadly encompassing introduction to this idea can be found in the edited volume *Principles of Ecosystem Stewardship: Resilience-Based Natural Resource Management in a Changing World* (Chapin *et al.* 2009). A key tenet in basing restoration objectives to build

resilience is the idea of adaptive management. A very succinct and applicable introduction to adaptive management can be found in Walters & Holling (1990), two of the originators of the concept. The open-access journal *Ecology and Society* is a particularly good source for current research on resilience and associated topics such as social learning and adaptive management.

4.7 Summary

Understanding the human dimension in stream restoration is necessary and, if done well, has the potential to accelerate the process of achieving stream restoration goals. It is all too easy to think that progress would be much faster without humans in the equation; not only is that unlikely, but explicitly recognizing the importance of understanding your partners' motivations and objectives will likely increase your potential for success in your restoration program.

We started this chapter by highlighting the value of understanding land ownership patterns and potential ways to approach various types of landowners. Ownership tends to be blocked into larger pieces in upper watershed areas; moving downstream generally increases the number of different landowners you will need to involve in your project, each one of which is likely to own a smaller piece of property. Further, upstream projects may have effects on downstream owners, just as projects may affect adjacent landowners. These potential effects need to be considered during project planning and design, not after implementation.

Stream restoration projects will be considered experimental or innovative or even hazardous to many of the landowners you approach. Understanding how people react to new ideas and the process of adoption will assist in targeting landowners who are more probable to be willing to accept the project, and whose acceptance may lead to their neighbors also joining in the program. Who will be willing to work with you at any given time may mean that the best location for a project may not be the most likely landowner to cooperate, and thus you may have to use other sites first while building acceptance from the landowner at the key location.

Most stream restoration projects involve the cooperation of a number of individuals and organizations. There will be landowners and managers, other organizations who provide funding and possibly technical assistance, and there will be governmental agencies with regulatory authority to assess and regulate the impacts of stream restoration projects. Organizations of all types (as well as the individuals associated with them) have norms and 'cultures' that typically emerge through their formation stages and direct their reactions. While individuals have their dominant behavioral type within organizations, these people influence, and are influenced by, their peers who they must work with daily and by the organization's culture itself. Understanding cultures and behavioral types will again increase your effectiveness and decrease frustration. Along these lines, although it is critical for all project members to understand the human dimension of their efforts, it can be helpful to ensure that the team includes members who are particularly strong in these social skills and employ techniques such as 'coffee klatches' and 'kitchen table' meetings as a strategy to make strong connections with potential project partners.

Finally, stream restoration should not be considered a single event, but a series of actions that take place over a period of time. Each interaction can be considered a negotiation: you want to do something and you want the other party to cooperate. We identified a game theory model within which to frame these negotiations – the Prisoner's Dilemma – a strategy that has proven to have the best outcomes over the long term. The key concept is that most relationships are based on reciprocity; while perhaps seeming self-serving, this is in fact a stronger approach than expecting continual altruistic behavior from your project partners.

The models, approaches and tools provided in this chapter are widely considered to be classic. Rather than being outmoded or dated, these approaches have stood the test of time across years and various cultures. Applying these concepts and techniques will assist in effectively implementing stream restoration efforts by increasing the effectiveness of getting projects implemented while, at the same time, decreasing the frustration that results when resistance or indifference is met.

4.8 References

Alexander, G.G. & Allan, J.D. (2007) Ecological success in stream restoration: case studies from the midwestern United States. *Environmental Management* **40**, 245–255.

Aronson, J., Milton, S.J. & Blignaut, J.N. (eds) (2007) *Restoring Natural Capital: Science, Business, and Practice.* The Science and Practice of Ecological Restoration

Series, Society for Ecological Restoration International, Island Press, Washington, DC.

Axelrod, R. (1984). *Evolution of Cooperation.* Basic Books, New York, New York.

Beechie, T.J., Sear, D.A., Olden, J.D. *et al.* (2010) Process-based principles for restoring river ecosystems. *BioScience* **60**, 209–222.

Bennett, A.F. & Saunders, D.A. (2010) Habitat fragmentation and landscape change. In: Sodhi, N.S. & Ehrlich, P.R. (eds) *Conservation Biology for All.* Oxford University Press, England, pp. 86–106.

Bernhardt, E.S., Palmer, M.A., Alexander, G. *et al.* (2005) Synthesizing U.S. river restoration efforts. *Science* **308**, 636–637.

Bouwen, R. & Taillieu, T. (2004). Multi-party collaboration as social learning for interdependence: Developing relational knowing for sustainable natural resource management. *Journal of Community & Applied Social Psychology* **14**, 137–153.

Boyce, J.K., Narain, S. & Stanton, E.A. (eds) (2007) *Reclaiming Nature: Environmental Justice and Ecological Restoration.* Anthem Press, London.

Brierley, G.J. & Fryirs, F.A. (2008) *River Futures: An Integrative Scientific Approach to River Repair.* The Science and Practice of Ecological Restoration Series, Society for Ecological Restoration International, Island Press, Washington, DC.

Chapin III, F.S., Kofinas, G.P. & Folke, C. (eds) (2009) *Principles of Ecosystem Stewardship: Resilience-Based Natural Resource Management in a Changing World.* Springer, New York, New York.

Cowx, I. & Welcomme, R. (eds) (1998) *Rehabilitation of Rivers for Fish.* Fishing News Books, Blackwell Science, Oxford, England.

Cronin, W. (1996) Introduction: In search of nature. In: Cronin, W. (ed.) *Uncommon Ground: Rethinking the Human Place in Nature.* Norton & Co., Inc., New York, New York, pp. 23–56.

Downs, A. (1967) *Inside Bureaucracy.* Little, Brown & Co., Boston, Massachusetts.

Elbourne, N. (2009) RRC's Houting Outing to Western Denmark. *River Restoration News* **34**, 1–3. Available at www.therrc.co.uk/newsletters/rrn_34.pdf

Elkington, J. (1998) *Cannibals with Forks: The Triple Bottom Line of 21st Century Business.* Conscientious Commerce Series, New Society Publishers, Gabriola Island, British Columbia, Canada.

Enserink, B., Patel, M., Kranz, N.F. & Maestu, J. (2007) Cultural factors as co-determinants of participation in river basin management. *Ecology and Society* **12**(2), 24.

European Commission (2000) European Water Framework Directive 2000/60/EC. *Official Journal of the European Communities L* **327**, 1–72. Available at: http://eur-lex.europa.eu/LexUriServ/LexUriServ.do?uri=OJ:L:2000:327:0001:0072:EN:PDF.

European Commission (2003) *Common Implementation Strategy for the Water Framework Directive* (2000/60/EC). Guidance Document No. 8: Public participation in relation to the Water Framework Directive. Produced by Working Group 2.9 – Public Participation.

Fairfax, S.K. (1978) A disaster in the environmental movement. *Science* **99**(4330), 743–748.

Fisher, R., Ury, W.L. & Patton, B. (eds) (1991) *Getting To Yes: Negotiating Agreement Without Giving In*, 2nd edition. Penguin Group, New York, New York.

Frikke, J. (1999) Operation Corncrake – An agricultural and environmental project for the Varde River valley and the meadows around Ho Bay in Denmark. *Wadden Sea Newsletter* **1999**(1), 29–31. Available at www.waddensea-secretariat.org/news/publications/Wsnl/Wsnl99-1/articles/11-frikke.pdf.

Gellis. A.C., Pavich, M.J., Bierman, P.R., Clapp, E.M., Ellevein, A. & Aby, S. (2004) Modern sediment yield compared to geologic rates of sediment production in a semi-arid basin, New Mexico: Assessing the human impact. *Earth Surface Processes and Landforms* **29**, 1359–1372.

Higgs, E. (2003) *Nature By Design: People, Natural Process, and Ecological Restoration*. MIT Press, Cambridge, Massachusetts.

Hillman, M. (2004) The importance of environmental justice in stream restoration. *Ethics, Place and Environment* **7**(1–2), 19–43.

Hillman, M. (2005) Justice in river management: Community perceptions from the Hunter Valley, New South Wales, Australia. *Geographical Research* **43**(2), 152–161.

Hillman, M., Aplin, G. & Brierley, G. (2003) The importance of process in ecosystem management: Lessons from the Lachlan Catchment, New South Wales, Australia. *Journal of Environmental Planning and Management* **46**(2), 219–237.

Houting Project (2011) River Varde Å LIFE Houting Project. Available at: www.snaebel.dk/English/Areas/Varde

Jensen, A.R., Nielsen, H.T. & Ejbye-Ernst, M. (2003) *National Management Plan for the Houting*. Published by the Danish Ministry of Environment, Forest and Nature Agency, The County of Ribe, and The County of Sønderjylland. April, 2003. 34 pp.

Jensen, J. (undated) Wadden Sea estuary, nature and environment improvement project LIFE99 NAT/DK/006456, Final Report. Available at: http://ec.europa.eu/environment/life/project/Projects/index.cfm?fuseaction=search.dspPage&n_proj_id=402&docType=pdf.

Kann, O. (2006) River Varde, Ribe county – re-meandering and laying out spawning grounds. In: Madsen, S. & Debois, P. (eds) *River Restoration in Denmark – 24 Examples*. Storstrøm County Technical and Environmental Division, pp. 67–70. Available at: http://www.snaebel.dk/NR/rdonlyres/815FB296-E869-4907-B5A0-CE2CE89DE441/0/Vandloebsbog_UK.pdf.

Lake, P.S., Bond, N. & Reich, P. (2007) Linking ecological theory with stream restoration. *Freshwater Biology* **52**, 597–615.

Leach, W.D. & Pelkey, N.W. (2001) Making watershed partnerships work: A review of the empirical literature. *Journal of Water Resources Planning & Management* **127**, 378–385.

Mabo v. Queensland (No 2) (1992) High Court of Australia 23, 175 CLR 1 F.C. 92/014.

Manøe, S. (2011) *Vandkraft i Vestjylland – Karlsgårde Vandkraftstation 1921–2011*. Esbjerg Byhistoriske Arkiv, Esbjerg, Denmark.

Moellenkamp, S., Lamers, M., Huesmann, C. *et al.* (2010) Informal participatory platforms for adaptive management: Insights into niche-finding, collaborative design and outcomes from a participatory process in the Rhine Basin. *Ecology and Society* **15**(4), 41.

Netting, R.McC. (1993) *Smallholders, Householders: Farm Families and the Ecology of Intensive, Sustainable Agriculture*. Stanford University Press, Palo Alto, California.

New Mexico Environment Department (NMED) (2010) *Water Quality Survey Summary for the Rio Puerco Watershed 2004*. Surface Water Quality Bureau, New Mexico Environment Department, Santa Fe, New Mexico.

Ostrom, E. (2005) *Understanding Institutional Diversity*. Princeton University Press, New Jersey.

Pahl-Wostl, C., Mostert, E. & Tàbara, D. (2008) The growing importance of social learning in water resources management and sustainability science. *Ecology and Society* **13**(1), 24.

Parkins, J.R. & Mitchell, R.E. (2005) Public participation as public debate: A deliberative turn in natural resource management. *Society and Natural Resources* **18**, 529–540.

Pedersen, A.B. (2009) The fight over Danish nature: Explaining policy network change and policy change. *Public Administration* **88**(2), 346–363.

Pedersen, A.B. (2010) Why David sometimes defeats Goliath: The power of actors in disprivileged land-use policy networks. *Land Use Policy* **27**, 324–331.

Pedersen, M.L., Andersen, J.M., Nielsen, K. & Linnemann, M. (2007) Restoration of Skjern River and its valley: Project description and general ecological changes in the project area. *Ecological Engineering* **30**, 131–144.

Pippen, S.J. & Wohl, E. (2003) An assessment of land use and other factors affecting sediment loads in the Rio Puerco watershed, New Mexico. *Geomorphology* **52**, 269–287.

Pollard, S. & du Toit, D. (2011) Towards adaptive integrated water resources management in southern Africa: The role of self-organization and multi-scale feedbacks for learning and responsiveness in the Letaba and Crocodile Catchments. *Water Resources Management* doi: 10.1007/s11269-011-9904-0.

Poundstone, W. (1992) *Prisoners' Dilemma: John von Neumann, Game Theory, and the Puzzle of the Bomb.* Doubleday, New York.

Putnam, R.D. (2000) *Bowling Alone: The Collapse and Revival of American Community.* Simon & Schuster, New York, New York.

Quinn, R.E. & Cameron, K. (1983) Organizational life cycles and shifting criteria of effectiveness: some preliminary evidence. *Management Science* **29**(1), 33–51.

Rio Puerco Management Committee (RPMC) (2001) *Watershed Restoration Action Strategy (WRAS) for the Rio Puerco Watershed of New Mexico.* Prepared by the WRAS Subcommittee of the Rio Puerco Management Committee. Submitted to the United States Environmental Protection Agency, Region 6, Dallas, Texas.

Rogers, E.M. (1995) *Diffusion of Innovations*, 4th edition. Free Press, New York.

Rosenberg, S. & Margerum, R.D. (2008). Landowner motivations for watershed restoration: Lessons from five watersheds. *Journal of Environmental Planning and Management* **51**(4), 477–496.

Sabatier, P.A., Focht, W., Lubell, M., Trachtenberg, Z., Vedlitz, A. & Matlock, M. (eds) (2005a) *Swimming Upstream: Collaborative Approaches to Watershed Management.* MIT Press, Cambridge, Massachusetts.

Sabatier, P.A., Weible, C. & Ficker, J. (2005b) Eras of water management in the United States: Implications for collaborative watershed approaches. In: Sabatier, P.A., Focht, W., Lubell, M., Trachtenberg, Z., Vedlitz, A. & Matlock, M. (eds). *Swimming Upstream: Collaborative Approaches to Watershed Management.* MIT Press, Cambridge, Massachusetts, pp. 23–52.

Scurlock, D. (1998) *From the Rio to the Sierra: An Environmental History of the Middle Rio Grande Basin.* General Technical Report RMRS-GTR-5. United States Department of Agriculture – Forest Service, Rocky Mountain Research Station, Ft. Collins, Colorado.

Shiva, V. (1991). *Ecology and the Politics of Survival.* Sage Publications Pvt. Ltd., London, England.

Skinner, A., Young, M. & Hastie, L. (2003) *Ecology of the Freshwater Pearl Mussel.* Conserving Natura 2000 Rivers Ecology Series No. 2. English Nature, Peterborough, United Kingdom.

Spink, A., Hillman, M., Fryirs, K., Brierley, G. & Lloyd, K. (2010) Has river rehabilitation begun? Social perspectives from the Upper Hunter catchment, New South Wales Australia. *Geoforum* **41**, 399–409.

Stegner, W. (1953) *Beyond the Hundredth Meridian: John Wesley Powell and the Second Opening of the West.* University of Nebraska Press, Lincoln.

Strategic Restoration and Management (STREAM) (2009) Report on Exchange Visit to the Houting Project (LIFE05 NAT/DK/000153), Denmark. Action E11, 29 June–3 July 2009. Available at http://www.snaebel.dk/NR/rdonlyres/AAEE714A-68E3-425C-8038-A0912941149C/92045/Houtingexchangevisit.pdf.

von Neumann, J. & Morgenstern, O. (1947) *Theory of Games and Economic Behavior.* Princeton University Press, New Jersey.

Walters, C.J. & Holling, C.S. (1990) Large-scale management experiments and learning by doing. *Ecology* **71**, 2060–2068.

Weber, M. (1962) Bureaucracy. *Essays in Sociology.* Oxford University Press, New York, New York.

Wiersum, F. (1995). 200 years of sustainability in forestry: Lessons from history. *Environmental Management* **19**, 321–329.

Wilkinson, C.F. (2000) *Messages from Frank's Landing: A Story of Salmon, Treaties, and the Indian Way.* University of Washington Press, Seattle.

Wondollek, J.M. & Yaffee, S.L. (2000). *Making Collaboration Work: Lessons from Innovation in Natural Resources Management.* Island Press, Washington, D.C.

5 Selecting Appropriate Stream and Watershed Restoration Techniques

Philip Roni[1], George Pess[1], Karrie Hanson[1] & Michael Pearsons[2]

[1]Northwest Fisheries Science Center, National Oceanic and Atmospheric Association, USA
[2]Pearson Ecological, Canada

5.1 Introduction

Selecting the appropriate restoration technique requires having clear restoration goals with specific objectives, an understanding of the disrupted processes and desired habitat conditions, and an understanding of which restoration techniques can achieve these objectives. Selection of restoration techniques logically follows a watershed assessment including identification of problems and specific objectives for areas or reaches in need of restoration (Chapter 3). It can occur before or in the early stages of the design phase (Chapter 7), but typically precedes the prioritization of specific restoration actions. Determining appropriate restoration techniques is not a trivial matter. Simply choosing the newest technique or one that has worked in other regions can lead to implementation of techniques that: (1) do not address the underlying problem; (2) are inappropriate to restore, improve, or create the habitats or desired conditions; (3) lead to short-term improvements in habitat or improvements in only a few functions or habitats; or (4) in the most extreme cases, lead to no improvement or further degradation of habitat conditions.

This chapter provides an overview of common restoration techniques to assist the reader in selecting an appropriate restoration technique that addresses underlying problems and helps to achieve restoration goals. We discuss the processes and habitats each major technique improves or creates, some common design considerations, effectiveness of various techniques, and other factors to consider when selecting a restoration technique including response time and longevity. We aim to provide the reader with the knowledge necessary to determine which technique or techniques are most appropriate to meet restoration objectives before entering into the design phase (Chapter 7). Because hundreds of different techniques exist, we strive to cover only the most widespread techniques. Detailed design, engineering, and construction considerations are often region- or project-specific and beyond the scope of this chapter and book. However, the reader can find details on construction and design in the existing regional manuals on restoration techniques and design for: North America (e.g. Hunter 1991; Slaney & Zaldokas 1997; Saldi-Caromile et al. 2004); Europe (e.g. RSPB et al. 1994; Petts & Calow 1996; Cowx & Welcomme 1998; O'Grady 2006); Australia (Rutherfurd et al. 2000); and east Asia (Parish et al. 2004). Similarly, while we provide overviews of the effectiveness of many common techniques, detailed information on the effectiveness of different restoration techniques can be found in Roni et al. (2002, 2008) and Smokorowski & Pratt (2007).

5.1.1 Common categories of techniques

Placing the numerous methods for restoring and improving habitat into distinct categories for purposes of discussion is challenging due to the sheer number of techniques and because some restore or improve more than one process or habitat. Moreover, many restoration projects employ more than one technique. Thus, any approach to classification will lead to some overlap or duplication. We

Stream and Watershed Restoration: A Guide to Restoring Riverine Processes and Habitats, First Edition. Philip Roni and Tim Beechie.
© 2013 John Wiley & Sons, Ltd. Published 2013 by John Wiley & Sons, Ltd.

Table 5.1 Common categories of stream and watershed restoration based on the processes restored and their objectives, examples of specific techniques and selected publications that provide detailed information on design. In addition to these, habitat protection through land acquisition, conservation easements, or laws to protect habitat can also be important measures for allowing passive recovery of habitat and limit further degradation, and should be a key part of a comprehensive restoration plan. Bank stabilization is a management technique that can lead to some improvements in habitat, although it is often implemented to protect infrastructure rather than improve habitat.

Process or habitat restored and typical objectives	Examples of techniques	Detailed *Design Guide*
Connectivity: Reconnect migration corridors, allow natural transport of sediment and nutrients, reconnect lateral habitats, allow natural migration of channel	• Dam removal or breaching • Culvert replacement • Fish passage • Levee setback or removal • Reconnection of sloughs and lakes	RSPB *et al.* 1994; Slaney & Zaldokas 1997; Cowx & Welcomme 1998; DVWK 2002; RRC 2002; O'Grady 2006
Sediment and hydrology: reduce or restore sediment supply, restore runoff and hydrology, improve water quality, provide adequate flows for aquatic biota and habitat, reduce sediment and runoff from farms	• Road removal or abandonment • Forest road improvements (e.g. resurfacing, stabilization, addition or removal of culverts, addition of cross-drains) • Urban road improvement (e.g. reduction of impervious surface, natural drainage systems) • Change agriculture practices • Increase instream flows and/or flood flows • Reconnect sediment sources	Bagley 1998; Waters 1995; ODF 2000; Ministry of Environment 2001; Novotny *et al.* 2010
Riparian: restore riparian zone, vegetation and processes, improve bank stability and instream conditions, increase or decrease shade	• Planting of trees and vegetation • Thinning or removal of understory or invasive species • Fencing and grazing reduction • Complete removal of grazing • Riparian buffers and protection	RSPB *et al.* 1994; FISRWG 1998; Izaak Walton League 2002; RRC 2002; Massingill 2003; O'Grady 2006
Habitat improvement and creation: improve instream habitat conditions for fish (pools, riffles, cover), improve spawning habitat (gravel addition), increase available habitat, increase cover and habitat complexity	• Placement of log or boulder structures • Engineered logjams • Placement of brush or other cover • Placement of spawning gravel • Remeandering a straightened stream • Excavation of new floodplain habitats • Reintroduce or protect beaver	Hunter 1991; Hunt 1993; Brookes & Shields 1996; Slaney & Zaldokas 1997; Cowx & Welcomme 1998; FISRWG 1998; RRC 2002; Saldi-Caromile *et al.* 2004; Brooks 2006; O'Grady 2006
Beaver reintroduction: increase pool habitat, trap sediment and aggrade channel, reconnect floodplain and restore riparian habitat	• Protect or reintroduce beaver • Provide food or habitat to enhance beaver populations	FISRWG 1998; Saldi-Caromile *et al.* 2004
Increase nutrients and productivity: boost productivity of system to improve biotic production, compensate for reduced nutrient levels from lack of anadromous fishes	• Addition of organic and inorganic nutrients • Addition of salmon carcasses	Slaney & Zaldokas 1997
Bank stabilization: stabilize banks to reduce erosion and improve riparian or instream habitat	• Revetments constructed of logs, branches, root wads, rock, and riparian cuttings, stakes and plantings	RSPB *et al.* 1994; Cowx & Welcomme 1998; Fischenich & Allen 2000; Izaak Walton League 2002; O'Grady 2006

have chosen to categorize restoration techniques by the processes they restore or the habitat they seek to improve (Table 5.1). This allows one to select the category of techniques that will best address objectives identified in the watershed assessment process, which should outline the process, function, or habitats to be restored (Chapter 3).

First, we will focus on techniques that restore watershed-scale processes (connectivity, sediment, and hydrology) and those that focus on riparian conditions and reach-scale processes. We then discuss those actions that aim to improve or create habitats, for example, wood placement, restoring sinuosity to straightened channels (remeandering), or increase productivity (nutrient addition). We focus on *active restoration* techniques: those that require some type of on-the-ground action designed to restore processes or improve habitats. We discuss *passive tech-niques* in less detail, not because they are not important but because they often require legal or regulatory frameworks to remove a pressure and allow the system to recover naturally, and creating or changing policies or laws is not the focus of this volume. Similarly, while we discuss many techniques that reduce runoff, fine sediment, and improve water quality, a detailed discussion of all techniques for restoring water quality and pollutants is beyond the scope of this chapter and volume.

5.1.2 Selecting the appropriate technique: What process or habitat will be restored or improved?

Before discussing the different techniques in detail, it is important to have some idea of what types of habitats or processes they restore. The appropriate technique to

Table 5.2 Restoration techniques and the major habitats and processes they typically restore. Processes: Con = connectivity, Sed = sediment, Hyd = hydrology, Rip = riparian and organic matter. Habitats: Flp = floodplain, Rif = riffle, Pl = pool, Spw = spawning, Cov = cover. Bank stabilization may improve sediment and riparian processes as well as cover, but is often implemented to protect infrastructure. Nutrient additions may lead to increased productivity, but does not directly affect physical processes or habitat conditions.

Technique	Process				Habitat				
	Con	Sed	Hyd	Rip	Flp	Rif	Pl	Spw	Cov
Dam removal	X	X	X	X	X		X		
Culvert replacement	X	X		X					
Fish passage structures	X								
Levee removal or setback	X	X		X	X				
Reconnection of floodplain habitats	X	X	X		X				
Road removal	X	X	X						
Road resurfacing		X							
Stabilization, upgrading stream crossings		X	X						
Reduce impervious surface			X						
Instream flows			X						
Agricultural practices		X							
Restore sediment sources		X				X	X		
Riparian replanting		X		X					
Thinning or removal of understory				X					
Removal or control of invasives				X					
Fencing		X		X					
Rest-rotation or grazing strategy		X		X					
Log or boulder structures						X	X	X	X
Natural LWD placement					X		X	X	X
Engineered logjams					X		X		X
Brush or other cover									X
Gravel addition					X	X	X	X	X
Remeandering of straightened channel	X				X	X	X	X	X
Creation of floodplain habitats					X			X	
Beaver reintroduction	X				X		X		
Nutrient additions									
Bank stabilization		X		X					X

address particular habitat degradation depends on the restoration goals for the site, which processes or habitat have been disrupted and are in need of restoration, key design considerations, project longevity, and other technical and socioeconomic factors discussed in previous chapters. At this point in the restoration process, it is assumed that the causes of degradation and the need for restoration have been identified (Chapter 3). The appropriate restoration action therefore should be based on addressing those factors identified in the assessment and depend on selecting techniques that adequately address these factors and meet the restoration goals and specific objectives. For example, if a particular reach was identified as having degraded riparian habitat, the overall restoration goal is to improve aquatic habitat, and one objective is to improve shade, then the potential techniques can be narrowed down to those riparian techniques that improve shade (Tables 5.1 & 5.2). If the objective is improving instream habitat complexity, then a suite of different instream habitat techniques may address those needs. The major processes or habitats restored for the major categories of restoration techniques are outlined in Table 5.2 to assist with selecting the appropriate technique.

5.2 Connectivity

The connectivity of watersheds and their habitats is critical for maintaining the flux of water, sediment, organic matter, nutrients, and the migration and movement of fish and other biota (Vannote *et al.* 1980; Fullerton *et al.* 2010). Three main types of connectivity exist and can be affected by anthropogenic alteration or restoration: longitudinal, lateral, and vertical. Longitudinal refers to upstream–downstream connectivity while lateral connectivity typically refers to the connection of the river to the floodplain and riparian area. Vertical connectivity is often thought of as the sole connection with the hyporheic zone and subsurface area, but also includes the incision of channels which can have a profound influence on lateral connections. For example, due to agricultural practices, urbanization, or other factors, many channels have become incised and isolated from their historic floodplains. Restoring vertical connectivity is often part of restoring lateral connectivity, restoring sediment transport, and placement of instream structures. To minimize overlap with other sections, we therefore focus our discussions on lateral and longitudinal connectivity and discuss vertical connectivity in Section 5.3.3.

5.2.1 Longitudinal connectivity

Restoring longitudinal connectivity often calls for the removal or modification of infrastructure that has caused the disconnection. Disruptions to upstream–downstream connectivity include dams, weirs, pipeline crossings, bridges, culverts, road or stream crossings, and other infrastructure. These structures disrupt transport of sediment, wood and organic matter, nutrients, and timing and volume of water and are often partial or complete barriers to migration of fish and other biota. Common approaches to addressing these include dam, culvert, or barrier removal and replacement and modification or installation of fish passage structures, each of which are described in detail below. In addition, water withdrawal and pollution can create barriers to movement of fish and other aquatic biota. Which technique for restoring connectivity is most appropriate obviously depends on the objectives of the restoration. For example, dam removal will restore all processes, while installation of fish passage structure will only meet objectives of restoring migratory pathways for some biota.

5.2.1.1 Dam removal and modification

Dam removal and the restoration of the hydrologic regime to more natural conditions are employed primarily to restore longitudinal connectivity of stream and floodplain habitats; improve fish access and migration; allow for natural transport of water, sediment, organic material, and nutrients; and maintain or restore natural riverine processes that create and maintain fish habitat (Pess *et al.* 2005b; Figure 5.1). Many dams in developed countries have reached the end of their useful life and are being considered for removal or restructuring because of ecological concerns, safety concerns, and the costs of continued operation or repair (Stanley & Doyle 2003). There are several alternatives to complete dam removal ranging from partial dam removal and restoration of former impoundment and stream channel to notching the dam to drain the reservoir but leaving most of the dam in place (Wunderlich *et al.* 1994; Shuman 1995; Quinn 1999). These also vary in the amount of sediment removal and reconstruction of the river channel in the former reservoir. Complete dam removal and restoration includes removal of all structures and the return of river and valley morphology and topography to a pre-construction state. This requires not only dam removal but also sediment removal, channel reconstruction, riparian planting, etc., and may require that suitable reference or potential channel conditions be determined (see Chapter 7). Partial dam removal or retention typically

Figure 5.1 Common approaches for restoring longitudinal and lateral connectivity including: (A) dam removal (Souhegan River, New Hampshire; from Pearson *et al.* 2011, reproduced by permission of American Geophysical Union, Copyright 2011 AGU); (B) a recently replaced culvert with a natural stream bottom (West Fork of Smith River tributary, Oregon, USA); (C) a restored (removal of bank protection) and (D) adjacent unrestored reaches of the River Aggar, Germany. (See Colour Plate 7)

involves the limited removal or modification of structures and an incomplete alteration of the topography to the pre-existing state (Quinn 1999). Depending on the alterations to the dam and former river channel, it may lead to full or partial restoration of hydrology, sediment delivery, floodplain function, and other processes. In contrast, partial retention of the dam and making no modifications to the former reservoir or channel would restore the hydrology, but not fully restore sediment and floodplain functions (Quinn 1999).

Determining whether or not to remove a dam, or undertake some form of flow or structural alteration, depends on the project objectives and incorporates the consideration of multiple factors such as stream channel morphology, river hydrology and hydraulics, sediment

properties (e.g. amount, size, toxicity) and transport, surface- and groundwater quality and quantity, instream habitat, floodplain vegetation, the aquatic community (i.e. fish, aquatic insects, and plants), and, of course, cost (Shuman 1995). A sediment management strategy in relation to which dam removal or modification technique is implemented is perhaps the most important consideration. Sediment management in dam removal projects can follow one of three major strategies including: (1) removal of all accumulated material behind a reservoir; (2) allowing a river to erode a new channel or channels through the accumulated sediment; or (3) removal of a specified area of the sediment deposit that is in the anticipated path of a river, and leaving the remaining sediment in place (Wunderlich *et al.* 1994; Quinn 1999).

A staged drawdown of impoundment water level may be needed several years prior to dam removal to determine sediment stability and accumulation in the reservoir and reservoir delta. This is critical information for selecting a dam removal alternative and sediment management strategy (Childers et al. 2000).

Geomorphic adjustment following dam removal typically leads to erosion of sediments accumulated in the reservoir and subsequent transport and deposition and bed aggradation downstream of the former dam structure (Doyle et al. 2005). The amount of sediment mobilized will be a function of several factors including the amount of stored sediment, sediment size composition, and the geomorphic characteristics of the former reservoir section, which can drive the rate and magnitude of downstream sediment response (Doyle et al. 2005). For example, Cheng & Granata (2007) found that following removal of a 2 m high dam, less than 1% of the sediment stored in the reservoir was transported downstream, resulting in minor channel adjustment downstream. In contrast, removal of the 14 m high Marmot dam on the Sandy River, Oregon, resulted in 50% of the reservoir sediment eroding by the end of the first year and prompted growth of new bars and enlargement of existing bars in the first 2 km below the dam site (Major et al. 2008; Podolak & Wilcock 2009). These examples highlight the variability in erosion and deposition of sediments depending on the amount and composition of the sediment, the stream energy and antecedent channel conditions (Doyle et al. 2005).

The newly exposed soils in the former reservoir are typically colonized rapidly by vegetation, although the type and diversity is related to the time since dam removal as well as soil, light, climate, and other conditions (Orr & Stanley 2006). Whether grasses or trees and shrubs colonize the surface can have a large effect on the persistence of erosion from the former reservoir site, with greater bank stability in tree-vegetated channels (Doyle et al. 2005). Often, planting and seeding of trees, shrubs, grasses, and other vegetation is included to help stabilize the soils in the former reservoir, reduce erosion and provide fish and wildlife habitat (see Section 5.4 on riparian restoration).

The benefits of dam removal to fish populations have been well documented in North America, Europe, and Asia and fish typically migrate upstream and colonize new habitats rapidly (Burdick & Hightower 2006; Stanley et al. 2007; Nakamura & Komiyama 2010). Dam removal often results in large shifts in the periphyton and macroinvertebrate community in the former reservoir area

as it is transformed from a lentic to lotic environment (Hart et al. 2002; Stanley et al. 2002; Maloney et al. 2008). Another important aspect to consider is the potential for the introduction or expansion of non-native species upstream of potential dam and migration barrier removal locations (Marks et al. 2010). This should be considered carefully, particularly if the removal or eradication of non-native species is unrealistic (Jansson et al. 2007).

5.2.1.2 Culvert and stream-crossing removal, replacement or modification

Many barriers or disruptions to longitudinal connectivity result from stream crossings infrastructure such as culverts and bridges that constrain the channel and disrupt or block movement of sediment, organic matter, fishes, and, in some cases, water. Common techniques to address this include removal of the stream crossing, replacing it with a larger bridge or culvert, or modifying the culvert or structure so that it will not be a barrier to fish migration (Table 5.3). In each case, the approach will vary depending on the specific objectives and cost of the modifications. For example, if the objective is to restore fish passage through an impassible culvert, then a suite of possible options are available to correct the situation. These range from replacing the culvert with a bridge to leaving the original culvert in place and modifying hydraulics by installing baffles within the culvert to lower water velocities and allow passage of target fish species. In some instances, if the channel below the culvert is incised or the gradient very steep, simply installing a series of weirs can lower the hydraulic head and velocities enough to allow fish passage though the culvert. In contrast, if the objectives are to restore channel migration, sediment, and organic transport, then complete removal or replacement with a bridge wider than the active channel or much larger culvert may be needed.

Complete elimination of the stream crossing would be ideal, but is often not feasible. In these cases, bridges are a preferred alternative because, depending on their width, they can allow the lateral and vertical adjustment of a stream to occur (Roni et al. 2008). However, a narrow bridge may only restore some processes and limit channel movement or lock the channel in place. There are numerous types of culverts including, but not limited to, smooth box or round, bottomless pipe arch, squash pipe or countersunk, round corrugated, baffled, and round corrugated with no baffles (Table 5.3). These are sometimes categorized either as 'stream simulation' or 'hydraulic design' based on whether they allow for the creation and maintenance of an actual streambed within the culvert

Table 5.3 Common types of stream crossings and the processes they restore or improve. Modified from Roni et al. 2002, 2005; Roni 2005.

Stream crossing type	Processes restored				Notes
	Fish passage	Sediment transport	LWD transport	Constrains channel?	
Bridge	Y	Y	Y	N	Assuming the bridge is as wide or wider than active channel
Stream simulation culverts (bottomless pipe arch)	Y	Y	Y/N	N	May allow for transport of LWD depending upon width, and height and size of LWD
Squash pipe or countersunk culvert	Y	N	N	Y	Designed so bottom of culvert is covered with gravel or substrate from stream
Hydraulic design culverts	Y	N	N	N	Can be a variety of culvert types but slope, grade, and material are designed to provide passage for specific fish species and life stage
Round corrugated with baffles culvert	Y	N	N	Y	Baffles designed to provide fish passage
Smooth (box or round) culvert	Y/N	N	N	N	Fish passage depends upon slope and length
Log, rock, or other weirs downstream of culvert	Y	N	N	Y	Designed to create pools or steps to allow fish passage through culvert; no modification of culvert
Ford crossing or low water crossing	Y	Y	Y	N	Bottom of stream is paved or armored so that road surface is underwater and traffic can cross stream at low to moderate flows

itself or are designed to meet specific criteria to allow migration of a particular species and life stage. The benefit of natural stream bottom culverts, particularly on smaller stream, is that they ideally do not create any more of a barrier to aquatic organisms than a natural stream channel (Price *et al.* 2010). Hydraulic designs are similar to fish passage structures in that they are typically engineered to meet the swimming performance of specific species and life stages; they therefore do not restore connectivity for all species and are typically not designed to restore transport of wood and sediment (Price *et al.* 2010).

In contrast to bridges or culverts which go over the stream, ford crossings, also known as shallow- or low-water crossings, simply consist of hardened crossings on the streambed that are either consistently submersed or submersed during higher-flow events. Ford crossings are typically placed in areas where the stream is intermittent, but can be used in perennial mountain streams where shallow rapid landslides and debris flows are common, or in agricultural settings to direct and allow livestock to cross the stream and limit damage to banks. In the case

of mountain streams prone to high sediment loads and debris flows, the objective is to prevent sediment or debris from accumulating behind and plugging the crossing which leads to road failure or diversion of the stream onto the roadbed. The ford crossing allows for greater movement of material, which can be readily excavated if it does accumulate on the crossing.

Efforts to restore longitudinal connectivity in the form of retrofitting or replacing culverts or other road crossings have been the focus of numerous restoration actions over the last several decades and have resulted in the reconnection of large amounts of habitat (Bernhardt *et al.* 2005; Price *et al.* 2010). For example, over the last 10 years in Washington State, approximately 3500 fish passage barrier culverts were replaced with fish-passable structures and an estimated 5990 km of fish habitat is now accessible to migratory fishes (Price *et al.* 2010). Studies evaluating effectiveness of replacement or removal of culverts in both Europe and North American have consistently shown rapid colonization by fishes with the increase a function of distance from the source population (Roni *et al.* 2008; Zitek *et al.* 2008). However, culverts often need maintenance and the success of replacement of impassable culverts at providing fish passage and restoring other processes depends on culvert design including slope, width, length, and the percent the culvert is countersunk (Price *et al.* 2010).

5.2.1.3 Fish passage structures
In many cases, dam or weir removal, culvert replacement, or other forms of reconnecting isolated habitats or stream reaches are not possible. In these cases, a variety of fish passage structures can be used to effectively restore migration corridors and longitudinal connectivity for fishes. These include various types of fish ladders, fish locks, addition of baffles or weirs, and bypass channels (Figure 5.2). These can generally be divided into two major types

of passage structures or fish passes: semi-natural and technical solutions (DVWK 2002). Semi-natural types attempt to create a semi-natural streambed and include bottom ramps, sills and slopes, bypass channels, and fish ramps. These can also include placement of boulders and substrate directly on a low head weir or sill to create suitable velocities and resting places for fish to swim over a weir or low-head obstruction, or the creation of semi-natural bypass channels around the obstruction. Technical solutions are highly engineered structures and include: pool, vertical slot, Denil, and eel passes or ladders; fish locks; fish lifts; and trap-and-haul facilities (Clay 1995; DVWK 2002).

Most of the fish passage structures previously identified are designed to provide for upstream migration of

Figure 5.2 Common types of structures used to pass fish at weirs and dams including: (A) a fish ramp Vara River, Italy (photo Enrico Pini-Prato); (B) a bypass channel for both fish passage and kayaking on Gave da Pau River, France (photo Enrico Pini-Prato); (C) pool and weir type pass at Lower Granite Dam, Snake River, Washington State, USA; and (D) a Denil steep pass at Bennett Dam, North Santiam River, Oregon (photo Ed Meyer). (See Colour Plate 8)

fish. In the case of fish ladders at large dams, water diversion, or abstraction weirs, the downstream migration of fish is also severely restricted and a suite of different types of screens, weirs, and other highly engineered structures are installed upstream or within the dam to divert fish away from water or turbine intake structures or direct fish into one of the bypass channels, pipes, or structures. These measures can generally be categorized as behavioral barriers (lights, sound, bubbles) and physical barriers including bar racks, fixed screens, traveling screens, fish diversion devices such as drum screens, screens that divert fish away from a structure, and fish collection devices such as traps and 'gulpers' (Clay 1995).

All fish passage structures require extensive engineering and design and several manuals and detailed text on design and engineering considerations exist (Clay 1995; DVWK 2002; see also *Bulletin Français de la Pêche et de la Pisciculture* 2002, Supplement 364). Key design parameters include (but are not limited to): dam, weir, or barrier height and width; river flow timing, magnitude, and velocity; bypass or ladder length, slope, substrate, and attraction flows; and species of interest. These factors also influence the effectiveness of these structures to pass various fish species both upstream and downstream. Periodic maintenance is needed on most fish passage structures to keep them clear of debris and ensure appropriate hydraulic conditions are met to maintain fish passage.

5.2.2 Techniques to restore lateral connectivity and floodplain function

Restoration of lateral connectivity typically seeks to restore connection of the river channel to its floodplain and to restore floodplain functions. The most appropriate approach will depend on the factors that have led to isolation of the channel from its floodplain. We discuss common approaches below based on whether they require modification or removal of levees or bank revetments, channel filling, or installation of passage structure to allow fish and other biota to access floodplain habitats (Figure 5.1).

5.2.2.1 Levee removal or setbacks

A levee removal or setback is the lowering, removal, or relocation of a levee adjacent to a mainstem river that allows a river to migrate or meander freely and maintain connection to the adjacent floodplain. Complete levee removal is usually used if the levee is damaged or obsolete and if there is little to no infrastructure or adjacent land

use that will be compromised. Theoretically, this method will eventually allow for full recovery of all or most riverine functions including: retention and natural exchange of water, wood, sediment, and nutrients between a floodplain and mainstem; fine sediment deposition; channel migration; the development of a greater diversity of riparian conditions; seed dispersal; and a greater diversity of habitat types (Pess *et al.* 2005a; Roni *et al.* 2008). However, levee removal may need to be coupled with other restoration techniques to encourage more rapid recovery of channel morphology. For example, if the channel has been straightened, remeandering or restructuring the channel may be used to create a sinuous channel (see section 5.5 below).

Partial levee removal is often used when the extent of restoration actions is limited by adjacent land-use or other resource constraints. One technique is a 'beaded approach' to floodplains (Cowx & Welcomme 1998), where alternating reaches of the levee are removed, allowing floodplain reconnection and restoration of functions in some reaches and not others. In the restored sections of floodplain, several habitat features such as floodplain channels and wetlands are restored or constructed. This technique allows portions of the floodplain to be inundated and encourages scour, erosion, and deposition in those areas. It is particularly suitable in tributary junctions or other areas where habitat diversity is high and the levee can only be removed or set back in some but not all reaches.

Rather than remove all or part of a levee, a levee setback is used to relocate the levee further back from the river bank, allowing the river to access more of the floodplain. The new wider river channel may also increase mainstem habitat and flood conveyance capacity. In incised channels, levee setbacks can be combined with floodplain excavation to lower the floodplain and for a higher inundation frequency (Williams *et al.* 2009a). In other incised channels, two-stage or multi-stage channels are constructed within the levee setback to create multiple floodplain terraces that are flooded at different flows (Cowx & Welcomme 1998; O'Grady 2006).

Another technique is intentional levee breach, which is simply the excavation of 'openings' in a levee system that allow for the accumulation and, if appropriately designed, conveyance of water during certain flow events. This technique can also lead to changes in floodplain deposition, erosion, and topography (Florsheim & Mount 2002). Alternatively, the height of the levee can be reduced to allow more frequent inundation of the floodplain during large flood events. This also requires providing channels, culverts, or drains in the lowered levee to ensure water does not become trapped or ponded behind the levee when flood flows subside.

In many cases, stream banks have been armored with cement, riprap, rock, or other material to prevent the stream from moving, limit erosion, or stabilize the bank. This can often lead to a loss of lateral connectivity. Removing revetments or bank armoring can be an effective method of restoring channel migration and other floodplain processes. In many cases, riprap or other bank protection is removed and replaced with logs, vegetation, fiber mats, or other 'soft' or 'bioengineering' approaches (see Section 5.6.2). We discuss these approaches in the section on habitat improvement because their objectives are to improve habitat and only partially restore channel processes.

Existing and ongoing evaluations of levee removal or modification indicate that these techniques lead to a wider active floodplain and allow for the re-establishment of connectivity between a river and its floodplain. They promote water retention in the floodplain, exchange between surface and subsurface flow, overbank fine sediment deposition, organic matter retention, and increased sinuosity, habitat diversity, and complexity (Jungwirth et al. 2002; Muhar et al. 2004; Konrad et al. 2008). Depending on residence time of water in the floodplain, increased primary productivity can occur in newly established or reconnected habitat, thus providing valuable food resources to the main channel (Junk et al. 1989; Schemel et al. 2004; Ahearn et al. 2006). Fish rearing in floodplain habitats created or reconnected following levee removal or setbacks often have higher growth rates than those in the mainstem. Levee setbacks and modifications have also been shown to benefit not only rheophilic fishes, but also amphibians and aquatic insects (Chovanec et al. 2002). Other studies have demonstrated improvements in physical habitat and restoration of natural erosional and channel migration processes as well as improvements in fish and riparian diversity and age structure (Jungwirth et al. 2002).

5.2.2.2 Reconnecting isolated floodplain wetlands, sloughs, and other habitats

Techniques similar to reconnecting longitudinal habitats are used when the objectives are to reconnect isolated floodplain water bodies or channels while maintaining the existing infrastructure. Culverts, notched or lowered levees or dikes, or controllable flow gates can be used to reconnect existing relict channels in areas where flow

needs to be regulated (Lister & Finnegan 1997; Cowx & Welcomme 1998). Two common problems that occur with such projects are: (1) elevation differences between the mainstem and relict channels due to mainstem channel or from natural sedimentation or aggradation in the relict feature; and (2) degraded water quality in the floodplain due to land use. For example, rehabilitation opportunities identified in the Stillaguamish River in Washington State include the reconnection of two former meander bends that are now floodplain sloughs (Pess et al. 1999). However, the river channel incised 1–2 m between the period when the bends where cutoff in 1929 and 1991, making reconnection of the former meanders difficult (Collins et al. 2003). This problem is common to many stream channels in Europe, North America, and throughout the world (Cowx & Welcomme 1998; Pess et al. 2005a).

One potential solution is the installation of submerged weirs (grade control structures) to raise the bed elevation of incised mainstem channels to reconnect them with their former channels and floodplain. In the Danube River in Slovakia, for example, check dams were installed in artificial channels and natural former side to aggrade the river to the point where older channels are now reconnected, increasing the retention time of water in the reach (Cowx & Welcomme 1998). In other cases, rather than aggrade the mainstem channel, the floodplain feature can be reconnected with a culvert or channel with a fish passage structure installed to allow for movement of fish in and out of reconnected habitat. These techniques vary greatly in success depending on the flow, level of connectivity of the habitats, and the fish species of interest.

Another approach used to reconnect floodplain habitats in rivers that have been dredged, channelized, and straightened includes filling in the straightened channel and forcing it back into its original meandering channel. One of the most high-profile cases of this is the Kissimmee River, Florida. Using material from levees constructed when the river was straightened and channelized in the 1960s, the channelized mainstem has been filled in at specific locations, forcing the river to flow through its original meandering channel. This approach, of course, is only feasible if much of the original meandering channel still exists and the floodplain or some of its key habitats remain intact (Cowx & Welcomme 1998). Studies on effectiveness of various floodplain reconnection methods have consistently shown rapid colonization and use of reconnected floodplain habitats (Schmutz et al. 1994; Grift et al. 2001; Roni et al. 2008).

5.3 Sediment and hydrology

Because so many human activities impact basic sediment and hydrologic processes, a suite of different approaches has been developed to restore more natural levels and timing of sediment and runoff. Many of these techniques are designed to address both sediment delivery and hydrology and to address a specific type of human infrastructure or land use, making it difficult to categorize them based on the process they restore. For example, efforts to minimize or eliminate the impacts of roads are intended to reduce sediment inputs and restore natural hydrologic processes. Other efforts such as reduction of water withdrawal and restoration of instream flows focus more on restoring the natural hydrologic system, but also result in a more natural transport and routing of sediment in the system. We therefore discuss the methods for restoring sediment and hydrology by technique and highlight which combination of processes they seek to restore.

5.3.1 Reducing sediment and hydrologic impacts of roads

Roads typically alter stream flows, increase fine and coarse sediment loads to stream channels, transport nutrients and other pollutants, and degrade water quality, all of which can negatively impact streams and their biota (Goudie 2000; Gucinski *et al.* 2001; Beechie *et al.* 2005). A variety of treatments have been developed to minimize these impacts, including road removal, decommissioning or closure, stabilizing or upgrading road crossings (culverts, bridges), increasing cross-drains, resurfacing, reducing traffic or tire pressure, or reducing impervious surface area (Beechie *et al.* 2005; Roni *et al.* 2008). These techniques can be categorized by whether their objective is to eliminate the impact of roads by partially or completely removing or abandoning the road or to minimize road impacts by improving the road or road maintenance. Most of these techniques have been developed on and applied to unpaved roads to minimize the impacts of forestry operations (timber removal, road construction), but they can also be applied in rural areas with similar unpaved road surfaces and drainage issues. Paved roads and impervious surfaces in urban and residential areas are discussed in Section 5.3.1.3.

5.3.1.1 Forest and unpaved road removal and restoration

Forest and other unpaved roads and associated systems of ditches and cross-drains capture, intercept, and con-centrate water and sediment from the road surface and hillslope that would have originally fallen on the forest floor or flowed downslope as groundwater, essentially increasing drainage density. The road and associated ditches and drains concentrate flow and deliver it to the stream at a faster rate and greater magnitude. This can lead to increased surface erosion rates and fine sediment loads, which increase stream channel sedimentation rates. The ideal approach to eliminating or reducing these impacts would be removal or decommissioning the road (deactivation) (Figure 5.3). This is an increasingly common approach on low-traffic forest roads in areas of the western United States and Canada where a dense infrastructure of unpaved roads was built in the 20th century for harvest and extraction of timber (Baker *et al.* 1988). The goal of road removal is to slow surface runoff, reduce fine sediment produced from the road surface, or reduce landslides from unstable roads or road crossings. Road removal can also restore hydrology because the road and its drainage network no longer intercept water and deliver it directly to the stream. Removal or decommissioning is not only done for environmental reasons, but often for economic reasons because annual maintenance and repair of roads can be costly. Complete road removal includes removing fill, recontouring the slope, and replanting or seeding so the footprint of the road is no longer visible and the natural contours of the land are restored (Beechie *et al.* 2005). In contrast, road closure or decommissioning ranges from simply preventing access to the road to allow it to be recolonized by native vegetation to removing culverts and stream crossings, installing water bars, and scarring the road surface so that it is more readily colonized by plants (Beechie *et al.* 2005).

Whether complete removal, closure, or decommissioning of the road is appropriate depends on a number of factors including condition of the road, soils, steepness of terrain, compaction of the road surface, and other factors (ODF 2000; Roni *et al.* 2008). Closing a road without stabilizing stream crossings or treating the road surface can lead to additional erosion in some soils and terrain. However, both simple road closure and complete removal have been demonstrated to greatly reduce sediment delivery (Hickenbottom 2000; Madej 2001; Switalski *et al.* 2004; see Roni *et al.* 2008 for a review). The type of treatment following road removal or closure (e.g. recontouring of slope, scarring or ripping of road surface, removal of stream crossings, placement of mulch, seeding, and planting) can influence the sediment production and infiltration capacity of the former road bed (Cotts *et al.* 1991; Maynard & Hill 1992; McNabb

Figure 5.3 Complete removal of a forest road in Salmon Creek, Humboldt County, California: (A) before; (B) during; (C) immediately after; and (D) 3 years after removal of stream crossing and road fill. Note that the stream side-slopes were laid back to a stable grade that mimics natural hillslope topography and all bare slopes were mulched with rice straw. The excavated stream crossing has also been sloped to a stable angle and mulched to control surface erosion. Photos courtesy of California Department of Fish and Game Fisheries, Bureau of Land Management, Pacific Coast Fish, Wildlife & Wetlands Restoration Association, and Pacific Watershed Associates. (See Colour Plate 9)

1994; Luce 1997; Elseroad *et al.* 2003). In general, these studies have found that recontouring slope and site preparation (ripping and mulching of former road bed) are most effective at inducing plant growth and reducing fine sediment production, although the position of the road on the slope and time since treatment also play a role in the effectiveness of these methods.

5.3.1.2 Road improvements

In many cases road removal, decommissioning or closure are not feasible and restoration efforts must focus on minimizing the impacts of roads on sediment, hydrology, and water quality by stabilizing the road and roadside ditch network, minimizing exposed soil, disconnecting road drainage from the stream, and improving stream crossings to minimize erosion. Many of these same tech-

niques are used during road removal or abandonment to ensure that the former road will not continue to lead to erosion or hydrologic impacts.

5.3.1.2.1 Reducing road-related mass wasting

Poorly designed roads or roads designed on unstable slopes can greatly increase mass wasting (e.g. avalanches, landslides, soil creep, debris flows) and delivery of coarse and fine sediment to stream channels (Gucinski *et al.* 2001). The objective of reducing or eliminating mass wasting caused by road construction and existence is primarily to prevent coarse and fine sediment delivery to the stream channel. This objective is achieved by removing, stabilizing, or reinforcing the unstable road and slope material and often by routing water away from unstable

Table 5.4 Summary of common methods for reducing impacts of forest and unpaved roads and improving or restoring natural sediment levels and hydrology (modified from Beechie *et al.* 2005). Note that complete road removal would address all of these objectives and utilize a number of the specific techniques including sidecast removal, removing fill, removing culverts and cross-drains, as well as recontouring, ripping, and replanting road surface.

General objective	Site-specific objective	Techniques
Reduce mass wasting	Remove unstable road material	• Sidecast removal
	Reinforce unstable material	• Buttress toe slope
		• Retaining walls or other geotechnical approaches
	Route water away from unstable material	• Enhance road drainage control
		• Subsurface drain pipes or other drainage modifications
Reduce surface erosion	Reduce traffic effects	• Block vehicle entry with gate
		• Block vehicle entry with barrier (boulders/tank traps)
	Armor running surface	• Surface road with gravel or crushed rock
		• Pave road
	Vegetate exposed soil surfaces	• Seeding and planting
		• 'Rip' (i.e. decompact) tread to improve growth of vegetation
	Armor exposed soil surfaces	• Cover with rock or other resistant material
		• Cover with matting
Disconnect road drainage from streams (reduce hydrologic impacts and surface erosion)	Disconnect road runoff from stream	• Add more cross-drains (e.g. culverts, water bars) between streams
		• Outslope tread
	Filter sediment from road runoff prior to stream entry	• Install settling ponds
		• Install slash filter windrows
Reduce erosion at stream crossings	Reduce potential for plugging or diversion	• Replace undersized culverts with adequate structure
		• Remove culvert or crossing structure
		• Construct drainage dip or hump over structure
	Reduce fill erosion at stream crossings	• Remove fill at crossing
		• Replace soil fill with rock or concrete
		• Armor fill surface with rip-rap
		• Plant woody species on fill
	Improve drainage between crossings	• Reshape tread (e.g. inslope, crown)
		• Repair or upsize cross-drains
		• Clear or enlarge ditches
	Improve cross-drainage to minimize diversion	• Replace culverts with water bars or dips
		• Construct dips or backup water bars over culverts

material or slopes. Common techniques include removal of sidecast material, stabilizing the toe of the slope through retaining walls or other geotechnical approaches, installing geotextiles or planting of vegetation, drainage systems to route water away from unstable area, or some combination of these approaches (Roni *et al.* 2002; Beechie *et al.* 2005; Miller *et al.* in press; Table 5.4). These techniques are generally effective at reducing landslides, but may require periodic maintenance (i.e. cleaning culverts and ditches) to maintain benefits (ODF 2000; Daniels *et al.* 2004).

5.3.1.2.2 *Reducing fine sediment from the road surface*

Techniques designed to reduce surface erosion and fine sediment production include resurfacing the road, reducing traffic or blocking vehicle entry, installing dips (low spots in the road), and reducing tire pressure of trucks traveling the road (Table 5.4). These approaches focus on reducing the amount of fine sediment produced from the road surface that enters the stream channel either through erosion during periods of precipitation or as dust from traffic. Increasing the thickness of gravel on the road surface or paving an unpaved road can significantly reduce fine sediment (Reid & Dunne 1984; Kochenderfer & Helvey 1987; Burroughs & King 1989). Similarly, on roads with heavy traffic from logging operations, reducing the tire pressure and/or the traffic levels can reduce fine sediment production (Reid & Dunne 1984; Bilby *et al.* 1989). The effectiveness of these approaches depends partly on the geology, the type of material used for the road surface, and the level and type of traffic. Geology is particularly important in determining natural sediment delivery; the sediment delivery from roads, side slopes, and drainage ditches; and the overall success of sediment reduction efforts (Bloom 1998).

5.3.1.2.3 Disconnecting road drainage from the stream

Interception of water by ditches and the road itself can lead to erosion of the road, upslope, and downslope areas, and to hydrologic changes by channeling water directly to the stream channel that would have fallen onto and infiltrated into the forest floor or continued downslope as groundwater. In addition, erosion of the road ditch can increase fine sediment, increase erosion of the road, and destabilize the slope above the road cut. Methods to disconnect the road drainage from the stream channel include water bars, dips, cross-drain culverts, and depressions designed to transport water off the road (Beechie *et al.* 2005; Table 5.4). These approaches are designed to prevent the concentration of flow on the road surface or in roadside ditches by directing and dispersing the water onto the forest floor, allowing it to infiltrate into the soil. Installation of slash filters (woody material) and settling ponds are also used to limit erosion and control runoff (Table 5.4). Similar to approaches to reduce road-related landslides, these techniques also depend on occasional maintenance to ensure that they continue to route water away from the road onto the forest floor.

5.3.1.2.4 Reducing erosion at stream crossings

Undersized stream crossings or crossings on steep slopes with considerable fill can become plugged with debris or overtopped at high flows, leading to erosion of fill material or diversion of the stream onto the road or into the roadside ditch (Furniss *et al.* 1991, 1997). The latter can result in the stream abandoning the former stream channel and flowing down the road. All can lead to large amounts of fine and coarse sediment being delivered to the stream channel and increase the likelihood of road failures (Furniss *et al.* 1991, 1997). Techniques to minimize this include replacing the culvert with a bridge,

drainage dip, or ford-type crossing or a larger culvert, reducing or stabilizing fill at the stream crossing to minimize erosion or potential for failure during high flows, reshaping road tread, repairing or upgrading cross-drains, and consistent maintenance to clear ditches and blockages. Undersized culverts can also prevent the transport of wood and sediment; when possible, bridges or other crossings that will allow for the natural transport of wood and sediment should be used (Roni *et al.* 2002). Stabilization of the stream crossing may only be a short-term solution if the culvert or crossing is undersized. As discussed in Section 5.2.1, culverts are often barriers to fish migration and this should be considered when upgrading a stream crossing to reduce erosion.

5.3.1.3 Reducing or eliminating impacts of paved roads and impervious surfaces

The previous road treatments focused primarily on dirt and forest roads. Paved roads and impervious areas in urban and suburban areas can cause even more severe impacts to aquatic ecosystems through dramatic increases in frequency and magnitude of peak flows, channel incision, simplification of instream habitat, reduced water quality and biotic diversity, and lower biotic production (Walsh *et al.* 2005). Several studies have shown that when roads and impervious surfaces approach 10% of the total watershed area, significant negative impacts on aquatic habitat and biota occur (Paul & Meyer 2001). There is considerable debate as to whether reductions in biotic diversity are the result of alterations in hydrology, sediment supply or the pollutants contained within the water or sediment (water quality). It is generally believed to be a combination of these factors and that the driving factor varies from one watershed to another (Paul & Meyer 2001). Most restoration efforts that focus on runoff from paved roads and urbanization aim to improve or restore natural hydrology, balance sediment supply, and lower pollutant levels by reducing, controlling, and treating stormwater runoff.

The continued growth of urban areas and increased understanding of resulting negative impacts on streams has led to a number of engineering techniques to address these impacts through storing stormwater or delaying its arrival into the stream, reducing the amount of impervious surface and increasing infiltration rates, or filtering or otherwise treating stormwater to improve water quality. Approaches can be divided into two categories: (1) those that attempt to remove sediment and pollutants and (2) those that are designed to affect the timing and volume of runoff, though many attempt to do both.

Traditional approaches to stormwater management such as dry stormwater retention ponds, which store water only during storm events, were designed to handle 2-year flood events. These have often proven too small to appreciably reduce peak flows or to deal with the pollutants and sediment produced from roads, parking lots, and other impervious surface areas. Many new detention ponds are wet retention ponds that include a design feature called extended detention, designed to retain water for longer periods and thus delay runoff and allow sediments and pollutants to settle out (Fischenich & Allen 2000). The success of both dry and wet detention ponds at reducing runoff depends largely on their size, design, and level of impervious surface area and maintenance. Because of limited retention time, dry ponds are typically not effective at removing pollutants.

A detailed discussion of methods to reduce pollutants is beyond the scope of this chapter and book, but it is important to note that a number of devices have been designed to remove pollutants. These include infiltration basins, trenches and dry wells, and bioretention ponds or swales, which can be constructed adjacent to or in some cases under parking lots and roads (Fischenich & Allen 2000; Novotny *et al.* 2010). Many of these are effective at removing pollutants and often coupled with stormwater detention ponds to address both pollutants and runoff (Davis & McCuen 2005). As with stormwater retention ponds, considerable maintenance is required.

In recent years, several new techniques have been employed to reduce runoff and increase infiltration of both new and existing infrastructure. These include a variety of detention, retention, settling, or seepage basins or ponds; overflow wetlands; rainwater cisterns; high-flow bypass channels; alternative drainage systems or low-impact development (LID) techniques; narrower roads; replacing impervious surfaces with porous paving or pervious pavers; removing hard surfaces; and planting vegetation (Figure 5.4) (Sieker & Klein 1998; Riley 1998). A variety of 'green' alternatives such as bioswales, rain gardens, and green roofs have been implemented and some have resulted in dramatic reductions in stormwater runoff (USEPA 2010). Evaluations of retrofitting existing housing developments in large cities with rain gardens, reducing street widths or addition of vegetated swales (bioswales) have also seen storm flows reduced by 90% in some cases (Nassauer 1999). In rapidly urbanizing areas, the emphasis has been on utilizing these and other LID techniques in new construction to minimize impacts; in existing urban centers, the emphasis has been on retrofitting existing infrastructure in hopes of restoring

Figure 5.4 Example of (A) before and (B) after narrowing a paved residential street to reduce impervious surface area and creation of bioswales for water retention. Runoff was reduced by more than 90% following reduction of impervious surface and addition of bioswales and vegetation (photos courtesy of City of Seattle Public Utilities).

natural hydrologic processes or preventing further degradation of streams and riparian areas (Horner & May 1999; Novotny *et al.* 2010). Additional information on these techniques can be found in texts on stormwater management and urban design (e.g. Davis & McCuen 2005).

5.3.2 Reducing sediment and pollutants from agricultural lands

Land cultivation can lead to a large increase in the amount of fine sediment from erosion by wind or water (Waters 1995; Ministry of Environment 2001). In addition, the conversion of forest or grassland to agricultural land can affect hydrology, the impacts of which can vary by crop and season (Goudie 2000). This increase in sediment and runoff is not only damaging to aquatic systems, but costly

Table 5.5 Common agricultural practices for reduction of erosion from cultivated lands (modified from Waters 1995).

Practice	Method	Purpose	Remarks
Contour planting	Tilling and seeding in rows that follow contours	Reduce rill erosion, promote infiltration of water	Oldest method of erosion control
Terracing	Grading to create areas of level or reverse slope	Promote infiltration	Steep, grassed backslopes; most useful on steep slopes
Terrace outlets	Grassed ditches or underground culverts	Provide for removal of excess water from terrace	Size or width based on discharge from 10-year storm
Strip cropping	Alternating strips of row crops and forage	Reduction of field erosion, promote infiltration	Runoff velocity reduced greatly on grass strip
Crop rotation	Annual alternation of row crops and forage	Long-term reduction of erosion	Allows time for improvement of soil structure
Spring plowing	Substitution for fall plowing	Avoid bare soil in winter	Reduces wind erosion and waters erosion in spring snowmelt
Reduced row crop spacing	Planting rows closer	Reduce erosion of bare soils between rows	Added benefit in increased crop yield
Crop residues	Leaving stubble and other crop material instead of plowing under	Reduce exposure of disturbed soil, filters sediment in surface runoff	A major element of conservation tillage
Conservation tillage	Avoidance of deep plowing; minimal or no tillage; one-operation tilling and planting	Almost total avoidance of exposed, disturbed soil, increased infiltration	Many variations in shallow plowing; chisel and point plowing; combined with crop residues and reduced irrigation, fertilization, and pesticide use
Buffer strips	Leaving grass or riparian vegetation along stream or ditches on farmlands	Prevents erosion and filters sediment and pollutants in runoff	These range from very narrow (<1 m) to tens of meters. Effectiveness at improving water quality depends upon width, vegetation, and amount of sediment and pollutants

in terms of loss and degradation of soil and land for agricultural production. Considerable agricultural research has focused on how to cultivate crops in a manner that greatly reduces the loss of topsoil and transport of fine sediment and nutrients to streams.

Common strategies to reduce water-related erosion include crop selection and crop rotation, contour planting, terracing, conservation tillage (tilling that leaves a portion of last year's crop on the surface to reduce erosion), leaving crop residue, or the addition of organic matter to reduce erosion potential of soil (Waters 1995; Toy *et al.* 2002; Table 5.5). These are often coupled with supportive practices such as contouring or orientating ridges or crop rows perpendicular to the runoff and wind direction to reduce erosion (Toy *et al.* 2002). Silt fences, vegetative strips or riparian buffers, straw bales, and other barriers attempt to minimize or filter sediment delivered

from overland flow particularly when soils are bare. Buffers with mixed natural vegetation appear to be the most effective at reducing erosion and filtering nutrients and pollutants from agricultural runoff or subsurface flow (Sather & May 2008). The appropriate width of buffers to stabilize banks, minimize bank erosion, provide shade, and intercept and filter sediment and pollutants depends on the objectives and can range from tens to hundreds of meters. In general, buffers of 15 m or more are recommended as a minimum to meet water quality and fish habitat objectives (Fischer & Fischenich 2000). The path of surface and subsurface water flow, buffer width, vegetation type, depth of rooted zone, and runoff are all factors that influence the effectiveness of riparian buffers. Other approaches, such as settling ponds or terraces, grade reduction structures, control of head-cut erosion, lining the channel, or protecting stream banks, attempt to reduce erosion in channels draining agriculture lands.

Methods for reducing wind erosion of sediments include many of the techniques used for water erosion control above, with a focus on trying to provide vegetative, canopy, and ground cover to reduce fetch and dissipate wind energy. Windbreaks, hedgerows, forest, and other vegetative buffers are also commonly used and effective methods to reduce erosion from wind as well as runoff. Which techniques are most appropriate for reducing wind- or water-related sediment supply to streams from agricultural lands is highly dependent on the soils, topography, climate, and crop (Waters 1995; Ministry of Environment 2001). The best approaches for a particular area can be found by consulting with local soil conservation or agricultural support services.

5.3.3 Increasing sediment supply, retention and aggrading incised channels

While considerable emphasis is placed on reducing sediment supply that has increased due to human activities, some streams have become sediment starved due to human activities that reduce sediment supply and natural rates of bed or bank erosion (e.g. dams, bank armoring, gravel mining, torrent control) (Surian & Rinaldi 2003; Rosenfeld *et al.* 2010). This has led to loss of vertical and lateral connectivity and often results in channel incision. Channel incision can also result from changes in hydrology due to land use, channelization and changes in channel slope and hydraulics, reduction in sediment supply, or a combination of these factors. A detailed basin-scale analysis of sediment supply, erosion, hydrology, and hydrologic modifications is needed to determine whether the lack of sediment supply or incision is a basin-scale or local reach-scale problem, what is causing the problem, and which type of intervention might be most appropriate (CIRF 2010).

Approaches for dealing with channel incision caused by hydrologic impacts include stormwater management, reduction of impervious surfaces, and changes in agricultural and land-use practices discussed in the previous sections. Change in channel slope, shape and hydraulics to reduce sediment transport may include restoring meanders to straightened channels, widening the channel, reintroducing beaver, or adding weirs or other instream structures to reduce velocities, trap sediment, and prevent further incision. Common approaches for increasing sediment supply or restoring access to sediment sources include: removal of levees; bank armoring and other structures that limit channel mobility or erosion (see Section 5.2.2 on lateral connectivity); removal of dams, weirs, and torrent control structures that trap sediment; or cessation of gravel mining from the active channel or adjacent floodplain (Cowx & Welcomme 1998; CIRF 2010). These are the same approaches often used to restore lateral connectivity or floodplain habitats. The addition of gravels, discussed in the habitat improvement section, is also used as a mitigation measure to aggrade channels or restore sediment supply that has been interrupted by dams or weirs.

5.3.4 Increasing instream flows and flood pulses

The diversion and abstraction of flow from rivers and water bodies is a prime cause of loss and degradation of aquatic habitats throughout the world (Goudie 2000; see also Chapter 1). Adequate instream or base flows (also environmental or limiting flows), flood pulses, or flushing flows are needed to maintain aquatic and riparian habitat and production of aquatic ecosystems and biota (Stanford *et al.* 1996; Annear *et al.* 2002; Arthington & Pusey 2003). Increasing instream flows is largely a legal or regulatory issue that is most often addressed through enforcing new regulations, reaching new agreements with water users, or, if possible, purchasing or leasing water rights. In the United States and many other countries, instream flows have often been set to ensure adequate flows for downstream agriculture, industrial and urban users, transportation, or, in some cases, fish migration. Increasing or augmenting flows can be considered a form of passive rather than active restoration, as the assumption is that once the flows are increased the channel and biota will recover to function more naturally. Increasing instream flows on rivers with water diversions or 'rewatering' stream reaches where most or all water has been abstracted are methods known to reduce water temperatures, improve water quality, increase aquatic habitat diversity, and generally benefit biota (Weisberg & Burton 1993; Ban *et al.* 2011).

Flood pulses, flushing flows, and a varied hydrograph are important for maintaining healthy aquatic and riparian ecosystems (Poff *et al.* 1997). In stark contrast to the need to eliminate unnatural peak flows in many urban streams, the restoration of flood flows or flood pulses is an increasingly common strategy on highly regulated rivers. When dam removal is not feasible, restoring a more natural flood regime by controlled dam releases can improve a wide array of processes such as connectivity of the floodplain, sediment transport, and restoration of

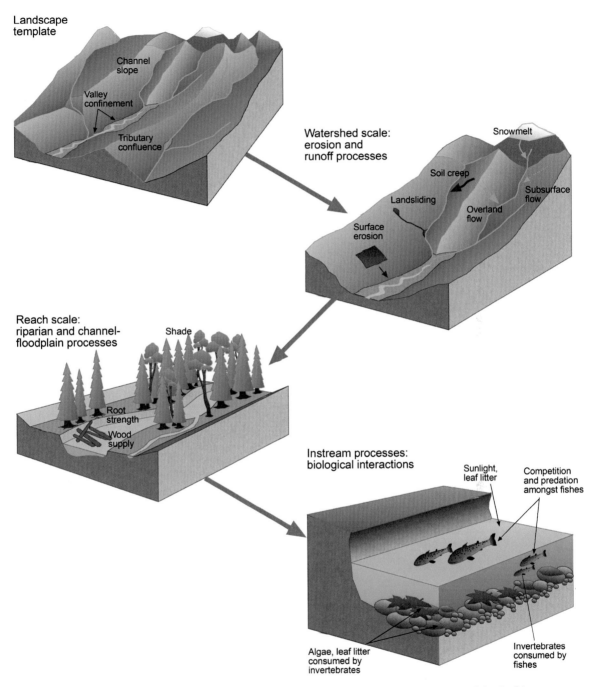

Plate 1 (Figure 2.2) Hierarchical nesting of processes controlling population and community responses of riverine biota. Higher-level controls set limits on the types of habitat features or ecosystem attributes that can be expressed at lower levels, and lower-level processes control the expression of attributes within those limits (based on Beechie *et al.* 2010).

Stream and Watershed Restoration: A Guide to Restoring Riverine Processes and Habitats, First Edition. Philip Roni and Tim Beechie.
© 2013 John Wiley & Sons, Ltd. Published 2013 by John Wiley & Sons, Ltd.

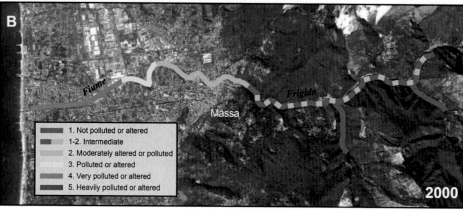

Plate 2 (Figure 3.22) Example of using multispecies metrics to display habitat and biological conditions within the watershed of the Fiume Frigido, Italy in (A) 1982 and (B) 2000. In 1982 the river was heavily impacted by sedimentation from marble quarries in the headwaters, as well as agriculture and development pressures in the lower watershed. Efforts to reduce sedimentation and other impacts have significantly improved biological and habitat status in the last two decades. (Adapted from Banchetti *et al.* 2004; CIRF 2006; background image copyright 2012 Google, 2012 TeleAtlas, 2012 Digital Globe.).

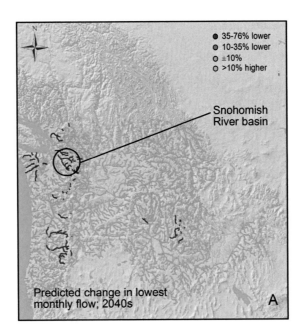

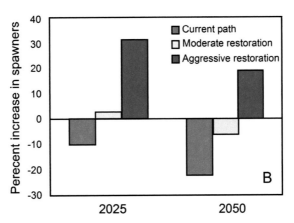

Plate 3 (Figure 3.24) Estimates of (A) change in lowest monthly stream flow by the 2040s in Pacific Northwestern USA (data from University of Washington Climate Impacts Group, Seattle, USA, http://www.hydro.washington.edu/2860/report/), and (B) change in Chinook salmon population size by 2025 and 2050 for the Snohomish River basin, Washington State, USA based on varying levels of restoration effort and a moderate CO_2 emissions scenario (adapted from Battin *et al.* 2007).

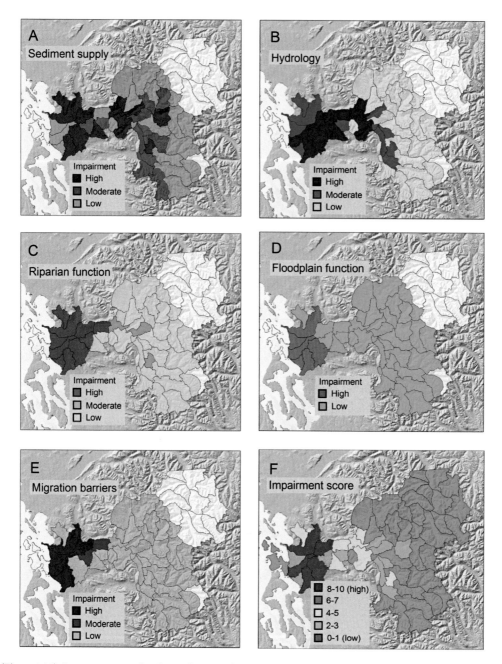

Plate 4 (Figure 3.25) Summary maps of analyses of causes of habitat change in the Skagit River basin, Washington State, USA (based on unpublished data from Skagit Watershed Council, Mount Vernon, Washington State, USA). Five panels illustrate altered watershed processes including (A) sediment supply; (B) hydrology; (C) riparian functions; (D) floodplain connectivity; and (E) amounts of habitat blocked to salmon migration by artificial barriers. (F) Impairment rating map illustrates the combined level of impairments across all five factors; each factor receives a score of 2 if impairment was rated high, a score of 1 of impairment was rated moderate, and a score of 0 if impairment was rated low. Score of 0 indicates no or low impairment to any of the five processes and a score of 10 indicates that all five processes are impaired.

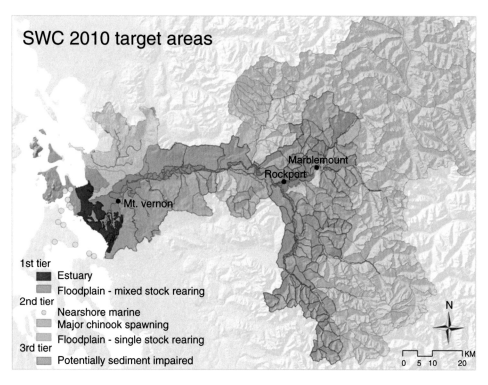

Plate 5 (Figure 3.26) Illustration of target restoration areas in the Skagit River basin, USA. Adapted from Skagit Watershed Council (2010). Descriptions of target areas are listed in Table 3.18.

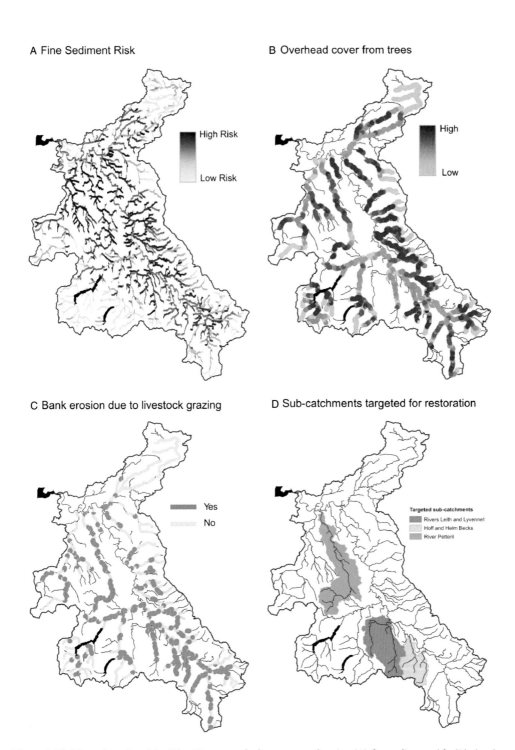

Plate 6 (Figure 3.28) Mapped results of the Eden River watershed assessment showing (A) fine sediment risk; (B) riparian overhead cover; (C) bank erosion due to livestock grazing; and (D) the three sub-catchments targeted for restoration efforts. See also Table 20 for description of targeted sub-basins shown in (D).

Plate 7 (Figure 5.1) Common approaches for restoring longitudinal and lateral connectivity including: (A) dam removal (Souhegan River, New Hampshire; from Pearson *et al.* 2011, reproduced by permission of American Geophysical Union, Copyright 2011 AGU); (B) a recently replaced culvert with a natural stream bottom (West Fork of Smith River tributary, Oregon, USA); (C) a restored (removal of bank protection) and (D) adjacent unrestored reaches of the River Aggar, Germany.

Plate 8 (Figure 5.2) Common types of structures used to pass fish at weirs and dams including: (A) a fish ramp Vara River, Italy (photo Enrico Pini-Prato); (B) a bypass channel for both fish passage and kayaking on Gave da Pau River, France (photo Enrico Pini-Prato); (C) pool and weir type pass at Lower Granite Dam, Snake River, Washington State, USA; and (D) a Denil steep pass at Bennett Dam, North Santiam River, Oregon (photo Ed Meyer).

Plate 9 (Figure 5.3) Complete removal of a forest road in Salmon Creek, Humboldt County, California: (A) before; (B) during; (C) immediately after; and (D) 3 years after removal of stream crossing and road fill. Note that the stream side-slopes were laid back to a stable grade that mimics natural hillslope topography and all bare slopes were mulched with rice straw. The excavated stream crossing has also been sloped to a stable angle and mulched to control surface erosion. Photos courtesy of California Department of Fish and Game Fisheries, Bureau of Land Management, Pacific Coast Fish, Wildlife & Wetlands Restoration Association, and Pacific Watershed Associates.

Plate 10 (Figure 5.6) Cartron Stream, County Sligo, Ireland (A) before and (B) three years after fencing (photos Martin O'Grady, Inland Fisheries Ireland) and Vernon Creek, Wisconsin (C) before and (D) 15 years after livestock exclusion (photos Ray J. White).

Plate 11 (Figure 5.7) Common instream habitat improvement techniques including (A) placing LWD in a small stream (photo M. Pearsons); (B) boulder weir on an Erriff River, Ireland (photo Dan Kircheis); (C) gravel and wood placement below a Howard Hansen Dam, Green River, Washington State, USA (photo Scott Pozarycki); and (D) engineered logjam Elwha River, Washington (photo M. McHenry).

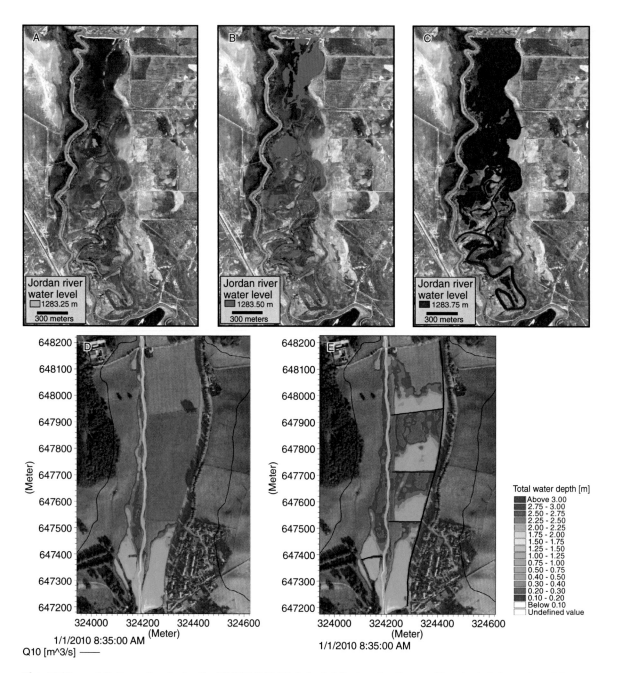

Plate 12 (Figure 7.6) Examples of using the 1D HEC-RAS (Hydrological Engineering Center – River Analysis System) model to predict inundation of wetlands at (A) low; (B) moderate; and (C) high flows as part of a playa wetland restoration project on the Jordan River, Utah, USA and the 2D MIKE21 model to predict flow depths, velocities, floodplain inundation patterns, and flood storage under existing and design scenarios for the Darnhall reach of the Edellston Water, UK. The 2D examples show (D) existing and (E) design configurations with the addition of cross-floodplain berms to increase flood storage in the design scenario. Figures (A–C) adapted from SWCA *et al.* (2008); Figures (D) and (E) courtesy of cbec, Inc.

riparian vegetation and associated processes (Rood & Mahoney 2000; Ellis *et al.* 2001; Stevens *et al.* 2001). Restoration of more natural flows, whether they are base flows or flood pulses, is critical for the success of many other restoration measures including riparian restoration and habitat improvement. Determining appropriate instream flows and flood pulses on regulated rivers is a complex process that includes not only detailed hydrologic analysis, but also economic and political discussions. It is important to consider the benefits and potential impacts to both natural resources and human infrastructure and population (Acreman 2000).

The understanding that the quantity, timing, and quality of flows is critical for sustaining key physical and ecological processes and ecosystem health has led to a move away from setting instream flows based on purely human needs or the needs of one or a few target species and towards setting 'environmental flows' needed to maintain and restore a variety of physical processes and ecological needs (Petts 2009). Determining appropriate environmental flows is site or river specific and is typically done by using sophisticated hydrodynamic and ecological models.

5.4 Riparian restoration strategies

Efforts to restore riparian vegetation and processes usually have multifaceted objectives such as increasing shade and reducing stream temperatures; restoring riparian forests and input of organic matter (e.g. large wood, leaf litter); reducing erosion and stabilizing stream banks; or improving water quality through filtration of fine sediment, nutrients, or pollutants from agriculture, roads, or urbanized landscapes. Because of the wide variety of land uses, restoration objectives, and riparian species present in different regions, a suite of different riparian restoration techniques have been developed. Regardless of technique, many riparian restoration projects are implemented on private land or on actively grazed or managed lands. It is therefore important that a long-term land-use agreement or contract with the land owner or owners is in place to ensure that the riparian area will remain protected once restored.

For ease of presentation and discussion, we divide riparian restoration into silviculture strategies, grazing management or animal exclusion, and riparian buffer protection strategies. Silviculture strategies seek to restore or improve riparian areas through replanting, thinning, or removal of trees and other vegetation. They are often paired with protection from further harvest or vegetation removal. Grazing management techniques are typically more passive and involve the removal or reduction of grazing pressure on riparian areas to allow vegetation to recover naturally, although they may be coupled with planting. Riparian buffer strips or zones include protecting a narrow buffer along streams to filter out sediment, pollutants, and runoff from agricultural, urban, or other land uses. They may be actively restored through replanting or passively left to regenerate naturally. Approaches for restoring floodplain areas and habitat connectivity, such as increasing instream flows, restoring seasonal floods, installing riparian large woody debris (LWD), or reintroducing beaver can also help restore riparian functions, but their primary objectives focus on restoring other processes. They are therefore treated in more detail in sections 5.2, 5.5 and 5.6. Nonetheless, they can be an important component of riparian restoration and are often implemented in conjunction with the riparian restoration techniques we discuss below.

5.4.1 Silviculture techniques

A variety of silvicultural strategies are used to actively restore riparian forests and improve riparian conditions. These include seeding, planting, and removal of trees, competing understory, or invasive plants (Pollock *et al.* 2005). These techniques can be divided based on whether the objective is to establish vegetation (planting, seeding, coppicing, staking, and layering) or remove or kill vegetation (thinning/harvest, girdling, competitive release, pruning, removal of invasive species with mechanical or herbicide application) (Table 5.6). Regardless of technique, riparian restoration involves initial site preparation, planting, and establishment – all of which need to be considered during project design.

5.4.1.1 Planting

The most widely used method for improving riparian habitats is to plant live trees, shrubs (potted or bare-root), live stakes or cuttings, forbes, grasses, or seeds obtained from a nursery or wild stock (Figure 5.5). Successful riparian planting and other silviculture treatments depend on a number of factors, many of which are specific to the geographic area and specific site being treated. Moreover, live plants, cuttings, or seeds of different species may require very different soils, water table levels, sun, shade, soil fungi, or rainfall patterns to germinate and thrive, and these must be considered regardless of what type of plantings are used. Information derived from a watershed assessment such as the current and

Table 5.6 Common techniques used in riparian rehabilitation. Pruning and harvest are also occasionally used to improve or restore riparian functions (modified from Pollock *et al.* 2005). Dam removal and restoration of floods, which also can improve riparian conditions, are discussed in sections 5.2.1 and 5.3.4.

Technique	Definition	Objectives/comments
Vegetation removal		
Thinning/ selective harvest	Cutting of a portion of the trees in a stand, usually the smaller, less vigorous trees	The goal of thinning is to increases the growth rate of remaining trees. Cut trees can be left on site to provide organic material to forest floor. Harvest is typically done to remove a portion of commercially valuable trees and can have positive or negative effects on riparian conditions
Girdling	Killing of trees by killing the cambial layer	Same as thinning, but trees are left standing to create snag habitat and (eventually) LWD
Competitive release	Killing of vegetation that is competing with desired species (e.g. hardwoods or shrubs competing with conifers)	Increases the growth rate of remaining trees. Can be labor intensive and needs to be repeated in riparian areas where growth rates are robust
Removal of invasive species	Using a variety of the above techniques or chemical or biological approaches to remove invasive exotic species	Prepare site for replanting or to improve growth and survival of desired species
Vegetation planting		
Planting	Placing live plants of target species into the ground	Standard technique for establishing plants. Cost of planting largely depends on size of plants. Site conditions, species and plant stock, protection from herbivores, watering, and planting depth can all influence success
Seeding	Planting of seeds of desired species	Establishes desired vegetation. Often unreliable, depending on species and weather
Coppicing	Regeneration from vegetative sprouts (stumps, limbs)	Easily establishes vegetation from stumps or limbs (usually certain deciduous trees) and creates increased shrub density or cover
Staking	Vertical insertion of live stems or branches partially into the ground to then take root	Used to quickly establish trees or bushes. In riparian settings, commonly used to establish willows and cottonwoods
Layering	Complete or partial horizontal burial of live stems that then take root	Useful for bank stabilization or other projects where rapid root growth is need. Also used to propagate some conifers that grow slowly
Grazing and passive riparian restoration		
Removal of grazing	Elimination of livestock from riparian area or entire watershed	To allow natural recovery of riparian area and stream channel, water quality and biota
Fencing	Placement of fencing perpendicular or adjacent to stream to exclude livestock from part or all of riparian area	To allow natural recovery of riparian area and stream channel, water quality and biota while allowing for livestock use outside of enclosure
Grazing system	Controlling the number and distribution of livestock, duration and/or season of grazing, or control of forage use	Minimize impacts to allow for recovery of riparian vegetation, stream channel, water quality, and biota while still allowing for livestock use
Riparian buffer protection	Protecting riparian areas or buffers and allowing natural recovery of riparian vegetation	Remove the human pressure to allow natural recovery; typically done by purchasing land or passing laws to protect riparian areas

Figure 5.5 Photos of (A) willow stakes to stabilize newly contoured bank following removal of bank armoring (Drau River, Austria) and (B, C) live plantings to restore riparian trees (Fraser River Valley, British Columbia, Canada). Note use of fencing to protect plantings from herbivores. Photos (B) and (C) M. Pearsons.

historic riparian, soil, and channel conditions as well as land use will be helpful to determine which riparian treatments and species are most appropriate.

The level of site preparation needed will depend on the specific goals and objectives of the project, riparian species to be used, and condition of the site. Site preparation is critical for plant survival and growth and may include removal of invasive species or other vegetation, grading, exposing soil, draining, irrigating, or adding fertilizer and other soil amendments. Site-specific factors such as soil type and condition, elevation, flood, low-flow and groundwater levels, competing and invasive species present, wildlife (herbivores such as deer and beaver), and disease or insects will also affect the success of planting and the appropriate species and plant stock (Massingill 2003).

Many species grow quickest in the open although others prefer shade, so one must carefully assess site conditions to understand which are most appropriate before deciding which species and types of plantings to use. Obtaining plants from either wild or nursery stock can be costly, but some species such as willows (*Salix spp.*), poplars (*Populus spp.*), and dogwood (*Cornus spp.*) are successfully established from cuttings. These cuttings can be used in staking and layering; although there are many methods using live stakes or layering cuttings, staking generally involves the vertical insertion of cuttings into the ground while layering generally involves layering or partially burying cuttings nearly horizontally on or near the soil surface. In cases where adequate seeds from native plant sources are not available to colonize a site naturally, seeding may be used following site preparation to establish desired riparian species.

With sufficient lead time (1–2 years), a local nursery or soil and water conservation organization can custom-grow native plant material from the site itself or a nearby site, usually at little or no additional cost. Local material will be best adapted to local conditions, which may result in higher survivorship and growth rates. The use of a nearby healthy riparian area with similar physical characteristics to the project site as a template or reference to help guide restoration efforts can often greatly increase success of riparian treatments. The use of appropriately sized plant stock is also a key consideration. Although it is often tempting to stretch a budget by planting larger numbers of smaller plants, survivorship of seedlings or small cuttings may be much lower.

The success of many riparian planting and other silviculture treatments depends on periodic maintenance (watering, invasive removal, protection from predators) until vegetation becomes well established. Considerable maintenance in the first 5–7 years after planting is often needed, which may include staking, pruning and replacing dead plants, irrigation, and additional invasive species removal (Massingill 2003). In addition, the protection of plantings from deer, voles, beaver, and other herbivores is often necessary to limit damage and mortality of plantings and improve growth of trees and other plantings (Sweeney et al. 2002; Pollock et al. 2005).

Studies evaluating riparian silviculture treatments have largely focused on short-term survival and growth of trees, shrubs and other plants and many have shown positive results (Pollock et al. 2005). The long-term goal of many silviculture treatments such as planting trees or removal of understory is to reduce water temperatures and improve instream habitat, particularly for fish. Because of the long lag time between treatment and changes in instream conditions or delivery of large wood or organic matter, only a few short-term studies have examined the response of fish or other instream biota to various riparian silviculture treatments; these have produced variable results depending on the region and treatment (Penczak 1995; Parkyn et al. 2003). In addition, most riparian treatments influence reach-scale conditions and processes, while in-channel conditions are often more affected by upstream or watershed-scale features which, if not addressed, will limit physical and biological response in the project area.

5.4.1.2 Thinning to promote tree and vegetation growth

Thinning or selective harvest is often used to promote more rapid growth of desirable species through increased light or reduced competition among trees and also to improve fish or wildlife habitat. Selective harvest usually involves removing a portion of the trees within the riparian forest and is often done as part of commercial forestry management to either remove trees that are commercially valuable or reduce tree density and promote rapid growth of commercially valuable species (Pollock et al. 2005). While thinning and selective harvest are sometimes used as an approach to improve or restore riparian conditions (e.g. establish late successional characteristics or increase stand diversity), it can reduce shade, inhibit the development of large snags, and reduce production of downed wood and recruitment of wood and organic matter to both the forest floor and stream channel.

Competitive release is similar to thinning or selective harvest, but often includes killing undesirable exotic or invasive understory species to prevent them from competing with desirable native or target species. This may include manual removal of species, mulching, or grubbing to dig up roots. Pruning can promote competitive release of desirable species and reduce windfall or fire in conifer forests that are managed for both ecological and commercial forestry production (Pollock et al. 2005). Pruning is sometimes used as a fisheries enhancement tool to decrease shade and increase light and production in small stream channels where dense hardwood shrub growth has led to 'tunneling' of streams and reduced primary and secondary production (O'Grady 2006).

Girdling involves removing the layer of bark and cambium around the circumference of the tree to induce mortality and result in standing dead trees (snags). It is used on large or small trees and more closely mimics the natural riparian processes of density-dependent mortality or fire or insect outbreak than thinning or selective harvest (Pollock et al. 2005). For example, girdling is often used to kill alder (Alnus spp.) in the Pacific Coast of the United States to encourage succession to conifer-dominated riparian stands (Pollock et al. 2005), conifers being the late successional species historically found in the region before extensive harvest of riparian forests.

5.4.1.3 Removal of exotic and invasive species

Many restoration efforts focus on controlling exotic or invasive species in the hope of restoring native vegetation and riparian conditions. The high rate of human and natural disturbance in riparian areas results in patches of exposed soil susceptible to colonization by many invasive, often exotic, plants (Pollock et al. 2005). Many of these exotic species pose little threat to riparian areas, but some are highly invasive and can form monocultures or become the dominant species excluding native species. For example, Tamarix spp., a native of Eurasia, has caused degradation of riparian habitat throughout the American Southwest, Australia and Argentina (Shafroth & Biggs 2008). Other problematic riparian invasive species common in North American and Europe include Japanese knotweed (Fallopia japonica), reed canary grass (Phalaris arundinacea), giant reed (Arundo donax), leafy spurge (Euphorbia esula), Himalayan blackberry (Rubus armeniacus), multiflora rose (Rosa multiflora), and many others. Oddly enough, some exotic species which are invasive outside their native range are the focus of restoration within their native range. For example, Spartina alterniflora (smooth cord grass) is the focus of restoration efforts in estuarine and saltmarsh habitats along the Atlantic coast of the United States, while S. alterniflora and its hybrids are considered invasive exotics in Europe and along the Pacific coast of the United States. Other exotic aquatic plants can choke stream channels and are often the subject of harvest or eradication, particularly in highly modified channels.

Control or removal of invasive riparian plants is frequently done before planting of native riparian vegetation as part of initial site preparation. Common methods of control include mechanical removal, chemical treatment, and biological or ecological manipulations (e.g. groundwater levels, increasing shade). Mechanical removal can include hand pulling, cutting or pulling with tools, mowing, grazing, covering, and bulldozing. Removal of aquatic or semiaquatic invasive species in managed landscapes through cutting or harvesting with machinery is also used to open up channels that have become choked with vegetation. For example, reed canary grass can completely block or choke small low-gradient streams, restricting flow and increasing flooding, and is often cut back to allow more natural stream flows. Application of herbicide is often done with a truck or backpack sprayer, but may include applying directly onto foliage or bark or injecting plant stems with a syringe (ACB 2003). Application of herbicides needs to be done accurately and with caution to avoid harming native flora and fauna or contaminate nearby streams. Both chemical and mechanical treatment can be very labor intensive and require repeated application before eradication or an acceptable level of mortality is achieved. Biological methods such as introduction of parasites or herbivores are sometimes used to control invasive species. For example, weevils, beetles, or moths have been used to control purple loostrife (*Lythrum salicaria*) and tansy ragwort (*Senecio jacobaeain*) in North America. Ecological approaches for invasive species control or removal may include manipulating groundwater or shade levels to create inhospitable conditions; for example, increasing shade either by planting or with shade cloth often controls reed canary grass.

The most appropriate techniques for invasive species control or removal are species and site specific and many local environmental agencies provide online guides, guidelines, and recommendations for control of different species (e.g. The Nature Conservancy: www.imapinvasives. org; Center for Invasive Plant Management: http://www. weedcenter.org/; UK Environment Agency: http://www. environment-agency.gov.uk/homeandleisure/wildlife/ 31350.aspx).

5.4.2 Fencing and grazing reduction

Intensive livestock grazing has led to degradation of riparian and stream habitats throughout the world (e.g. Platts 1991; NRC 1992; Robertson & Rowling 2000;

Jansen & Robertson 2001). Impacts on riparian areas and streams from excessive grazing include: changing community composition and reducing amounts or density of vegetation; destabilizing banks through denuding or hoof shear; increasing sediment; channel widening; and bed aggradation (Armour *et al.* 1991; Platts 1991; Belsky *et al.* 1999). Extensive grazing throughout a basin can also lead to changes in hydrology due to soil compaction and loss of vegetation and result in channel incision and lowering of the water table (Belsky *et al.* 1999). Grazing often affects fish and aquatic habitat by reducing shade, cover, water quality, terrestrial food supply, pool area and depth; altering stream morphology; and increasing water temperature (Armour *et al.* 1991; Platts 1991; Belsky *et al.* 1999). The objectives of riparian grazing management and the processes and habitats they restore can vary by site and region, but most focus on fencing to allow passive recovery of streamside vegetation which increases shade and reduces bank erosion, channel width, and stream temperature. In other cases, a combination of fencing, planting, and other riparian techniques are used.

Techniques for recovering riparian areas and streams impacted by grazing include complete cessation or removal of grazing, excluding livestock or other wildlife from part of the riparian zone with fences or other structures, or implementing a grazing management system that enables riparian vegetation to recover (Elmore 1992; Medina *et al.* 2005; Figure 5.6; Table 5.6). Excluding livestock through fencing is relatively straightforward, although many different materials (e.g. wood, barbed wire) and types of fences (single wire, multiple wire, electric) are used depending upon the site and animals present. In contrast, numerous different grazing management systems have been implemented throughout the world to reduce impacts to riparian areas, but all involve controlling the period and duration of grazing and the number of animals. For example, Platts (1991) described 17 different approaches used in the American west alone. Rest-rotation grazing management, which includes alternating periods of grazing and non-grazing, often rotated on multiple pastures, is one of the more common approaches used. Other approaches for limiting livestock grazing in riparian areas include providing water sources away from the stream (i.e. nose pumps) or providing salt away from the stream (Platts & Nelson 1985).

Excluding livestock may be followed by an increase in native ungulates, which can also negatively impact efforts for recovery of riparian areas and the stream channel (Medina *et al.* 2005). This can be exacerbated by the elimination of natural predators or implementation of

Figure 5.6 Cartron Stream, County Sligo, Ireland (A) before and (B) three years after fencing (photos Martin O'Grady, Inland Fisheries Ireland) and Vernon Creek, Wisconsin (C) before and (D) 15 years after livestock exclusion (photos Ray J. White). (See Colour Plate 10)

hunting restrictions. The recent reintroduction of wolves, which reduce native ungulate populations, led to natural control of grazing and the recovery of riparian areas in Yellowstone National Park (William & Beschta 2003). Limiting excessive grazing by native herbivores may be important for the recovery of riparian vegetation (Opperman & Merenlender 2000). Regardless of the technique, if grazing by native ungulates or domestic livestock is considerably reduced or completely eliminated, existing research suggests that riparian vegetation will recovery fairly quickly (5–10 years; Roni et al. 2008). However, because so many site-specific conditions come into play, recovery of a natural riparian forest and all functions may take many decades depending on the level of impacts from grazing, the forest type, and the geographic location. In heavily modified landscapes, exclusion of grazing can result in monocultures of invasive species (e.g. reed canary grass, Himalayan blackberry) if no active restoration is implemented concurrent with grazing removal.

Improvement in riparian vegetation cover and functions including increased shade, sediment storage, water storage, and aquifer recharge following livestock exclusion or dramatic reductions in grazing intensity have been documented in several studies (Elmore & Beschta 1987; Myers & Swanson 1995; Clary et al. 1996; Kauffman et al. 1997; Clary 1999; O'Grady et al. 2002; Carline & Walsh 2007). While fencing narrow buffers of 5–10 m wide can lead to recovery of some riparian processes, wider buffers are more effective at recovering the full suite of riparian and channel processes (Fischenich & Allen 2000). The success of rest-rotation grazing systems in inducing vegetation recovery is influenced by many factors including the number of days of grazing, season, seasonal livestock dispersal, and the level of compliance with the grazing system (whether livestock are moved as planned) (Myers 1989).

Most common riparian and livestock exclusion techniques, if implemented correctly, are effective at allowing passive riparian habitat recovery, although recovery of some riparian functions and instream habitat may lag far behind the recovery of vegetation. Response of aquatic biota to grazing removal or reduction appears to be less clear with only a few studies reporting increased fish abundance or macroinvertebrate diversity, although few studies have been long term or at a scale to measure population response (Rinne 1999; Roni et al. 2008). Factors limiting the success of grazing removal or reduction projects include the scale of the project, livestock densities, plant community, the width of riparian areas fenced, and the ability to address other management actions and upstream watershed processes that can effect sediment supply (Medina et al. 2005; Roni et al. 2008). Rest-rotation systems have shown promise under proper management and the right physical, morphological, and climatic conditions. The success of using fences for removal of grazing is dependent on periodic fence maintenance to ensure that livestock are excluded and the riparian area protected. Fencing and grazing projects are often implemented at a reach scale and the results are often localized, which can limit the physical and biological effectiveness of the restoration measures.

5.4.3 Riparian buffers and protection

A variety of other passive restoration approaches are used to protect riparian areas including leasing or purchasing land or rights/restrictions on its use(sometimes called conservation easements), or the implementation of regulations and laws that protect riparian areas and require leaving vegetated buffers along streams and other water bodies. In some instances, passive restoration (protection of riparian areas) may allow for sufficient recovery such that planting or other silviculture treatments are not necessary (Briggs 1996). Similar to riparian planting and grazing strategies, the success of passive restoration through protected riparian buffers depends on the disturbance being removed and ensuring that invasive species do not compete with native or desired vegetation. Moreover, the wider and longer the riparian area protected or restored, the more likely it is to see improvements in water temperature, water quality, macroinvertebrates and other biota (Roni et al. 2008). Periodic monitoring ensures that riparian buffers continue to recover from disturbance and that laws protecting these areas are being enforced (Lucchetti et al. 2005). While a detailed discussion of these regulatory or legal mechanisms is beyond the scope of this chapter or book, they are an important part of protecting and restoring watersheds. These passive approaches have been shown to be effective at reducing sediment, nutrient, and pesticide concentrations delivered to streams and improving bank stability and water quality (e.g. Osborne & Kovacic 1993; Barling & Moore 1994; Dosskey et al. 2005; Mayer et al. 2005; Puckett & Hughes 2005; Vought & Loucosiere 2010).

5.5 Habitat improvement and creation techniques

In many cases, direct intervention in the form of improving or creating new habitats is needed to create rapid changes in habitat or as part of the restoration processes discussed above. In fact, many restoration techniques developed over the last 75–100 years have been designed to improve and even create new riverine habitats that have been lost or degraded from human activities. Common habitat improvement techniques include various instream habitat techniques (e.g. log weirs, boulders structures, logjams), remeandering of straightened channels, and creation of new floodplain habitats including the construction of new side channels, ponds, wetlands, or other floodplain habitats (Figures 5.7 & 5.8; Table 5.7). We discuss instream habitat techniques and creation of floodplain habitat in Sections 5.5.1 and 5.5.2, respectively.

Because the restoration activities discussed in this section directly modify habitat rather than improve an underlying process (e.g. riparian, hydrology, sediment transport) and typically do not attempt to return habitats to some pre-disturbance state, we refer to them as habitat improvement or enhancement. Understanding watershed processes and reach characteristics are, however, critical to selecting and designing the correct habitat improvement techniques. When designed and implemented with considerations of current and historical site potential and watershed processes, these methods often result in relatively rapid improvements in habitat quality and quantity (physical area) and often produce increases in fish and other biota within a few years (Roni et al. 2002, 2008).

5.5.1 Instream habitat improvement techniques

The channelization of streams, removal of riparian vegetation and the removal of boulders and logs, most often carried out to facilitate navigation and transportation or to reduce flood risk, have greatly simplified stream channels and degraded instream habitat. This has been exacerbated by a suite of other human land-use activities in riparian

Figure 5.7 Common instream habitat improvement techniques including (A) placing LWD in a small stream (photo M. Pearsons); (B) boulder weir on an Erriff River, Ireland (photo Dan Kircheis); (C) gravel and wood placement below a Howard Hansen Dam, Green River, Washington State, USA (photo Scott Pozarycki); and (D) engineered logjam Elwha River, Washington (photo M. McHenry). (See Colour Plate 11)

and upland areas resulting in low habitat diversity, reduced cover, and degraded spawning and rearing habitat for many lotic fishes. Most instream habitat restoration techniques have been designed to mitigate these degraded conditions by improving or recreating some of the structure lost from both the historic and current human activities described above and in Chapter 2.

5.5.1.1 Structures to create pools, riffles, and cover and improve complexity

The placement of wood, boulders, gravel, and other structures such as check dams, deflectors and logjams into streams to create pools and riffles to alter channel morphology and provide cover and habitat for fish and other biota are among the oldest instream restoration techniques, with many methods pioneered in the early

20th century (see Section 1.5). Some of these same structures can be used to trap gravel or to encourage aggradation in channels that have been incised or scoured to bedrock, or to prevent further downcutting. Once the channel has aggraded, this is sometimes followed by the addition of logs and other natural materials to improve habitat complexity and cover (Roni *et al.* 2006). In some areas such as the UK, Ireland, and other parts of Europe rubble mats and other boulder and cobble placement are used to create riffle habitats, which are rare in many channelized or constrained channels (O'Grady 2006). Similarly, random or boulder clusters can be used to recreate the natural heterogeneity of streams where boulders were removed for transport of logs, as seen in many streams in Northern Europe and North America (Slaney & Zaldokas 1997; Roni *et al.* 2005; Rosenfeld *et al.* 2010).

Figure 5.8 Example of (A) remeandering of a formerly channelized stream (River Duhn, Germany); (B) construction of a groundwater channel to create spawning and rearing habitat, Skagit River, Washington State, USA (photo Sarah Morley); (C) reclaimed gravel pit (Rhine River, Germany); and (D) reconnected and reconstructed side channel (River Drau, Austria).

The placement of root wads, deflectors, brush bundles, and single or multiple log cover and boulder structures are generally designed to provide fish cover, but also create pools by concentrating flow (Roni *et al.* 2005; Table 5.7). In addition, pools are sometimes excavated as part of stream channel realignment or remeandering or excavated or blasted into bedrock in places where channels are scoured to bedrock (O'Grady 2006).

Most of these techniques are effective at improving or creating the desired habitat and can result in rapid changes in instream habitat conditions (Roni *et al.* 2002, 2008) as well as increased organic matter retention, habitat complexity, and flow heterogeneity (Jungwirth *et al.* 1995; Pretty *et al.* 2003). The ability of these techniques to improve physical habitat is partly determined by the durability or longevity and continuing interaction

with the active channel. The longevity and durability of the various techniques vary widely and are dependent on the structure type, design and construction, stream size and power, sediment characteristics, and whether they are designed to be fixed structures. Many fixed structures such as log weirs and deflectors were originally designed for use in small (<12 m bankfull width), low-gradient (<2%) streams. In recent years, techniques that mimic natural wood or boulder accumulations such as logjams or boulder clusters are used in place of permanent static structures. These more natural techniques require little maintenance and are less likely to 'fail' as they are not designed to be static; rather, they are intended to adjust position naturally as part of a dynamic stream setting. In contrast, traditional static habitat improvement structures such as weirs, gabions, deflectors, and other

Table 5.7 Common types of instream habitat improvement techniques (modified from Roni 2005).

Type of instream rehabilitation	Definition	Typical purpose
Log structures (e.g. weirs, sills, deflectors, logs, wing deflectors)	Placement of logs or log structures into active channel	Create pools and cover for fish, trap gravel, confine channel, or create spawning habitat
Logjams (multiple log structures, engineered logjams)	Multiple logs placed in active channel to form a debris dam and trap gravel	Create pools and holding and rearing areas for fish, trap sediment, prevent channel migration, and restore floodplain and side channels
Boulders structures (weirs, clusters, deflectors)	Single or multiple boulders placed in wetted channel	Create pools and cover for fish, trap gravel, confine channel, or create spawning habitat
Gabions	Wire mesh baskets filled with gravel and cobble	Trap gravel and create pools or spawning habitat
Rubble matts or boulder additions to create riffles	Addition of boulders and cobble to create riffles	Increase riffle diversity (velocity and depth), create shallow water habitat
Cover structures (lunker structures, rock, or log shelters)	Structures embedded in stream bank	Provide fish cover and prevent erosion
Brush bundles/root wads	Placement of woody material in pools or slow water areas	Provide cover for juvenile and adult fish, refuge from high flows, substrate for macroinvertebrates
Gravel additions and spawning pads	Addition of gravels or creation of riffles	Provide spawning habitat for fishes
Channel remeandering	Alter channel morphology by excavating new channel to restore meander patterns or return to historic channel	Restore meander patterns, increase habitat complexity and pool-riffle ratio, reduce channel width, reconnect floodplain

structures often require maintenance and have high failure rates, especially in higher-energy streams (Roni et al. 2002). Many failures of instream structures are due to a lack of understanding of larger hydrologic, geomorphic and geologic (erosion and deposition) factors that need to be considered during the design phase (see Chapter 7). To limit the mobility and maximize interaction with the channel and changes in habitat, the size of LWD and boulders should be scaled to channel size and stream power and the orientation (perpendicular to flow or angled downstream) considered in relation to natural wood and boulder accumulations.

Despite questions about their durability, the placement of instream structures has a long and popular history partly because a number of studies have demonstrated not only improvements in habitat but increases in fish numbers. Recent reviews and meta-analysis have consistently shown that placement of instream structures can lead to localized increases in fish production, particularly for salmonid fishes (Roni et al. 2002, 2008; Whiteway et al. 2010). In particular, large wood placement projects have been shown to lead to increases in local fish density, particularly for salmonids (Roni & Quinn 2001; Smoko-

rowski & Pratt 2007; Whiteway et al. 2010). Which species of fish respond positively to habitat improvement largely depends on the type of habitat created and the habitat preferences of the species in question (Roni et al. 2008). Species diversity can respond positively to increased pool habitat or habitat complexity (e.g. Shields et al. 1993, 1995; Jungwirth et al. 1995; Pretty et al. 2003) although, in many cases, factors other than habitat complexity such as water quality or productivity may limit species diversity (Roni et al. 2008).

Instream habitat improvement techniques are designed primarily to create or enhance fish habitat rather than to change macroinvertebrate abundance or diversity, and several existing studies show only relatively small changes in macroinvertebrate diversity due to instream structures (Roni et al. 2008; Miller et al. 2010). However, substantial increases in aquatic macroinvertebrate populations have been noted following enhancement programs in Ireland to stabilize banks (reduce erosion) and following installation of rubble mats to channelize streams (Lynch & Murray 1992; O'Grady et al. 2002). There is also recent evidence that constructed logjams in large river systems that typically lack wood can support a unique community of invertebrates that are often overlooked in lotic systems (i.e. meiofauna), but may be contributing substantially to overall biological diversity (Coe *et al.* 2009). There may be some short-term negative impacts on macroinvertebrates due to disturbance associated with construction of instream structures and other restoration techniques, but recolonization is very rapid (Roni *et al.* 2008). The differing response of macroinvertebrates to instream habitat improvement is not altogether surprising, given that the primary habitat improvement goal is often to enhance fish habitat rather than increase primary or secondary production.

A key factor to consider when designing and implementing instream habitat improvements is the intensity of treatment (number and complexity of structures per kilometer); those projects that produce the largest change in physical habitat produce the largest increases in fish abundance (Kennedy & Johnston 1986; Roni & Quinn 2001; Roni *et al.* 2006). The success or failure of particular techniques at improving habitat or increasing fish abundance is largely dependent on the geomorphic context (i.e. slope and confinement) and whether other larger factors or processes such as water quality or riparian conditions have been addressed.

5.5.1.2 Gravel addition and creation of spawning habitat

Adding gravel is most often used to temporarily improve or create spawning habitat for salmon, trout and other species that require gravel substrate for reproduction. It is particularly common where dams or other obstructions have cut off gravel transport and sediment supply, resulting in coarsening or armoring of the substrate. It is also sometimes used in areas where agriculture or other activities have led to increased fine sediment or where dredging or straightening of channels has removed or disconnected the channel from the floodplain and gravel deposits. Particle size is a key consideration in gravel placement. If stream slope (and power) is high relative to gravel size, placed gravel may be quickly transported downstream or out of the project area. Log or boulder retention structures are often used to trap gravel and reduce stream gradient and power in the study site or reach.

The addition of gravel or placement of structures to trap gravel has been shown to lead to suitable habitat for salmonids and other fishes and, in some cases, increases in number of redds (nests) or spawners (e.g. Merz *et al.* 2004; Roni *et al.* 2008; McManamay *et al.* 2010). Studies of gravel additions below dams suggest that salmon successfully reproduce in placed gravels,

although additional studies are needed (Merz *et al.* 2004). In addition, macroinvertebrates quickly colonize placed gravels and achieve density and diversity similar to native instream gravel deposits (Merz & Chan 2005). The success of gravel or structures to improve spawning habitat, spawning numbers and reproductive success of salmonids is less likely in areas with low velocities and high levels of fine sediment (Iversen *et al.* 1993; Roni *et al.* 2005). Other approaches to improving spawning habitat such as sediment traps and gravel cleaning have largely been abandoned because these areas quickly fill with fine sediment and thus provide little improvement in spawning habitat.

5.5.1.3 Recreating meanders

Restoring the sinuosity or *remeandering* straightened and narrowed channels is a common technique to restore channelized streams (Pess *et al.* 2005a; Vought & Lacoursiere 2010; Figure 5.8). Several large high-visibility restoration projects costing tens of millions of dollars or euros have been or are being implemented to remeander or reconnect short sections of rivers that were straightened and channelized only 50 years previously (e.g. Kissimmee River in the United States and Skerne, Skjern, Brede and Cole projects in the United Kingdom and Denmark; Brookes & Shields 1996; Pedersen *et al.* 2007). Similarly, 'daylighting' (exposing and remeandering streams that have been piped and buried under parking lots, residential developments or fields) is also an increasingly common technique, particularly in urban and agriculture areas (Nielsen 1996; Riley 1998; Vought & Lacoursiere 2010).

The objectives of remeandering channels are typically to: (1) restore the natural sinuosity and meander patterns to streams that have been straightened from human impacts; (2) create a new floodplain at a lower level; (3) increase stream length, habitat complexity and diversity; and (4) allow the river to migrate or move naturally across the floodplain. In contrast to techniques such as levee setbacks or removal of bank armoring, which remove constraints and largely allow the river to create its own meanders or side channels through natural hydrologic and geomorphic processes, remeandering is a highly engineered approach whereby a new sinuous channel is carefully designed and constructed. If adequate space is available, this may include widening and remeandering the straightened channel or diverting the straightened channel into an adjacent newly constructed meandering channel. In cases where the channel is deeply incised from the historic floodplain or where only a limited corridor is available to remeander the channel, a multi-stage or two-stage channel may be created at the new base level (Rosgen 1994; Cowx & Welcomme 1998; O'Grady 2006). In other cases where a stream has been straightened but is not deeply incised, rather than actual excavation of a new channel, a series of alternating deflectors can be used to create meanders (O'Grady 2006). Channel remeandering also typically includes the placement of boulders, wood and other instream structures to create habitat complexity and cover, as well as riparian replanting to restore the riparian area and protect and stabilize soil exposed during construction.

In some cases, passive restoration such as cessation of bank maintenance (vegetation cutting) and channel dredging can lead to natural recovery of meandering, although studies have generally shown that active channel remeandering produces more rapid results than passive approaches (Friberg et al. 1998). The ability and length of time required for a channel to naturally redevelop its former sinuosity and geometry is largely dependent on stream power, sediment load, and channel and bank stability (Brookes 1992). Natural channel recovery in some low-gradient low-energy channelized streams with stable hydrology and banks with cohesive soils could take centuries, while recovery in some higher-energy and hydrologically dynamic streams may be of the order years or decades (Brookes 1992).

Remeandering straightened or incised channels leads to dramatic increases in total stream length and habitat areas. For example, in a review of Danish stream rehabilitation, Iversen et al. (1993) reported increases in stream length ranging from 17% to more than 60% for five river meander reinstatement projects. Intensive evaluation of the Skjern, Brede, Cole, and other remeandering projects in Denmark and the UK demonstrate obvious improvement in habitat complexity and channel morphology, flood frequency, amount of water passing onto the floodplain, and nutrient retention as well as an increase in sediment deposition, sediment-associated phosphorous, and a decrease in erosion (Kronvang et al. 1998; Sear et al. 1998; Pedersen et al. 2007). Monitoring of other Danish remeandered rivers has demonstrated some small increases in macroinvertebrates, fish fauna, and aquatic vegetation (Iversen et al. 1993; Hansen 1996; Friberg et al. 1998). Improvements in both physical habitat and fish species diversity have also been reported from Austrian streams and in some streams in the northeastern United States (Jungwirth et al. 1995; Baldigo et al. 2010). In contrast, other studies in both North America and Europe have reported little fish response to remeandering (Moerke & Lamberti 2003; Cowx & Van Zyll De Jong

2004; Pedersen et al. 2007). The lack of response of fish and some other biota has largely been attributed to water quality and other broader or upstream problems that had not been addressed (Cowx & Van Zyll De Jong 2004), the stocking of fish or continuation of other management activities (Pedersen et al. 2007), or attempting to design a static meandering channel in a highly dynamic stream reach (Miller & Kochel 2009).

Designing channel remeander projects without considering the historic and current potential of the site, or current sediment and hydrologic regimes, can lead to total failure as can attempts to use standardized designs or channel types incorrectly or out of context (Kondolf 2006). The methodology discussed in Chapter 7 was developed partly to increase the success of channel remeandering and floodplain restoration projects. It is also important to realize that because remeandering projects involve disturbing the soil and recreating habitats, they typically incorporate many other techniques discussed in this chapter such as riparian planting, soft revetments (bank protection), and placement of instream structures.

5.5.2 Creation of floodplain habitats

Creation of new floodplain habitats is often used where side-channel, alcove, and pond habitats have been lost due to a variety of human activities. The most common habitat creation efforts on floodplains are constructed side channels and a variety of backwaters, off-channel ponds, and wetlands (Roni et al. 2005). Side channels are most often either groundwater or surface-water fed and typically designed to create spawning and rearing habitat for fishes. Surface-water channels are connected at both the upstream and downstream ends to the main river channel; their flow is therefore determined by conditions in the main river channel. However, they can be constructed with an intake structure or weir at the upstream end to control flow and sediment entering the channel. In contrast groundwater channels, which are generally only feasible to construct in floodplain areas with a high groundwater table, are connected to the main river channel only at the downstream end and have very stable flow and temperature. Because of differences in flow, temperature, and substrate size of surface versus groundwater channels, they typically provide habitat for different fish species (Pess et al. 2005a).

Creation and construction of off-channel ponds, wetlands, backwaters, or alcoves are designed to replace, recreate, or reclaim lost floodplain habitats. Construction methods include excavation, blasting, damming, or flow control structures on an existing perennial or intermittent stream, pond, or wetland (Cederholm *et al.* 1988; Slaney & Zaldokas 1997; Cowx & Welcomme 1998). The construction of alcoves or backwaters directly connected and adjacent to the river is another technique that is popular in both North America and Europe (Slaney & Zaldokas 1997; Cowx & Welcomme 1998). Regardless of how ponds or channels are constructed, it is important to ensure that these constructed habitats are consistently connected to the river throughout the year or at least seasonally.

The restructuring and connection of former gravel pits, borrow areas, millponds, and dredged ponds with the river channel provides opportunities for additional floodplain habitats (Norman 1998). These areas can provide important rearing habitats for a variety of fishes that prefer slow-water habitats, although they sometimes provide refuge for predators or invasive species (Roni *et al.* 2008). The successful conversion of gravel pits, millponds, and other constructed floodplain lakes depends on several factors including depth, morphology, availability of prey resources, cover, and water quality (Norman 1998; Roni *et al.* 2008). Assuming water temperature and quality are not issues, constructing shallow ponds (<2 m deep) with complex morphology and shoreline provide the best opportunities for successful use by juvenile fishes; large and deep ponds will provide less prey and pose a higher risk of main channel avulsion into the pond (Peterson 1982; Norman 1998).

As with reconnected habitats, the connectivity of constructed habitats to the main river channel plays a large role in the physical and biological effectiveness of these habitats and is a key design consideration. Much of the information on effectiveness of constructed habitats comes from studies on constructed floodplain ponds and wetlands, which have been shown to be effective at both providing habitat for juvenile salmonids and increasing their survival and abundance (Sheng *et al.* 1990; Raastad *et al.* 1993; Lister & Bengeyfield 1998; Solazzi *et al.* 2000; Giannico & Hinch 2003; Roni *et al.* 2006). Excavation of groundwater channels is a particularly popular and successful technique for creating spawning and rearing habitats for salmonid fishes (Bonnell 1991; Cowan 1991; Hall *et al.* 2000). These techniques are effective for many other species of fishes including, but not limited to, northern pike (*Esox lucius*) (Cott 2004), nase (*Chondrostoma nasus*), other rheophilic fishes (Chovanec *et al.* 2002) and age 0 coarse fishes (Langler & Smith 2001). Constructed side channels can also have higher diversity of habitats, substrates and macroinvertebrates and higher densities of fish than in mainstem river reaches

(Habersack & Nachtnebel 1995). These and other studies demonstrate that properly constructed floodplain habitats can provide important spawning and rearing areas for a variety of fishes. The effectiveness of constructed floodplain habitats is closely tied to their connection with the main channel and other floodplain habitats and their morphology, depth and complexity with shallower and more complex habitats often more productive (Roni *et al.* 2008).

5.5.3 When are habitat improvement techniques appropriate?

Despite their long history and widespread use, instream habitat improvement projects can be controversial because they often do not address the underlying cause of degradation. In cases where rapid improvement or creation of habitat is needed to recover important species or restoration of processes may not be possible due to land-use constraints, these techniques may be reasonable short-term options. Because they often produce rapid changes, habitat improvement techniques can be coupled with restoration of natural processes which may take longer to recover, thus providing a combination of short-term and long-term restoration strategies. An example of this is the Elwha River in the United States, where engineered logjams were placed below two large dams to improve habitat complexity and trap sediment that is being released during dam removal.

Clearly, the most appropriate technique or approach depends on the restoration objectives identified in the watershed analysis, the types of habitat improved or restored locally, the dynamics of the channel, and logistical and social constraints. Some instream habitat improvement techniques try to create static habitats in dynamic stream channels and, for this reason, require periodic maintenance or may only last a few decades and need to be repeated if underlying causes of habitat degradation have not recovered or been restored during the life of the project. In other cases, habitat improvement may be implemented as an interim measure until more comprehensive measures can be implemented.

5.6 Miscellaneous restoration techniques

A few common approaches for habitat restoration or improvement do not fit neatly into one category of restoration. These include beaver restoration, soft engineering approaches for bank stabilization, nutrient addition, and vegetation management.

5.6.1 Beaver restoration or control

Beaver were historically abundant throughout the northern hemisphere, but have been eliminated from a large part of their range or are at lower than historic population levels due to extensive harvest for the fur trade (Pollock *et al.* 2003). More recently their numbers have been controlled in some areas because of the damage their tree cutting and dam building causes to private property and infrastructure. Beavers are, however, key riparian species that provide many important ecological functions. They create pools, wetlands, ponds, and other floodplain features that are critical lentic habitats for fish, birds, amphibians, and other biota (Collen & Gibson 2000; Pollock *et al.* 2004). The removal of beaver has had severe impacts on streams and floodplains in some areas and has greatly reduced the amount of fish habitat (Pollock *et al.* 2003). Reintroduction of beaver is increasingly being used to restore floodplain habitats and riparian processes and increase fish numbers in Europe, Russia, Mongolia and North America. Beaver restoration can be active or passive: active efforts including not only reintroducing them but, in some cases, providing food sources (willow *Salix* spp. or cottonwood *Populus* spp. branches) or instream structures to provide the basis for dam construction. Passive approaches include restricting or banning commercial or recreational harvest and allowing beaver numbers to naturally recover. Studies in both the USA and Europe demonstrate that beaver rapidly recolonize and modify habitats following reintroduction as long as the animals are not harvested or consumed by predators (Zurowski & Kasperczyk 1988; Roni *et al.* 2008). Because beaver dams aggrade channels, reintroduction and protection of beaver is an effective strategy for aggrading incised stream channels and restoring floodplain habitat; in highly incised channels, however, the addition of wood and boulders may initially be needed to provide suitable dam sites for beaver (Michael Pollock, NOAA Fisheries, personal communication 2011).

Because they actively cut down trees and modify and flood habitats, beaver can dramatically alter riparian areas and their reintroduction in urban, suburban, and other populated areas can create considerable damage to parks, gardens, and infrastructure (e.g. roads and culverts). Similarly, beavers can negatively impact a riparian replanting project if plants are not protected. In cases where beaver have been introduced outside their natural range, such as in southern Argentina and Chile, they have caused considerable damage and removal or control of beaver is seen as a method of restoration. In other cases, the removal of beaver has been used as a fisheries management tool to eliminate ponds and improve the number of catchable trout (Avery 2004). While beaver reintroduction has many long-term benefits, several socioeconomic factors need to be considered when using them as a restoration strategy (Collen & Gibson 2000).

5.6.2 Bank stabilization

The hardening of banks with riprap, cement, and other hard structures is a major cause of habitat degradation and eliminating these structures is the focus of many restoration efforts (Cowx & Welcomme 1998; Schmetterling *et al.* 2001). While these hard engineering approaches may stabilize banks and protect infrastructure, they do not improve habitat or restore processes. In contrast, many restoration practitioners use soft or bioengineering approaches to stabilize and create more natural banks, particularly to assist in recovery of riparian vegetation in newly constructed projects with disturbed soils. For example, remeandering a channelized stream typically includes excavating an entirely new channel and banks and often bioengineering approaches using a combination of wood (e.g. logs, trees, branches, cuttings, fascines), rock, and fiber mat to protect the bank until riparian vegetation grows or naturally colonizes and stabilizes the bank (RSPB *et al.* 1994; Cowx & Welcomme 1998; O'Grady 2006). Similarly, a levee setback project may include pulling the levee back and recontouring the bank with vegetation and rock to create a sloped and stable bank for the new channel, or a fencing project in a heavily grazed meadow may include the placement of trees and brush along exposed banks to help stabilize the bank until vegetation recovers following grazing removal. Not all bank protection should therefore be thought of as detrimental to the system.

Most beneficial strategies for bank protection are bioengineering approaches designed to stabilize the bank and provide ecological benefits. Common approaches include placing root wads, trees or logs, live stakes, fascines, brush layering, fiber mats, coir logs, turf reinforcement mats, or other rolled erosion control products (Figure 5.9). The effectiveness of these structures at improving habitat depends partly on how similar they are in material to natural stream banks and the amount of wood and vegetation incorporated (Sudduth & Meyer 2006). Their effectiveness at bank stabilization varies greatly depending on local hydrologic, hydraulic, and other site conditions or other restoration measures. For example, Christmas tree revetments in Irish streams show high success rates in stabilizing banks and reducing

Figure 5.9 Examples of bioengineering approaches for bank protection using brush or logs. (A) Christmas tree revetments to help limit erosion while riparian zone recovers following fencing to exclude livestock on an Irish river immediately after placement and (B) 17 years later (photos Martin O'Grady, Inland Fisheries Ireland). (C, D) Bioengineering bank revetments using large woody debris (photos Herrera Inc.).

erosion when coupled with grazing reduction (98% success rate; O'Grady 2006). While bioengineering approaches are often an improvement over traditional hard armor bank protection and can be used as part of many restoration activities, it is important to realize that if they prevent the channel from moving they likely do little to restore natural channel processes. Additional information on the design of bank protection can be found in RSPB *et al.* (1994), Cowx & Welcomme (1998), FISRWG (1998), Fischenich & Allen (2000), O'Grady (2006), and other texts focusing on restoration design.

There is a gray area between bank stabilization to restore or improve habitat and hard armoring or bank protection designed to protect property and infrastructure. It is largely dependent on the project objectives, as both bioengineering approaches and bank protection may use rock or wood. If the main objective is rather to

protect property and infrastructure, it is probably not restoration or habitat improvement. In contrast, if one of the primary objectives is to stabilize the bank to minimize erosion and improve aquatic and riparian habitat, it is more likely habitat improvement or partial restoration.

5.6.3 Nutrient additions

In some oligotrophic waters in temperate and arctic latitudes, the addition of organic and inorganic nitrogen (N) or phosphorus (P) has been used as a method to increase primary and secondary productivity with the intention of increasing fish growth and abundance (Roni *et al.* 2008; Janetski *et al.* 2009). This would seem contrary to efforts in streams in most parts of the world where efforts to restore water quality focus on reducing nutrient inputs (N and P) from agricultural runoff (both fertilizers and animal waste), stormwater inputs and treated and untreated sewage (Kiffney *et al.* 2005). Moreover, techniques that focus on restoring sediment, hydrology, and riparian processes discussed earlier are designed to reduce sediment and nutrients transported from urban or rural areas. In many oligotrophic streams in the northern United States and Canada, the reduction in the number of returning salmon and other anadromous fishes has however led to a substantial reduction in the transport of marine-derived nutrients (N and P) to freshwater habitats (Gresh *et al.* 2000). It is estimated that reductions in returning Pacific salmon has reduced marine derived N and P delivered to streams in western Canada and the United States to less than 10% of historic levels (Gresh *et al.* 2000). Carcasses of Pacific salmon have been shown to both directly and indirectly provide food for macroinvertebrates, fishes, mammals, and birds and the nutrients from these carcasses increase primary and secondary production and growth of riparian vegetation (Krokhin 1975; Cederholm *et al.* 1999; Kiffney *et al.* 2005). The importance of other anadromous fishes such as smelt (*Osmeridae*), shad (*Alosa* spp.), Atlantic salmon (*Salmo salar*), and sea trout (*Salmo trutta*) provide similar benefits (Durbin *et al.* 1979; Garman & Macko 1998; Lyle & Elliott 1998).

In an effort to compensate for reductions in N and P historically transported to freshwater ecosystems by anadromous fishes, nutrient enrichment techniques have been developed to boost the productivity of streams and lakes. These include the addition of organic (salmon or other fish carcasses) and inorganic (typically N and P) nutrients (Ashley & Slaney 1997). It is important to emphasize that nutrient addition is an enhancement or mitigation technique designed to compensate for over-harvest or other reductions in anadromous fishes and to provide a temporary boost in productivity and hopefully increase fish production while other efforts are underway to address reductions in anadromous fish numbers (Ashley & Slaney 1997). If these factors are not addressed, any increases in productivity will require continued addition of nutrients (Guthrie & Peterman 1988; Gross *et al.* 1998). Moreover, additions of fish carcasses or inorganic nutrients often have relatively localized effects. Naturally reproducing anadromous fish carcasses are more effectively consumed, and naturally spawning salmon are more efficient at transferring nutrients to an aquatic ecosystem than inorganic nutrients or even artificially placed salmon carcasses (Claeson *et al.* 2006; Kohler *et al.* 2008; Shaff & Compton 2009).

The recent popularity of nutrient enrichment in North America has led to some new methods for applying these nutrients including 'fish blocks', carcass analogs or 'silver bullets'; these are blocks or pellets produced using fish feed or bone and flesh waste from fish processing plants to create time-release fertilizers specially designed for stream applications (Sterling *et al.* 2000; Wipfli *et al.* 2004; Kiffney *et al.* 2005; Pearsons *et al.* 2007). Regardless of the application or nutrient type, these approaches all attempt to increase nutrient levels, boost primary and secondary production, and hopefully increase fish production even without addressing other factors that may have led to reductions in returning anadromous fishes. Research has consistently demonstrated that adding organic or inorganic nutrients to oligotrophic streams is successful at increasing P, cholorophyll-a, periphyton and macroinvertebrate growth and numbers (Roni *et al.* 2008; Janetski *et al.* 2009). Studies in North America and the United Kingdom have reported increased growth of salmonid fishes (Roni *et al.* 2008; Williams *et al.* 2009b). This increased growth appears to lead to increased survival and abundance, but only a few studies have examined or reported this (Slaney *et al.* 2003).

Because the addition of nutrients can lead to degradation of water quality or to introduction of pathogens if organic nutrients are used (Compton *et al.* 2006), this technique needs to be applied with caution and only to oligotrophic streams whose production is limited by N or P. Determining whether nutrient addition is a suitable short-term method to enhance the productivity of a system is therefore critical. Criteria and specific analyses recommended to determine whether a stream is N or P limited and a candidate for nutrient addition include: assess status of fish stocks; determine whether physical habitat conditions are limiting fish production; examine current sources of organic or inorganic N and P; collect water samples to determine N and P levels; and estimate potential impacts of added nutrients on the water body including uptake rates (Slaney & Zaldokas 1997; Kiffney *et al.* 2005).

5.6.4 Vegetation management

While many restoration efforts focus on planting and recovering riparian vegetation, in some highly managed channels such as those on many agricultural lands where riparian forest has largely been removed, the periodic trimming or removal of riparian or aquatic vegetation is used to improve habitat for birds, fish, and wildlife, and sometimes to minimize potential for local flooding (RSPB *et al.* 1994; Cowx & Welcomme 1998). For example, cutting aquatic and riparian vegetation is used in some managed channels to improve flow, create fish habitat, and allow for fish migration (Roni *et al.* 2005). Similarly, in some small channelized streams completely covered with dense riparian vegetation, removal of brush and riparian vegetation is used to increase light, primary production, and fish numbers (O'Grady 2006). Many of these techniques are effective at improving conditions for specific species or conditions. However, they do not attempt to address the underlying problem that has led to excess vegetation or shade, but rather manage for specific habitats. They may therefore be appropriate for creating specific conditions in channels that are managed for multiple human uses, but are typically not appropriate for channels where the objective is to restore natural riparian and channel processes.

5.6.5 Other factors to consider when selecting restoration techniques

In addition to processes, habitat restored, or effectiveness, it is important to consider several other factors when selecting a technique. These include response time, design and maintenance, longevity, landowner cooperation, other restoration actions needed, and potential influence of climate change on the project. How quickly the action will lead to improvements in watershed processes or conditions and achieving objectives can be an important consideration (Table 5.8). The lag time between the completion of a restoration action and the changes in habitat conditions or watershed processes can take many years. For example, reducing landslides or sediment delivery from forest roads may reduce fine and coarse sediment delivery, but it may take several years for the channel to recover from the excess sediment that is already in the channel. A riparian fencing project may lead to increases in bank stability and shade within the first five years of project implementation, but a mature forest and delivery of LWD to a channel may take numerous decades. These factors need to be considered when planning projects and worked into setting appropriate and achievable goals and objectives.

The effectiveness of many restoration techniques is dependent on a variety of design considerations. Many of these considerations are the same regardless of the technique, including addressing the root cause of degradation, incorporating watershed and geomorphic processes into the design, and accounting for socioeconomic, political, or other logistical constraints. Key design and planning considerations are discussed in detail in Chapter 7 and key resources for design of different project types are

found in numerous texts often specific to the category of restoration technique (Table 5.1). It is important to realize that many projects such as riparian restoration or road improvements require periodic maintenance to ensure their effectiveness. Finally, since most techniques have not been thoroughly evaluated, implementation and design of rigorous monitoring and evaluation are needed to keep improving on existing and newly developing restoration techniques (Chapter 8).

While the processes and habitats are restored, response time and longevity are the all-important considerations when selecting a restoration technique; the 'human dimension' of restoration discussed in Chapter 4 also plays a major role in the types of restoration action that might be feasible. If landowners are unwilling to allow certain types of restoration actions on their land, this will not only greatly influence the selection of restoration actions but also design alternatives, which we discuss in the following chapters.

We addressed different types of actions separately, but it is unrealistic to think that only one type of project will be implemented at all locations. In reality, restoration in a particular reach or location usually includes multiple actions. For example, in reaches lacking habitat complexity and LWD, riparian replanting is often coupled with placement of instream log structures to provide both short-term improvements in habitat and a long-term source of large wood and habitat complexity. In fact, riparian planting or other restoration is typically used simultaneously with many other restoration actions including road removal or abandonment, instream restoration, habitat creation, restoration of lateral connectivity, and others. Restoration of a location may be intended to achieve multiple objectives and may therefore employ more than one technique. Coupling habitat improvements with restoration of processes is recommended in many cases to provide both relatively rapid improvements in habitat, while providing for long-term recovery of processes. This strategy has been used frequently when dealing with endangered species, where both short-term protection and improvement in habitat is needed to prevent further declines and long-term more holistic restoration is needed to maintain the species into the future.

A final factor to consider when selecting a restoration technique and prioritizing restoration actions is the potential influence of climate change. Most climate change models predict large changes in water temperature and runoff patterns (e.g. low flow, peak flows) for many streams (Arnell 1999). Some techniques such as riparian planting, reconnection of floodplain habitats

Table 5.8 Response time, longevity and maintenance needs for common restoration techniques discussed in this chapter. Maintenance is rated as high (H) if it is needed annually or throughout the life of the project, medium (M) if it is needed less than annually or just in initial years of project, or low (L) if little or no maintenance is required. Asterisk (*) indicates that the longevity of the project depends on periodic maintenance or the length of the agreement with landowners, water users, or natural resource agency. Some techniques have variable response times because some processes recover quickly (1–5 years), but others may take much longer (5–20 years).

Technique	Response time in years	Longevity in years	Maintenance
Dam removal	1–5, 5–20	>50	L
Culvert replacement	1–5	>50	M
Fish passage	1–5	>50*	M
Levee removal or setback	5–20	>50	M
Reconnection of floodplain habitats	1–5	>50*	M
Road removal	5–20	>50	L
Road resurfacing	5–20	10–50	M
Stabilization, upgrading stream crossings	5–20	10–50	M
Reduce impervious surface	1–5	>50	M
Instream flows	1–5	>50*	L
Agricultural practices	1–5, 5–20	10–50*	H
Restore sediment sources	1–5	>50	L
Riparian replanting	>50	>50	M
Thinning or removal of understory	5–20	>50	L
Removal or control of invasives	1–5	<10	M
Fencing	1–5, 5–20	>50*	M
Rest-rotation or grazing strategy	1–5, 5–20	>50*	H
Log or boulder structures	1–5	10–50	H
Natural LWD placement	1–5	10–50	L
Engineered logjams	1–5	>50	L
Brush or other cover	1–5	10–50	L
Gravel addition	1–5	10–50	L
Remeandering of straightened channel	1–5	>50	L
Creation of floodplain habitats	1–5	>50	L
Beaver reintroduction	1–5	10–50	L
Nutrient addition	1–5	<10*	H
Bank stabilization	1–5	10–50	L

and restoration of connectivity are known to reduce water temperatures or provide summer refuge for many cold water fishes (Roni *et al.* 2008). In contrast, methods such as placement of instream structures typically do not ameliorate impacts of climate change and, in fact, need to be planned and designed with consideration of expected changes in hydrology due to climate change. In stream reaches where climate impacts are expected to be substantial, techniques that ameliorate changes in flow and temperature should therefore be either given higher priority or implemented simultaneously with actions that improve habitat but do not necessarily directly influence temperature and hydrology.

5.7 Summary

Determining the appropriate watershed or stream restoration technique requires a clear understanding of the goal of the restoration program, the restoration objectives for a site or reach, and an understanding of which restoration technique or techniques can achieve those objectives. In this chapter we described restoration techniques and the processes and habitat they restore including techniques that restore connectivity (i.e. barrier removal, levee setbacks, fish passage), sediment and hydrologic processes (i.e. road removal or improvement, instream flows, agricultural practices), riparian conditions and processes (i.e. planting, thinning, fencing, grazing removal) and instream habitat conditions (log and boulder placement, channel remeandering, nutrient addition). Techniques that focus on restoring connectivity, sediment and hydrology, and to a lesser extent riparian conditions, are often used to restore watershed-scale and reach-scale processes, while habitat improvement techniques are often used to improve more localized conditions and processes. Instream habitat improvement and restoration of connectivity usually results in rapid improvements, while techniques such as road removal or riparian planting may take many years to detect benefits. These long-term and short-term strategies can be implemented simultaneously to provide initial benefits while setting the system on a path to long-term recovery. Most techniques are effective if designed properly and other watershed processes are addressed or incorporated into the design. Many restoration actions require periodic maintenance, particularly during the first 5 to 10 years after completion and occasionally throughout the life of the project, to ensure that they are leading to the desired benefits. Additional factors to consider when selecting appropriate restoration techniques include sociopolitical constraints such as land ownership (see Chapter 4), restoration response time, and longevity of the restoration action.

5.8 References

ACB (Alliance for Chesapeake Bay) (2003) *Citizen's guide to the control of invasive plants in wetland and riparian areas*. Alliance for Chesapeake Bay, Camp Hill, Pennsylvania.

Acreman, M. (2000) *Managed Flood Releases from Reservoirs: Issues and Guidance*. Center for Hydrology and Ecology, Wallingford, UK.

Ahearn, D.S., Viers, J.H., Mount, J.F. & Dahlgren, R.A. (2006) Priming the productivity pump: flood pulse driven trends in suspended algal biomass distribution across a restored floodplain. *Freshwater Biology* **51**, 1417–1433.

Annear, T., Chisholm, I., Beecher, H. *et al.* (2002) *Instream Flows for Riverine Resource Stewardship*. Instream Flow Council, Cheyenne, Wyoming.

Armour, C.L., Duff, D.A. & Elmore, W. (1991) The effects of livestock grazing on riparian and stream ecosystems. *Fisheries* **16**(1), 7–11.

Arnell, N.W. (1999) Climate change and global water resources. *Global and Environmental Change* **9**, 31–49.

Arthington, A.H. & Pusey, B.J. (2003) Flow restoration protection in Australian rivers. *River Research and Applications* **19**(5–6), 377–395.

Ashley, K.I. & Slaney, P.A. (1997) Accelerated recovery of stream, river and pond productivity by low level nutrient placement. In: Slaney, P.A. & Zaldokas, D. (eds) *Fish Habitat Rehabilitation Procedures*, Ministry of Environment, Lands and Parks, Vancouver, Canada, pp. 13.1–13.24.

Avery, E.L. (2004) *A Compendium of 58 Trout Stream Habitat Development Evaluations in Wisconsin – 1985–2000*. Wisconsin Department of Natural Resources, Bureau of Integrated Science Services, Waupaca, Wisconsin.

Bagley, S. (1998) *The Road–Rippers Guide to Wildland Road Removal*. Wildlands CPR, Missoula, Montana.

Baldigo, B.P., Ernst, A.G., Warren, D.R. & Miller, S.J. (2010) Variable responses of fish assemblages, habitat, and stability to natural-channel-design restoration in Catskill Mountain streams. *Transactions of the American Fisheries Society* **139**, 449–467.

Baker, R., Burt, L., Maxwell, R.S., Treat, V.H. & Dethloff, H.C. (1988) *Roads in the National Forests*. United States Department of Agriculture, Forest Service, Washington, DC.

Ban, X., Du, Y., Liu, H.Z. & Ling, F. (2011) Applying instream flow incremental method for the spawning habitat protection of Chinese sturgeon (*Acipenser sinensis*). *River Research and Applications* **27**, 87–98.

Barling, R.D. & Moore, I.D. (1994) Role of buffer strips in management of waterway pollution: a review. *Environmental Management* **18**, 543–558.

Beechie, T.J., Veldhuisen, C.N., Schuett-Hames, D.E., DeVries, P., Conrad, R.H. & Beamer, E.M. (2005) Monitoring treatments to reduce sediment and hydrologic effects from roads. In Roni, P. (ed.) *Monitoring Stream and Watershed Restoration*. American Fisheries Society, Bethesda, Maryland, pp. 35–66.

Belsky, A.J., Matzke, A. & Uselman, S. (1999) Survey of livestock influences on stream and riparian ecosystems in the western United States. *Journal of Soil and Water Conservation* **54**, 419–431.

Bernhardt, E.S., Palmer, M.A., Allan, J.D. *et al.* (2005) Synthesizing U.S. river restoration efforts. *Science* **308**, 636–637.

Bilby, R.E., Sullivan, K. & Duncan, S.H. (1989) The generation and fate of road-surface sediment in forested watersheds in southwestern Washington. *Forest Science* **35**, 453–468.

Bloom, A.L. (1998) *An Assessment of Road Removal and Erosion Control Treatment Effectiveness: A Comparison of 1997 Storm Erosion Response Between Treated and Untreated Roads in Redwood Creek Basin, Northwestern, California.* Master of Science in Environmental Systems: Geology. Humboldt State University, Arcata, California.

Bonnell, R.G. (1991) Construction, operation, and evaluation of groundwater-fed side channels for chum salmon in British Columbia. In: Colt, J. & White, R.J. (eds) *Fisheries Bioengineering Symposium*. American Fisheries Society, Bethesda, Maryland, pp. 109–124.

Briggs, M.K. (1996) *Riparian Ecosystem Recovery in Arid Lands: Strategies and References*. University of Arizona Press, Tuscon, Arizona.

Brookes, A. (1992) Recovery and restoration of some engineered British River Channels. In: Boon, P.J., Calow, P. & Petts, G.E. (eds) *River Conservation and Management*, John Wiley & Sons, Chichester, England, pp. 337–352.

Brooks, A. (2006) *Design Guideline for the Reintroduction of Wood into Australian Streams*. Land & Water Australia, Canberra.

Brookes, A. & Shields, F.D. (1996) *River Channel Restoration: Guiding Principles for Sustainable Projects*. John Wiley & Sons, Chichester, England.

Burdick, S.M. & Hightower, J.E. (2006) Distribution of spawning activity by anadromous fishes in an Atlantic slope drainage after removal of a low-head dam. *Transactions of the American Fisheries Society* **135**, 1290–1300.

Burroughs Jr., E.R. & King, J.G. (1989) *Reduction of Soil Erosion on Forest Roads*. United States Department of Agriculture, Forest Service, Intermountain Research Station, General Technical Report INT-264, Ogden, Utah.

Carline, R.F. & Walsh, M.C. (2007) Responses to riparian restoration in the Spring Creek watershed, central Pennsylvania. *Restoration Ecology* **15**(4), 731–742.

Cederholm, C.J., Scarlett, W.J. & Peterson, N.P. (1988) Low-cost enhancement technique for winter habitat of juvenile coho salmon. *North American Journal of Fisheries Management* **8**, 438–441.

Cederholm, C.J., Kunze, M.D., Murota, T. & Sibatani, A. (1999) Pacific salmon carcasses: Essential contributions of nutrients and energy for aquatic and terrestrial ecosystems. *Fisheries* **24**(10), 6–15.

Cheng, F. & Granata, T. (2007) Sediment transport and channel adjustments associated with dam removal: field observations. *Water Resources Research* **43**, W03444.

Childers, D., Kresch, D.L., Gustafson, S.A., Randle, T.A., Melena, J.T. & Cluer, B. (2000) *Hydrologic Data Collected during the 1994 Lake Mills Drawdown Experiment, Elwha River, Washington*. United States Geological Survey, Water Resources Investigations Report 99-4215, Tacoma, Washington.

Chovanec, A., Schiemer, F., Waidbacher, H. & Spolwind, R. (2002) Rehabilitation of a heavily modified river section of the Danube in Vienna (Austria): biological assessment of landscape linkages on different scales. *International Review of Hydrobiology* **87**, 2–3.

CIRF (Centro Italiano per la Riqualificazione Fluviale) (2010) *Riqualificazione Fluviale Numero 3.2/2010*. Centro Italiano per la Riqualificazione Fluviale, Mestre, Italy. Available at: www.cirf.org.

Claeson, S.M., Li, J.L., Compton, J.E. & Bisson, P.A. (2006) Response of nutrients, biofilm, and benthic insects to salmon carcass addition. *Canadian Journal of Fisheries and Aquatic Sciences* **63**, 1230–1241.

Clary, W.P. (1999) Stream channel and vegetation responses to late spring cattle grazing. *Journal of Range Management* **52**(3), 218–217.

Clary, W.P., Shaw, N.L., Dudley, J.G., Saab, V.A., Kinney, J.W. & Smithman, L.C. (1996) *Response of a Depleted Sagebrush Steppe Riparian System to Grazing Control and Woody Plantings*. INT-RP-492. United States Department of Agriculture Forest Service, Intermountain Research Station, Ogden, Utah.

Clay, C.H. (1995) *Design of Fishways and Other Fish Facilities*. Lewis Publishers, Boca Raton, Florida.

Coe, H.J., Kiffney, P.M, Pess, G., Kloehn, K & McHenry, M. (2009) Periphyton and invertebrate response to wood placement in large Pacific coastal rivers. *River Research and Applications* **25**, 1025–1035.

Collen, P. & Gibson, R.J. (2000) The general ecology of beavers (*Castor spp.*), as related to their influence on stream ecosystems and riparian habitats, and the subsequent effects on fish – a review. *Reviews in Fish Biology and Fisheries* **10**(4), 439–461.

Collins, B.D., Montgomery, D.R. & Sheikh, A.J. (2003) Reconstructing the historic riverine landscape of the Puget Lowland. In: Montgomery, D.R., Bolton, S.M., Booth, D.B. & Wall, L. (eds) *Restoration of Puget Sound Rivers*. University of Washington Press, Seattle, pp. 79–128.

Compton, J.E., Andersen, C.P., Phillips, D.L. *et al.* (2006) Ecological and water quality consequences of nutrient addition for salmon restoration in the Pacific Northwest. *Frontiers in Ecology and the Environment* **4**, 18–26.

Cott, P.A. (2004) Northern pike (*Esox lucius*) habitat enhancement in the Northwest Territories. *Canadian Technical Report of Fisheries Aquatic Sciences No. 2528.* Department of Fisheries and Oceans Canada, Yellowknife, Northwest Territories, Canada.

Cotts, N.R., Redente, E.F. & Schiller, R. (1991) Restoration methods for abandoned roads at lower elevations in Grand-Teton National-Park, Wyoming. *Arid Soil Research and Rehabilitation/Arid Soil Research and Rehabilitation* 5(4), 235–249.

Cowan, L. (1991) Physical characteristics and intragravel survival of chum salmon in developed and natural groundwater channels in Washington. In: Colt, J. & White, R.J. (eds) *Fisheries Bioengineering Symposium.* American Fisheries Society, Bethesda, Maryland, pp. 125–131.

Cowx, I.G. & Welcomme, R.L. (1998) *Rehabilitation of Rivers for Fish.* Food and Agriculture Organization of the United Nations, Fishing News Books, Oxford, UK.

Cowx, I.G., & Van Zyll De Jong, M. (2004) Rehabilitation of freshwater fisheries: tales of the unexpected? *Fisheries Management and Ecology* 11(3–4), 243–249.

Daniels, B., McAvoy, D., Kuhns, M., & Gropp, R. (2004) *Managing Forests for Water Quality: Forest roads.* NR/FF/010, Utah State University, Logan, Utah.

Davis, A.P. & McCuen, R.H. (2005) *Stormwater Management for Smart Growth.* Springer, New York.

Dosskey, M.G., Eisenhauer, D.E. & Helmers, M.J. (2005) Establishing conservation buffers using precision information. *Journal of Soil and Water Conservation* 60, 349–353.

Doyle, M.W., Stanley, E.H., Orr, C.O., Selle, A.R., Sethi, S.A. & Harbor, J.M. (2005) Stream ecosystem response to small dam removal: lessons from the heartland. *River Research and Applications* 71, 227–244.

Durbin, A.G., Nixon, S.W. & Oviatt, C.A. (1979) Effects of the spawning migration of the Alewife, *Alosa Pseudoharengus*, on freshwater ecosystems. *Ecology* 60(1), 8–17.

DVWK (Deutscher Verband für Wasserwirtschaft und Kulturbau e.V.) (2002) *Fish Passes, Design, Dimensions and Monitoring.* Food and Agriculture Organization of the United Nations, Rome, Italy.

Ellis, L.M., Crawford, C.S. & Molles Jr., M.C. (2001) Influence of annual flooding on terrestrial arthropod assemblages of a Rio Grande riparian forest. *Regulated Rivers: Research & Management* 17(1), 1–20.

Elmore, W. (1992) Riparian responses to grazing practices. In: Naiman, R.J. (ed.) *Watershed Management,* Springer-Verlag, New York. pp. 442–457.

Elmore, W. & Beschta, R.L. (1987) Riparian areas: Perceptions in management. *Rangelands* 9(6), 260–265.

Elseroad, A.C., Fule, P.Z. & Covington, W.W. (2003) Forest road revegetation: effects of seeding and soil amendments. *Ecological Restoration* 21(3), 180–185.

Fischenich, J.C. & Allen, H. (2000) *Stream Management.* United States Army Corps of Engineers, Water Operations Technical Support Program, Report ERDC/EL SR-W-00-1. Fort Worth, Texas.

Fischer, R.A. & Fischenich, J.C. (2000) *Design Recommendations for Riparian Corridors and Vegetated Buffer Strips.* Ecosystem Management and Restoration Research Program Technical Note EMRRP-SR-24, United States Army Corps of Engineers, Fort Worth, Texas. Available at: http://el.erdc.usace.army.mil/elpubs/pdf/sr24.pdf.

FISRWG (Federal Interagency Stream Restoration Working Group) (1998) *Stream Corridor Restoration: Principles, Processes, and Practices.* GPO Item No. 0120-A, United States Department of Agriculture, Washington, DC.

Florsheim, J.L. & Mount, J.F. (2002) Restoration of floodplain topography by sand splay complex formation in response to intentional levee breaches, lower Cosumnes River, California. *Geomorphology* 44(1–2), 67–94.

Friberg, N., Kronvang, B., Hansen, H.O. & Svendsen, L.M. (1998) Long-term, habitat-specific response of a macroinvertebrate community to river restoration. *Aquatic Conservation: Marine and Freshwater Ecosystems* 8(1), 87–99.

Fullerton, A.H., Burnett, K.M., Steel, E.A. *et al.* (2010) Hydrological connectivity for riverine fish: measurement challenges and research opportunities. *Freshwater Biology* 55, 2215–2237.

Furniss, M.J., Roelofs, T.D. & Yee, C.S (1991) Road construction and maintenance. In: Meehan, W.R. (ed.) *Influences of Forest and Rangeland Management on Salmonid Fishes and Their Habitats, Special Publication 19.* American Fisheries Society, Bethesda, Maryland, pp. 297–324.

Furniss, M.J., Love, M. & Flanagan, S.A. (1997) *Diversion Potential at Road-Stream Crossings.* Report 9777 1806-SDTDC, United States Department of Agriculture Forest Service, Pacific Southwest Region, San Dimas, California.

Garman, G.C. & Macko, S.A. (1998) Contribution of marine-derived organic matter to an Atlantic coast, freshwater tidal stream by anadromous clupeid fishes. *Journal of North American Benthological Society* 17, 277–285.

Giannico, G.R. & Hinch, S.G. (2003) The effect of wood and temperature on juvenile coho salmon winter movement, growth, density and survival in side-channels. *River Research and Applications* **19**(3), 219–231.

Goudie, A. (2000) *The Human Impact on the Natural Environment*. MIT Press, Cambridge, Massachusetts.

Gresh, T., Lichatowich, J. & Schoonmaker, P. (2000) An estimation of historic and current levels of salmon production in the Northeast Pacific ecosystem: Evidence of a nutrient deficit in the freshwater systems of the Pacific Northwest. *Fisheries* **25**(1), 15–21.

Grift, R.E., Buijse, A.D., Densen, W. & Klein Breteler, J.G.P. (2001) Restoration of the river-floodplain interaction: benefits for the community in the River Rhine. *Archiv fuer Hydrobiologie Supplement, Large Rivers* **12**(2–4), 173–185.

Gross, H.P., Wurtsbaugh, W.A. & Luecke, C. (1998) The role of anadromous sockeye salmon in the nutrient loading and productivity of Redfish Lake, Idaho. *Transaction of American Fisheries Society* **127**(1), 1–18.

Gucinski, H., Furniss, M., Ziemer, R. & Brookes, M.H. (2001) *Forest Roads: A Synthesis of Scientific Information*. General Technical Report PNW-GTR-509 United States Department of Agriculture Forest Service, Portland, Oregon.

Guthrie, I.C. & Peterman, R.M. (1988) Economic evaluation of lake enrichment strategies for British Columbia sockeye salmon. *North American Journal of Fisheries Management* **8**, 442–454.

Habersack, H. & Nachtnebel, H.P. (1995) Short-term effects of local river restoration on morphology, flow field, substrate and biota. *Regulated Rivers: Research and Management* **10**, 291–301.

Hall, J.L., Timm, R.K. & Wissmar, R.C. (2000) Physical and biotic factors affecting use of riparian ponds by sockeye salmon (*Oncorhynchus nerka*). In: Wigington, P.J. & Beschta, R.L. (eds). *Proceedings of the American Water Resources Association International Conference on Riparian Ecology and Management in Multi-use Land Watersheds*. American Water Resources Association, Herndon, Virginia, pp. 89–94.

Hansen, H.O. (1996) *River Restoration – Danish Experience and Examples*. National Environmental Research Institute, Silkeborg, Denmark.

Hart, D.D., Johnson, T.E., Bushaw-Newton, K.L. *et al.* (2002) Dam removal: challenges and opportunities for ecological research and river restoration. *BioScience* **52**, 669–682.

Hickenbottom, J. (2000) *A Comparative Analysis of Surface Erosion and Water Runoff from Existing and Recontoured Forest Roads: O'Brien Creek Watershed, Lolo National Forest, Montana*. Masters of Science. University of Montana, Missoula, Montana.

Horner, R.R. & May, C.W. (1999) Regional study supports natural land cover protection as leading best management practice for maintaining stream ecological integrity. In: New Zealand Water and Wastes Association (ed.), *Comprehensive Stormwater & Aquatic Ecosystem Management*. The New Zealand Water and Wastes Association Inc., New Zealand, pp. 233–247.

Hunt, R.L. (1993) *Trout Stream Therapy*. University of Wisconsin Press, Madison, Wisconsin.

Hunter, C.J. (1991) *Better Trout Habitat: A Guide to Stream Restoration and Management*. Washington DC, Island Press.

Iversen, T.M., Kronvang, B., Madsen, B.L., Markmann, P. & Nielsen, M.B. (1993) Re-establishment of Danish streams: Restoration and maintenance measures. *Aquatic Conservation: Marine and Freshwater Ecosystems* **3**(2), 73–92.

Izaak Walton League (2002) A Handbook for Stream Enhancement & Stewardship. The Izaak Walton League of America, Gaithersberg, Maryland.

Janetski, D.J., Chaloner, D.T., Tiegs, S.D. & Lamberti, G.A. (2009) Pacific salmon effects on stream ecosystems: a quantitative synthesis. *Oecologia* **159**, 583–595.

Jansen, A. & Robertson, I.A. (2001) Relationships between livestock management and the ecological condition of riparian habitats along an Australian floodplain river. *Journal of Applied Ecology* **38**, 63–75.

Jansson, R., Nilsson, C. & Malmqvist, B. (2007) Restoring freshwater ecosystems in riverine landscapes: the roles of connectivity and recovery processes. *Freshwater Biology* **52**, 589–596.

Jungwirth, M., Muhar, S. & Schmutz, S. (1995) The effects of recreated instream and ecotone structures on the fish fauna of an epipotamal river. *Hydrobiologia* **303**, 195–206.

Jungwirth, M., Muhar, S. & Schmutz, S. (2002) Reestablishing and assessing ecological integrity in riverine landscapes. *Freshwater Biology* **47**, 867–887.

Junk, W.J., Bayley, P.B. & Sparks, R.E. (1989) The flood pulse concept in river-floodplain systems. In: Dodge, D.P. (ed.) *Proceedings of the International Large River Symposium*. Canadian Special Publications of Fisheries and Aquatic Sciences 106, pp. 110–127.

Kauffman, J.B., Bechta, R.L., Otting, N. & Lytjen, D. (1997) An ecological perspective of riparian and stream restoration in the western United States. *Fisheries* **22**(5), 12–24.

Kennedy, G.J.A. & Johnston, P.M. (1986) A review of salmon (*Salmo salar L.*) research on the River Bush. In: Crozier, W.W. & Johnson, P.M. (eds) *Proceedings of the 17th Annual Study Course*. Institute of Fisheries Management, University of Ulster at Coleraine, Coleraine, United Kingdom, pp. 49–69.

Kiffney, P.M., Bilby, R.E. & Sanderson, B.L. (2005) Monitoring the effects of nutrient enrichment to freshwater ecosystems. In: Roni, P. (ed.) *Monitoring stream and watershed restoration*, American Fisheries Society, Bethesda, Maryland, pp. 237–266.

Kochenderfer, J.N. & Helvey, J.D. (1987) Using gravel to reduce soil losses from minimum-standard forest roads. *Journal of Soil and Water Conservation* **42**, 46–50.

Kohler, A.E., Rugenski, A. & Taki, D. (2008) Stream food web response to salmon carcass analogue addition in two central Idaho, USA streams. *Freshwater Biology* **53**, 446–460.

Kondolf, G.M. (2006) River restoration and meanders. *Ecology and Society* **11**, 42. Available at: http://www.ecologyandsociety.org/vol11/iss2/art42/

Konrad, C.P., Black, R.W., Voss, F. & Neale, C.M.U. (2008) Integrating remotely acquired field data to assess effect of setback levees on riparian and aquatic habitat in glacial-melt water rivers. *River Research and Applications* **24**, 355–372.

Krokhin, E.M. (1975) Transport of nutrients by salmon migrating from the sea into lakes. In: Hasler, A.D. (ed.) *Coupling of Land and Water Systems*, Springer-Verlag, New York, pp. 153–156.

Kronvang, B., Svendsen, L.M., Brookes, A. *et al.* (1998) Restoration of the rivers Brede, Cole and Skerne: a joint Danish and British EU-LIFE demonstration project, III – Channel morphology, hydrodynamics and transport of sediment and nutrients. *Aquatic Conservation-Marine and Freshwater Ecosystems* **8**(1), 209–222.

Langler, G.J. & Smith, C. (2001) Effects of habitat enhancement on 0-group fishes in a lowland river. *Regulated Rivers – Research & Management* **17**, 677–686.

Lister, D.B. & Finnigan, R.J. (1997) Rehabilitating off-channel habitat. In: Slaney, P.A. & Zaldokas, D. (eds) *Fish Habitat Rehabilitation Procedures. Watershed Restoration Technical Circular No. 9*, Watershed Restoration Program, Ministry of Environment, Lands and Parks and Ministry of Forests, Victoria, British Columbia, Canada, pp. 7.1–7.29.

Lister, D.B. & Bengeyfield, W. (1998) An assessment of compensatory fish habitat at five sites in the Thompson River system. *Canadian Manuscript Report of Fisheries and Aquatic Sciences* No. 2444, Department of Fisheries and Oceans, Vancouver, Canada.

Lucchetti, G., Richter, K.O. & Schaefer, R.E. (2005) Monitoring acquisitions and conservation easements. In: P. Roni (ed.) *Monitoring Stream and Watershed Restoration*, American Fisheries Society, Bethesda, Maryland, pp. 277–312.

Luce, C.H. (1997) Effectiveness of road ripping in restoring infiltrating capacity of forest roads. *Restoration Ecology* **5**(3), 265–270.

Lyle, A.A. & Elliott, J.M. (1998) Migratory salmonids as vectors of carbon, nitrogen and phosphorus between marine and freshwater environments in north-east England. *Science of the Total Environment* **210**(1–6), 457–468.

Lynch, J.M. & Murray, D.A. (1992) Fishery rehabilitation and habitat enhancement following arterial drainage in Ireland. *Verhandlungen Internationale Vereinigung für Theoretische und Angewandte Limnologie* **25**(3), 1502–1503.

Madej, M.A. (2001) Erosion and sediment delivery following removal of forest roads. *Earth Surface Processes and Landforms* **26**(2), 175–190.

Major, J.J., O'Connor, J.E., Grant, G.E. *et al.* (2008) Initial fluvial response to the removal of Oregon's Marmot Dam. *EOS Transactions of the American Geophysical Union* **89**, 241–242.

Maloney, K.O., Dodd, H.R., Butler, S.E. & Wahl, D.H. (2008) Changes in macroinvertebrate and fish assemblages in a medium-sized river following a breach of a low-head dam. *Freshwater Biology* **53**, 1055–1068.

Marks, J.C., Halden, G.A., O'Neil, M. & Pace, C. (2010) Effects of flow restoration and exotic species removal on native fish: Lessons from a dam decommissioning. *Restoration Ecology* **18**(6), 934–943.

Massingill, C. (2003) *Coastal Oregon Riparian Silviculture Guide*. Coos Watershed Association. Charleston, Oregon.

Mayer, P.M., Reynolds Jr., S.K., Canfield, T.J. & McCutcheon, M.D. (2005) *Riparian Buffer Width, Vegetative Cover, and Nitrogen Removal Effectiveness*. EPA/600/R-05/118, United States Environmental Protection Agency, Cincinnati, Ohio.

Maynard, A.A. & Hill, D.E. (1992) Vegetative stabilization of logging roads and skid trails. *Northern Journal of Applied Forestry* **9**, 153–157.

McManamay, R.A., Orth, D.J., Dolloff, C.A. & Cantrell, D.A. (2010) Gravel addition as a habitat restoration for tailwaters. *North American Journal of Fisheries Management* **30**, 1238–1257.

McNabb, D.H. (1994) Tillage of compacted haul roads and landings in the boreal forests of Alberta, Canada. *Forest Ecology and Management* **66**, 179–194.

Medina, A.L., Rinne, J.N. & Roni, P. (2005) Riparian restoration through grazing management: considerations for monitoring project effectiveness. In: Roni, P. (ed.) *Monitoring Stream and Watershed Restoration*, American Fisheries Society, Bethesda, Maryland, pp. 97–126.

Merz, J.E. & Chan, L.K.O. (2005) Effects of gravel augmentation on macroinvertebrate assemblages in a regulated California river. *River Research and Applications* **21**, 61–74.

Merz, J.E., Setka, J.D., Pasternack, G.B. & Wheaton, J.M. (2004) Predicting benefits of spawning-habitat rehabilitation to salmonid (*Oncorhynchus spp.*) fry production in a regulated California river. *Canadian Journal of Fisheries and Aquatic Sciences* **61**(8), 1433–1446.

Miller, J.R. & Kochel, R.C. (2009). Assessment of channel dynamics in-stream structures and post-project channel adjustments in North Carolina and its implications to effective stream restoration. *Environmental Earth Sciences* **59**, 1681–1692.

Miller, S.J., Fischenich, J.C. & Thornton, C. (In press) *Stability Thresholds and Performance Standards for Flexible Lining Materials in Channel Restoration Applications*. U.S. Army Corps of Engineers, Engineering Research & Development Program, Vicksburg, Mississippi.

Miller, S.W., Budy, P. & Schmidt, J.C. (2010) Quantifying macroinvertebrate responses to in-stream habitat restoration: applications of meta-analysis to river restoration. *Restoration Ecology* **18**, 8–19.

Ministry of the Environment (2001) *Soil Conservation Technical Handbook*. Ministry of the Environment, Wellington, New Zealand.

Moerke, A.H., & Lamberti, G.A. (2003) Responses in fish community structure to restoration of two Indiana streams. *North American Journal of Fisheries Management* **23**, 748–759.

Muhar, S., Unfer, G., Schmutz, S., Jungwirth, M., Egger, G. & Angermann, K. (2004) Assessing river restoration programmes: habitat conditions, fish fauna and vegetation as indicators for the possibilities and constraints of river restoration. In: Garcia de Jalon, D. & Martinez, P.V. (eds) *Proceedings of the Fifth International Conference on Ecohydraulics – Aquatic Habitats: Analysis and Restoration*, International Association of Hydraulic Engineers, Madrid, Spain, pp. 300–305.

Myers, L.H. (1989) Grazing and riparian management in southwestern Montana. In: Gresswell, R.E., Barton, B.A. & Kershner, J.L. (eds) *Practical Approaches to Riparian Resource Management: An Education Workshop*, United States Department of the Interior, Bureau of Land Management, Billings, Montana, pp. 117–120.

Myers, T.J. & Swanson, S. (1995) Impact of deferred rotation grazing on stream characteristics in central Nevada: A case study. *North American Journal Fisheries Management* **15**(2), 428–439.

Nakamura, F. & Komiyama, E. (2010) A challenge to dam improvement for the protection of both salmon and human livelihood in Shiretoko, Japan's third Natural Heritage Site. *Landscape and Ecological Engineering* **6**, 143–152.

Nassauer, J.I. (1999) Ecological retrofit. *Landscape Journal* **17**(2), 15–17.

Nielsen, M.B. (1996) Lowland stream restoration in Denmark. In: Brookes, A. & Shields, F.D. (eds) *River Channel Restoration: Guiding Principles for Sustainable Projects*. John Wiley & Sons, Chichester, England, pp. 271–289.

Norman, D.K. (1998) Reclamation of flood-plain sand and gravel pits as off-channel salmon habitat. *Washington Geology* **26**(2–3), 21–28.

Novotny, V., Ahern, J. & Brown, P. (2010) *Water Centric Sustainable Communities: Planning, Retrofitting, and Building the Next Urban Environment*. John Wiley & Sons, Hobokan, New Jersey.

NRC (National Research Council) (1992) *Restoration of Aquatic Ecosystems; Science, Technology, and Public Policy*. National Academy Press, Washington, DC.

ODF (Oregon Department of Forestry) (2000) *Forest Engineering Roads Manual*. Oregon Department of Forestry, Salem, Oregon.

O'Grady, M. (2006) *Channels & Challenges. Enhancing Salmonid Rivers*. Irish Freshwater Fisheries Ecology & Management Series: Number 4, Central Fisheries Board, Dublin, Ireland.

O'Grady, M., Gargan, P., Delanty, K., Igoe, F. & Byrne, C. (2002) Observations in relation to changes in some physical and biological features of the Glenglosh River following bank stabilisation. In: Grady, M.O. (ed.) *Proceedings of the 13th International Salmonid Habitat Enhancement Workshop*, Central Fisheries Board, Dublin, Ireland, pp. 61–77.

Opperman, J.J. & Merenlender, A.M. (2000) Deer herbivory as an ecological constraint to restoration of degraded riparian corridors. *Restoration Ecology* **8**(1), 41–47.

Orr, C.H. & Stanley, E.H. (2006) Vegetation development and restoration potential of drained reservoirs following dam removal in Wisconsin. *River Research and Applications* **22**, 281–295.

Osborne, L.K. & Kovacic, D.A. (1993) Riparian vegetated buffer strips in water-quality restoration and stream management. *Freshwater Biology* **29**, 243–258.

Parish, F., Mohktar, M., Abdullah, B. & May, C.O. (2004) *River Restoration in Asia. Proceedings of the East Asia Regional Seminar on River Restoration.* Global Environmental Centre and Department of Irrigation and Drainage, Kuala Lumpur, Malaysia.

Parkyn, S.M., Davies-Colley, R.J., Halliday, N.J., Costley, K.J. & Croker, G.F. (2003) Planted riparian buffer zones in New Zealand: Do they live up to expectations? *Restoration Ecology* **11**(4), 436–447.

Paul, M.J. & Meyer, J.L. (2001) Streams in the urban landscape. *Annual Review of Ecology and Systematics* **32**, 333–365.

Pearson, A.J., Snyder, N.P., & Collons, M.J. (2011) Rates and processes of channel response to dam removal with a sand-filled impoundment. *American Water Resources Research* **47**, W08504, doi:10.1029/2010WR 009733.

Pearsons, T.N., Roley, D.D. & Johnson, C.L. (2007) Development of a carcass analog for nutrient restoration in streams. *Fisheries* **32**(3), 114–124.

Pedersen, M.L., Andersen, J.M., Nielsen, K. & Linnemann, M. (2007) Restoration of Skjern River and its valley: Project description and general ecological changes in the project area. *Ecological Engineering* **30**, 131–144.

Penczak, T. (1995) Effects of removal and regeneration of bankside vegetation on fish population-dynamics in the Warta River, Poland. *Hydrobiologia* **303**(1–3), 207–210.

Pess, G.R., Collins, B.D., Pollock, M., Beechie, T.J., Haas, A. & Grigsby, S. (1999) *Historic and Current Factors that Limit Coho Salmon Production in the Stillaguamish River Basin, Washington State: Implications for Salmonid Habitat Protection and Restoration.* Tulalip Tribes, Marysville, Washington.

Pess, G.R., Morley, S.A., Hall, J.L. & Timm, R.K. (2005a) Monitoring floodplain restoration. In: Roni, P. (ed.) *Monitoring Stream and Watershed Restoration.* American Fisheries Society, Bethesda, Maryland, pp. 127–166.

Pess, G.R., Morley, S.A. & Roni, P. (2005b) Evaluating fish response to culvert replacement and other methods for reconnecting isolated aquatic habitats. In: Roni, P. (ed.) *Monitoring Stream and Watershed Restoration.*

American Fisheries Society, Bethesda, Maryland, pp. 267–276.

Peterson, N.P. (1982) Population characteristics of juvenile coho salmon (*Oncorhynchus kisutch*) overwintering in riverine ponds. *Canadian Journal of Fisheries and Aquatic Sciences* **39**, 1303–1307.

Petts, G. (2009) Instream flow science for sustainable river management. *Journal of the American Water Resources Association* **45**, 1071–1086.

Petts, G. & Calow, P. (1996) *River Restoration: Selected Extracts from the Rivers Handbook.* Blackwell Science, Oxford, U.K.

Platts, W.S. (1991) Livestock grazing. In: Meehan, W.R. (ed.) *Influences of Forest and Rangeland Management on Salmonid Fishes and Their Habitats – Special Publication 19.* American Fisheries Society, Bethesda, Maryland, pp. 389–423.

Platts, W.S. & Nelson, R.L. (1985) Impacts of rest-rotation grazing on stream banks in forested watersheds in Idaho. *North American Journal of Fisheries Management* **5**, 547–556.

Podolak, C.J. & Wilcock, P.R. (2009) The formation and growth of gravel bars in response to increased sediment supply following the Marmot Dam removal. *Geological Society of America, Abstracts with Programs* **41**(7), 573.

Poff, N.L., Allan, J.D., Bain, M.B. *et al.* (1997) The natural flow regime. *BioScience* **47**, 769–784.

Pollock, M.M., Heim, M. & Werner, D. (2003) Hydrologic and geomorphic effects of beaver dams and their influence on fishes. In: Gregory, S.V., Boyer, K.L. & Gurnell, A.M. (eds) *The Ecology and Management of Wood in World Rivers,* American Fisheries Society, Bethesda, Maryland, pp. 213–233.

Pollock, M.M., Pess, G.R. & Beechie, T.J. (2004) The importance of beaver ponds to coho salmon production in the Stillaguamish River basin, Washington, USA. *North American Journal of Fisheries Management* **24**(3), 749–760.

Pollock, M.M., Beechie, T.J., Chan, S.S. & Bigley, R. (2005) Monitoring of restoration of riparian forests. In: Roni, P. (ed.) *Monitoring Stream and Watershed Restoration.* American Fisheries Society, Bethesda, Maryland, pp. 67–96.

Pretty, J.L., Harrison, S.C., Shepherd, D.J. *et al.* (2003) River rehabilitation and fish populations: assessing the benefit of instream structures. *Journal of Applied Ecology* **40**(2), 251–265.

Price, D.M., Quinn, T. & Barnard, T.J. (2010) Fish passage effectiveness of recently constructed road crossing culverts in the Puget Sound Region of Washington State.

North American Journal of Fisheries Management **30**, 1110–1125.

Puckett, L.J. & Hughes, W.N. (2005) Transport and fate of nitrate and pesticides: hydrogeology and riparian zone processes. *Journal of Environmental Quality* **24**, 2278–2292.

Quinn, E.M. (1999) Dam removal: a tool for restoring riverine ecosystems. *Restoration and Reclamation Review* **5**, 1–12.

Raastad, J.E., Lillehammer, A. & Lillehammer, L. (1993) Effect of Habitat Improvement on Atlantic Salmon in the Regulated River Suldalslagen. *Regulated Rivers – Research & Management* **8**(1–2), 95–102.

Reid, L.M. & Dunne, T. (1984) Sediment production from forest road surfaces. *Water Resources Research* **20**(11), 1753–1761.

Riley, A.L. (1998) *Restoring Streams in Cities*. Island Press, Washington DC.

Rinne, J.N. (1999) Fish and grazing relationships: the facts and some pleas. *Fisheries* **24**(8), 12–21.

Robertson, A.I. & Rowling, R.W. (2000) Effects of live-stock on riparian zone vegetation in an Australian dryland river. *Regulated Rivers: Research and Management* **16**(5), 527–541.

Roni, P. (2005) *Monitoring Stream and Watershed Restoration*. American Fisheries Society, Bethesda, Maryland.

Roni, P. & Quinn, T.P. (2001) Density and size of juvenile salmonids in response to placement of large woody debris in western Oregon and Washington streams. *Canadian Journal of Fisheries and Aquatic Sciences* **58**, 282–292.

Roni, P., Beechie, T.J., Bilby, R.E., Leonetti, F.E., Pollock, M.M. & Pess, G.P. (2002) A review of stream restoration techniques and a hierarchical strategy for prioritizing restoration in Pacific Northwest watersheds. *North American Journal of Fisheries Management* **22**, 1–20.

Roni, P., Hanson, K., Beechie, T.J., Pess, G.R., Pollock, M.M. & Bartley, D.M. (2005) *Habitat Rehabilitation for Inland Fisheries: Global Review of Effectiveness and Guidance for Rehabilitation of Freshwater Ecosystems*. FAO Fisheries Technical Paper, No. 484, Food and Agriculture Organization of the United Nations, Rome, Italy.

Roni, P., Pess, G., Bennett, T., Morley, S. & Hanson, K. (2006) Rehabilitation of stream channels scoured to bedrock: the effects of boulder weir placement on aquatic habitat and biota in southwest Oregon, USA. *River Research and Applications* **22**, 962–980.

Roni, P., Hanson, K. & Beechie, T.J. (2008) Global review of the physical and biological effectiveness of stream habitat rehabilitation techniques. *North American Journal of Fisheries Management* **28**, 856–890.

Rood, S.B. & Mahoney, J.M. (2000) Revised instream flow regulation enables cottonwood recruitment along the St. Mary River, Alberta, Canada. *Rivers* **7**(2), 109–125.

Rosenfeld, J., Hogan, D., Palm, D. *et al.* (2010) Contrasting landscape influences on sediment supply and stream restoration priorities in Northern Fennoscandia (Sweden and Finland) and Coastal British Columbia. *Environmental Management* doi:10.1007/s00267-010-9585-0.

Rosgen, D.L. (1994) A classification of natural rivers. *Catena* **22**, 169–199.

RRC (River Restoration Centre) (2002) *Manual of River restoration Techniques for the Web*. River Restoration Centre, Silsoe, U.K. Available at: http://www.therrc.co.uk/rrc_manual.php.

RSPB, NRA & RSNC (Royal Society for Protection of Birds, National Rivers Authority & Royal Society for Nature Conservation) (1994) *The New Rivers and Wildlife Handbook*. Royal Society for Protection of Birds, Bedfordshire, UK.

Rutherfurd, I.D., Jerie, K. & Marsh, N. (2000) *A Rehabilitation Manual for Australian Streams*. Land and Water Resources Research and Development Corporation, Canberra, Australia.

Saldi-Caromile, K., Bates, K.K., Skidmore, P., Barenti, J. & Pineo, D. (2004) *Stream Habitat Restoration Guidelines Final Draft*. Washington Department of Fish and Wildlife, Olympia, Washington.

Sather, N.K. & May, C.W. (2008) *Riparian Buffer Zones in the Interior Columbia River Basin: A Review of Best Available Science*. Battelle Memorial Institute, Marine Sciences Laboratory, Sequim, Washington.

Schemel, L.E., Sommer, T.R., Muller-Solger, A.B. & Harrell, W.C. (2004) Hydrologic variability, water chemistry, and phytoplankton biomass in a large floodplain of the Sacramento River, CA, USA. *Hydrobiologia* **513**, 129–139.

Schmetterling, D.A., Clancy, C.G. & Brandt, T.M. (2001) Effects of riprap bank reinforcement on stream salmonids in the western United States. *Fisheries* **26**(7), 6–23.

Schmutz, S., Matheisz, A., Pohn, A., Rathgeb, J. & Unfer, G. (1994) Colonisation of a newly constructed canal by fish (Marchfeldkanal, Austria). *Osterreichs Fischerei Salzburg* **47**(7), 158–178.

Sear, D.A., Briggs, A. & Brookes, A. (1998) A preliminary analysis of the morphological adjustment within and downstream of a lowland river subject to river restoration. *Aquatic Conservation: Marine and Freshwater Ecosystems* **8**(1), 167–183.

Shaff, C.D. & Compton, J.E. (2009) Differential incorporation of natural spawners vs. artificially planted salmon carcasses in a stream food web: evidence from delta N-15 of juvenile coho salmon. *Fisheries* **34**(2), 62–72.

Shafroth, P.B. & Biggs, M.K. (2008) Restoration Ecology and Invasive Riparian Plants: An Introduction to the Special Section on *Tamarix spp.* in Western North America. *Restoration Ecology* **16**, 94–96.

Sheng, M.D., Foy, M & Fedorenko, A.Y. (1990) Coho salmon enhancement in British Columbia using improved groundwater-fed side channels. Department of Fisheries and Oceans, *Canadian Manuscript Report of Fisheries and Aquatic Sciences* No. 2071, Vancouver, Canada.

Shields, F.D., Cooper, C.M. & Knight, S.S. (1993) Initial habitat response to incised channel rehabilitation. *Aquatic Conservation of Marine and Freshwater Ecosystems* **3**, 93–10.

Shields, F.D., Cooper, C.M. & Knight, S.S. (1995) Experiment in stream restoration. *Journal of Hydraulic Engineers* **121**(6), 494–503.

Shuman, J.R. (1995) Environmental considerations for assessing dam removal alternatives for river restoration. *Regulated Rivers: Research and Management* **11**, 249–261.

Sieker, H & Klein, M. (1998) Best management practices for stormwater runoff with alternative methods in a large urban catchment in Berlin, Germany. *Water Science & Technology* **38**, 91–97.

Slaney, P.A. & Zaldokas, Z. (1997) Fish habitat rehabilitation procedures. *Watershed Restoration Program, Watershed Restoration Circular No. 9*. Ministry of Environment, Lands and Parks and Ministry of Forests, Vancouver, Canada.

Slaney, P.A., Ward, B.R. & Whiteman, C.J. (2003) Experimental nutrient addition to the Keogh River and application to the Salmon River in coastal British Columbia. *American Fisheries Society Symposium* **34**, 111–126.

Smokorowski, K.E. & Pratt, T.C. (2007) Effect of a change in physical structure and cover on fish and fish habitat in freshwater ecosystems – a review and meta-analysis. *Environmental Review* **15**, 15–41.

Solazzi, M.F., Nickelson, T.E., Johnson, S.L. & Rodgers, J.D. (2000) Effects of increasing winter rearing habitat on abundance of salmonids in two coastal Oregon streams. *Canadian Journal of Fisheries and Aquatic Sciences* **57**, 906–914.

Stanford, J.A., Ward, J.V., Liss, W.J. *et al.* (1996) A general protocol for restoration of regulated rivers. *Regulated Rivers: Research & Management* **12**, 391–413.

Stanley, E.H. & Doyle, M.W. (2003) Trading off: the ecological effects of dam removal. *Frontiers in Ecology and the Environment* **1**, 15–22.

Stanley, E.H., Leubke, M.A., Doyle, M.W. & Marshall, D.W. (2002) Short-term changes in channel form and macroinvertebrate communities following low-head dam removal. *Journal of the North American Benthological Society* **21**, 172–187.

Stanley, E.H., Catalano, M.J., Mercado-Silva, N. & Orr, C.H. (2007) Effects of dam removal on brook trout in a Wisconsin stream. *River Research and Applications* **23**, 792–798.

Sterling, M.S., Ashley, K.I. & Bautista, A.B. (2000) Slow-release fertilizer for rehabilitating oligotrophic streams: a physical characterization. *Water Quality Research Journal of Canada* **35**(1), 73–94.

Stevens, L.E., Ayers, T.J., Kearsley, M.J.C. *et al.* (2001) Planned flooding and Colorado River riparian trade-offs downstream from Glen Canyon Dam, Arizona. *Ecological Applications* **11**(3), 701–710.

Sudduth, E.B. & Meyer, J.L. (2006) Effects of bioengineered streambank stabilization on bank habitat and macroinvertebrates in urban streams. *Environmental Management* **38**(2), 218–226.

Surian, N. & Rinaldi, M. (2003) Morphological response to river engineering and management in alluvial channels in Italy. *Geormorphology* **50**, 307–326.

Sweeney, B.W., Czapka, S.J. & Yerkes, T. (2002) Riparian forest restoration: Increasing success by reducing plant competition and herbivory. *Restoration Ecology* **10**(2), 392–400.

Switalski, T.A., Bissonette, J.A., DeLuca, T.H., Luce, C.H. & Madej, M.A. (2004) Benefits and impacts of road removal. *Frontiers in Ecology and the Environment* **2**(1), 21–28.

Toy, T.J., Foster, G.R. & Renard, K.G. (2002) *Soil Erosion: Processes, Prediction, Measurement, and Control.* John Wiley & Sons, New York.

USEPA (United States Environmental Protection Agency) (2010) *Green Infrastructure Case Studies*. EPA-841-F-10-004, United States Environmental Protection

Agency, Office of Wetlands and Oceans, Washington, DC.

Vannote, R.L., Minshall, G.W., Cummins, K.W., Sedell, J.R. & Cushing, C.E. (1980) The river continuum concept. *Canadian Journal of Fisheries and Aquatic Sciences* **37**, 130–137.

Vought, L.B.M. & Locoursiere, J.O. (2010) Restoration of streams in the agricultural landscape. In: Eiseltova, M. (ed.) *Restoration of Lakes, Streams, Floodplains, and Bogs in Europe*. Springer, New York, pp. 225–242.

Walsh, C.J., Roy, A.H., Feminella, J.W., Cottingham, P.D, Groffman, P.M. & Morgan, R.P. (2005) The urban stream syndrome: current knowledge and the search for a cure. *Journal of the North American Benthological Society* **24**(3), 706–723.

Waters, T.F. (1995) *Sediment in Streams. Sources, Biological Effects, and Control*. American Fisheries Society, Bethesda, Maryland.

Weisberg, S.B. & Burton, W.H. (1993) Enhancement of fish feeding and growth after an increase in minimum flow below the Conowingo Dam. *North American Journal of Fisheries Management* **13**, 103–109.

Whiteway, S.L., Biron, P.M., Zimmermann, A. *et al.* (2010) Do in-stream restoration structures enhance salmonid abundance? A meta-analysis. *Canadian Journal of Fisheries and Aquatic Sciences* **67**, 831–841.

William, J.R. & Beschta, R.L. (2003) Wolf reintroduction, predation risk, and cottonwood recovery in Yellowstone National Park. *Forest Ecology and Management* **184**, 299–313.

Williams, P.B., Andrews, E., Opperman, J.J., Bozkurt, S. & Moyle, V. (2009a) Quantifying activated floodplains on a lowland regulated river: Its application to floodplain restoration in the Sacramento Valley. *San Francisco Estuary and Watershed Science* **7**(1), 1–25. Available at: http://repositories.cdlib.org/jmie/sfews/vol7/iss1/art4.

Williams, K.L., Griffiths, S.W., Nislow K.H. *et al.* (2009b) Response of juvenile Atlantic salmon, *Salmo salar*, to the introduction of salmon carcasses in upland streams. *Fisheries Management and Ecology* **16**(4), 290–297.

Wipfli, M.S., Hudson, J.P. & Caouette, J.P. (2004) Restoring productivity of salmon-based food webs: contrasting effects of salmon carcass and salmon carcass analog additions on stream-resident salmonids. *Transactions of the American Fisheries Society* **133**, 1440–1454.

Wunderlich, R.C., Winter, B.D. & Meyer, J.H. (1994) Restoration of the Elwha River Ecosystem. *Fisheries* **19**, 11–19.

Zitek, A., Schmutz, S. & Jungwirth, M. (2008) Assessing the efficiency of connectivity measures with regard to the EU-Water Framework Directive in a Danube-tributary system. *Hydrobiologia* **609**, 139–161.

Zurowski, W. & Kasperczyk, B. (1988) Effects of reintroduction of European beaver in the lowlands of the Vistula Basin. *Acta Theriologica* **33**(12–25), 325–338.

6 Prioritization of Watersheds and Restoration Projects

Philip Roni[1], Tim Beechie[1], Stefan Schmutz[2], and Susanne Muhar[2]

[1]Fish Ecology Division, National Oceanic and Atmospheric Administration, USA
[2]University of Natural Resources and Life Sciences, Vienna, Austria

6.1 Introduction

In the context of restoration, prioritization is the process of ranking projects, habitats, or watersheds to determine their sequencing for funding and implementation. It is an important component of restoration planning that logically follows identification of restoration actions and opportunities (see Chapters 3–5). The need for prioritizing restoration actions largely stems from the need to make the best use of limited resources (i.e. funding, people, equipment, materials, and time) and the need to protect or restore lands before further degradation occurs (Noss *et al.* 2009). In cases where only a few restoration projects are identified, the prioritization may be relatively simple. In most cases, however, dozens of restoration actions need to be prioritized and these actions differ in costs, benefits, constraints, supporters, opponents, feasibility, and other factors. In some instances, the sequencing of projects is important because the success of one project may depend upon another. For example, the success of restoration of habitat for migratory fish above a man-made fish-migration barrier is dependent upon dam removal or installation of fish passage facilities. In other situations, watersheds or habitats may first be ranked and then actions within those locations prioritized for restoration. Whether determining the best sequence of watersheds, habitats, or restoration actions is desired, a consistent, repeatable, systematic, and well-documented

approach for ranking projects and determining priorities is needed.

A suite of different approaches can be used to rank restoration and conservation actions, ranging from simple scoring procedures to complex computer models. Fortunately, the steps needed to prioritize both areas for restoration and restoration actions are similar. These steps include identifying the goals and scale, determining who will prioritize actions, selecting the prioritization approach and criteria, collecting the data to be used, and analyzing and ranking projects or watersheds (Figure 6.1). In this chapter we discuss each of these and describe the most common approaches to prioritizing watersheds or restoration projects at a variety of scales. Examples for ranking restoration projects at different scales are provided to demonstrate the differences in approaches and assist the practitioner or student in developing a sound strategy. Documentation of each step and the decisions made is critical for providing a transparent prioritization process. This is particularly important for modern programs that often cover large geographic areas that include, and may affect, many stakeholders and user groups.

The reader should keep in mind that the prioritization process and approaches described in this chapter can assist restoration practitioners and managers with making decisions, but not take the place of it. Moreover, priorities should be periodically revisited as new information becomes available, projects are completed, funding source or level changes, or new restoration and protection opportunities are identified.

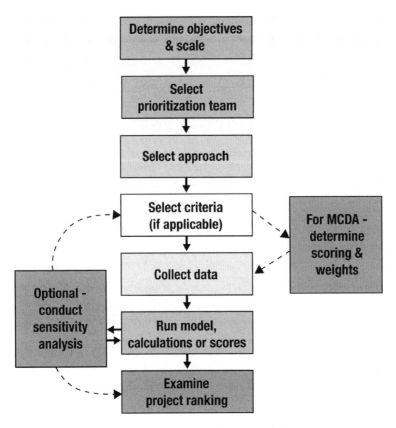

Figure 6.1 Key steps in prioritization process. MCDA – multi-criteria decision analysis.

6.2 Determine overall goals and scale

Before discussing what type of prioritization approach to use, a clear goal for restoration and specific objectives for prioritization must be defined. As discussed in previous chapters, the restoration goal or goals are typically set before assessing watershed conditions, disrupted processes, lost and degraded habitats, and identification of restoration actions. The restoration goal must be defined prior to commencing with prioritization. A detailed discussion of goal setting can be found in Chapter 3, but well-defined watershed restoration goals typically (1) identify ecological or biological objectives; (2) address underlying causes of ecosystem degradation; and (3) acknowledge social, economic and land-use constraints (Beechie *et al.* 2008). These goals can serve as a basis for the team to define the specific objectives and criteria used in the prioritization process. Restoration goals initially defined during the assessment phase should be revisited to ensure that they include adequate detail on spatial and

temporal scales to guide ranking of restoration actions. During subsequent phases of the prioritization process, these goals can be translated into specific objectives that reflect the values and methods selected by the project team.

Clearly defining goals and specific objectives is crucial to selecting an approach and specific criteria for prioritizing projects. The more clearly and specifically objectives are defined, the easier it will be to select appropriate criteria for ranking projects. For example, if the specific objectives are to restore watershed processes, biological diversity, increase migratory fish numbers, and providing recreational opportunities, then appropriate criteria would likely be processes restored or type of restoration, expected change in biological diversity, and predicted increase in fish and recreational opportunities. Examples of how the goals of a restoration program can be translated into prioritization objectives and criteria are provided in Table 6.1. If the restoration goals have been well defined, the prioritization objectives may be almost identical to the restoration goals.

Table 6.1 Examples of national, regional or watershed restoration goals and how they could be translated into specific prioritization objectives and criteria. In some cases prioritization objectives are identical to restoration goals, while in others the restoration goals are very general and need to be translated into more specific objectives to be useful in the prioritization process.

General restoration goal	Prioritization	
	Goals, objective and scale	Examples of criteria
To conserve, protect, rehabilitate, and improve the rivers, streams, watercourses, and water impoundments of the catchment including its estuary and adjacent coastal area	Determine high-priority projects in the basin for protection and rehabilitation based on their ecological value and total area restored	Number of species present, area and number of unique habitats, index of biological integrity, habitat quality, sensitivity of habitat or area to be restored
Re-establish a watershed and ecosystem that is capable of supporting and maintaining a balanced, integrated, adaptive community of organisms having a species composition, diversity, and functional organization comparable to the natural habitat of the region	Identify and rank projects within the watershed that will provide the range of habitats needed to maintain and support diverse species and provide the largest benefit for the resources available within the next five years	Number of habitats restored, area of habitat restored, expected increase in species diversity, number of years to complete project
Protect and restore the ecological integrity of the river and its streams to enhance aquatic diversity, increase recreational use, and provide for a quality urban fishery	Rank projects within the watershed based on their ability to restore ecological integrity and aquatic diversity, provide recreation, and increase catch for recreational anglers	Estimated improvements in IBI or other metrics of integrity, species diversity, angler effort and catch rates
Reduce erosion, fine sediment, and nutrients; restore natural river flows, floodplains, and meadow; expand habitat corridor to strengthen natural ecosystem; maintain recreational and economic benefits in the watershed	Same as watershed goal	Erosion, flow, water quality (fine sediment or nutrients), area of meadow or floodplain restored, habitat corridor length, impact on recreation, cost-benefit or cost-effectiveness
Restore and protect watersheds to assist in recovery of threatened and endangered anadromous fishes	Determine high-priority watersheds within the region for protection, restoration and reintroduction of endangered fishes based on habitat quality, historical use by endangered species, land use, and susceptibility to climate change	Percent of watershed occupied by species, population density, genetic integrity, watershed condition and connectivity, water quality and flow, introduced species, susceptibility to climate change
Maximize jobs created or maintained through implementation of shovel-ready coastal and marine habitat restoration projects, and improve the short- and long-term economic condition of an area (e.g. increased fisheries benefits, increased tourism and recreation) based on the significance of the anticipated outcomes of the project	Prioritize restoration projects nationwide based on their ability to maximize benefits to coastal and marine habitats, fisheries resources, and economic benefits including tourism, recreation, and direct and indirect employment (jobs)	Area restored, jobs created, indirect economic benefits, recreation opportunities created, regional importance of project, increase in number or density of fish
Inspire and sustain local communities to restore coastal and riverine habitats	Determine high-priority watersheds in the region to fund restoration that will provide the most efficient use of resources to protect and recovery anadromous fishes	Number of anadromous species present, number of endangered species present, watershed condition, current and projected level of urbanization
Restoration of fish migration routes throughout the Danube Basin (see Box 6.2 for additional detail)	The objective is to provide a stepwise and ecologically efficient approach for implementation of barrier removal projects at a basin-wide scale	Number of species that are migratory, location of barrier (river kilometer), length of habitat opened, protected status of river segment

We discuss and present the goals and selection of a prioritization team sequentially (Figure 6.1); however, determining who will prioritize the projects and the goals and objectives of the process are intertwined, and selecting a team may occur before determining goals. In some cases, the broad restoration goals may be predetermined by an agency or watershed plan and then these goals translated into more specific objectives by the prioritization team. Regardless of the order in which they occur, it is important to define the objectives, the scale, and who will do the prioritizing before launching into detailed discussions of what approach or criteria should be used to prioritize projects or watersheds. The setting of goals and objectives can be an iterative process where the specific objectives are further refined by the prioritization team after the method and criteria have been decided upon.

6.2.1 Legal frameworks, funding, and goals

Legal frameworks such as the Clean Water Act and Endangered Species Act in the United States; the Environmental Protection Act, Species at Risk Act, Fisheries Act, and Fish Habitat Policy in Canada; the Environment Protection and Diversity Act in Australia; or the European Union's (EU) Water Framework Directive (WFD) and similar legislation in these and other countries and states have been the drivers for many river and watershed restoration efforts (see also Chapter 3). For example, the WFD requires that member states establish a program of basic and supplementary restoration measures in each river basin district (based on watersheds), with the aim of moving progressively towards achieving the environmental objective of protecting and restoring habitats. Each country's river basin management plans in the EU therefore provide a legal framework for stepwise implementation of restoration measures that follow individual prioritization schemes. In the Austrian National River Basin Management Plan, for example, priority areas and their respective water bodies have been designated for restoration using the following criteria:

• rivers and mouths of major tributaries (catchment area >500 km²) with medium-distance migrating fish species;
• Natura 2000 areas (sites by EU Birds and Habitats directives);
• rivers or reaches with urgent management issues (e.g. rivers completely dewatered due to abstraction).

The funding source, which is often tied to legislative goals, will also influence restoration priorities. For example, river restoration projects funded in the United States under the 2009 American Recovery and Reinvestment Act (commonly called the Stimulus Package) had two main goals: (1) create environmental jobs and (2) stimulate the economy. These served as both goals and criteria for the project selection and prioritization. This also demonstrates how funding goals may conflict with ecological goals, as some projects with high ecological value were not funded because they provided limited economic benefits (i.e. jobs). It is therefore important to understand the legal and funding framework under which restoration is occurring when determining the goals as well as the scale at which restoration will be prioritized.

6.2.2 Spatial and temporal scale

The goal should also clearly state the spatial scale at which the restoration and prioritization will occur and whether the objective is to rank or select watersheds, reaches, projects within or across watersheds, or some combination of these. Restoration actions can be ranked at national, state, regional, provincial, watershed, subwatershed or stream-reach scales (Roni et al. 2003; Williams et al. 2007; Beechie et al. 2008; Noss et al. 2009; Figure 6.2). The ranking of areas for restoration (watersheds, reaches, or habitat types) rather than individual projects is often done before or as part of the project prioritization process. For large regional programs, a multilevel approach may be necessary where first watersheds or reaches within watersheds are ranked and selected for restoration, then individual projects or actions within these watersheds are prioritized (Figure 6.2). In other cases, the types of habitats that need to be restored or protected might first be rated based on assessments discussed in Chapter 3. Even if the goal is to rank individual actions across several watersheds, we may initially want to rank watersheds for restoration to assist with prioritization and selection of projects across a region. In the United States for example, the National Oceanic and Atmospheric Administration (NOAA) Restoration Center is developing a prioritization strategy for watersheds based on physical and biological as well as legal and socioeconomic factors to help determine where to focus their efforts for their Community Based Restoration Program, which provides funding to local communities for restoration across the country (J. Steger, personal communication, 2011). Key features of this strategy are that it considers the restoration potential of fish populations, watershed condition, land-use constraints, and number of protected or sensitive species. In this case, the funding source and agency goals helped define the scale at which prioritization would occur.

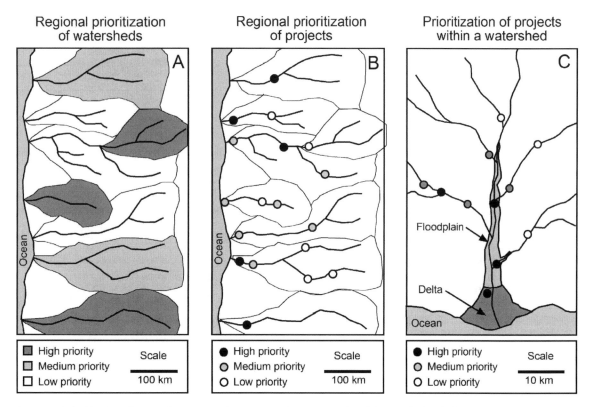

Figure 6.2 Prioritization of restoration actions often occurs at (A, B) regional or (C) watershed scale, but may also include prioritization of sub-watersheds or reaches within a watershed. Ideally, watersheds would be prioritized for restoration, then sub-watersheds or reaches within watersheds, and finally projects within those sub-watersheds, but this may be difficult in practice.

The appropriate scale is often driven by legal mandates that set ecological targets. In Europe for example, the WFD calls for basin-scale management plans to help achieve ecological targets. Similarly, for endangered fishes in the western United States, the focus of recovery and restoration is on distinct fish populations and the watersheds they inhabit. This has led to the focus on restoration of whole watersheds, although removing a particular species from the United States Endangered Species Act requires that a large number of well-distributed populations (and presumably watersheds) achieve specified recovery goals for abundance, population growth rate, spatial distribution, and diversity. Recent modeling efforts suggest that to produce measureable increases in salmon and trout abundance at a watershed or population scale, a significant amount of habitat within a watershed needs to be restored (>20 to 100%); this suggests the need to focus limited resources on watersheds that have the largest potential for salmon

recovery and restoration (Roni *et al.* 2010). However, while focused effort at the watershed scale is necessary to achieve measurable gains in abundance of a salmon population, successful restoration of the species will require focused effort in many watersheds across the range of a species. Regardless of the driver, it is important that the goals clearly state whether the focus is on prioritizing areas for restoration or actions (or both) and at what spatial scale that will be done. In addition, if the goal is recovery of endangered species, biological goals should be clearly defined.

It is also important to define the temporal scale within which projects may be prioritized, as many projects may take several years to develop and implement. The main factors that influence temporal scale are the logistic and funding sources, the size and complexity of the project, and when the project will produce the desired restoration or reach its full potential. A large project such as a dam removal may have the highest ranking for a watershed or region, but technical, logistic, and legal hurdles may take years to overcome. Funding sources may also influence the temporal scale if funds have to be distributed or spent by the end of a calendar or fiscal year, or require that project construction be completed within a set time frame. In some cases, projects may need to be excluded from a list of potential projects because they cannot be implemented during the required funding and time constraints. Projects are often completed in phases to work around funding constraints. For example, project proponents may first ask for funding for the initial phase of a project to complete the design or to demonstrate the feasibility of the project by restoring only a portion of the area, then ask for additional funding once the design or feasibility has been demonstrated. Another temporal aspect is the lag time between project implementation and the actual response. For example, projects such as fish passage or migration barrier removal often have quick responses, while a riparian planting project may take several years before all the benefits of the project are realized. In this case, the number of years until full project benefits are realized would likely be used as a criterion for prioritizing projects.

6.3 Who will prioritize projects? Selecting the team

The selection of the decision makers or technical team who will take part or assist in the prioritization process is critical because it can affect the credibility and acceptance of the approach by managers, stakeholders, and the local community (Preister & Kent 1997; Turner 1997). Getting input from the community is therefore critical to gaining public support and educating landowners who might initially oppose restoration on their lands (see Chapter 4). As discussed previously, national legislation often outlines not only the overall goals for restoration and prioritization but also identifies key stakeholders or team members who should be involved in development of priorities. The broader the geographic area covered and the more diverse the stakeholders (those who may be positively or negatively impacted by restoration), the greater the need for a carefully selected, representative, and diverse group to serve on the prioritization team. The most successful teams generally include 5–10 individuals, which is the optimal working group size in a variety of fields (Jay 1971; Mears & Voehl 1997). Groups of a slightly larger size may work, but larger groups are more

difficult to coordinate and manage and have difficulty reaching consensus or making decisions. Additional detail on working with diverse groups in a watershed restoration context or building consensus can be found in Chapter 4.

6.4 Prioritization approaches and criteria

The appropriate strategy or approach depends upon the restoration goals; the criteria the team defines; the type, quality, and detail of data available; and the amount of time and resources available to complete the prioritization. Ideally, the approach should initially be discussed and considered before assessing watershed conditions and defining restoration projects, as much of the information needed to rank restoration actions will or can be collected during the assessment stage (Beechie *et al.* 2008). In this section we discuss the factors to consider when selecting both the prioritization approach and criteria. We then review common approaches, their strengths and weaknesses, and provide examples for many of them. It is important to keep in mind that there is no ideal approach, and the selection of criteria and the approach are closely connected.

The criteria and data needed vary from one prioritization approach to another, and should be considered when selecting an approach. Criteria used in various approaches include biological (e.g. fish density, fish or macroinvertebrate diversity, presence or absence of key species), physical (e.g. instream habitat area, type or condition, riparian condition, fine sediment), economic (e.g. cost, cost-effectiveness), sociopolitical (e.g. ownership, protected areas), and others that are difficult to categorize such as restoration type or when the project will or can be implemented. The restoration goals will help determine the criteria to include in the prioritization approach. For example, if the goal of the watershed restoration program is to increase migratory fish abundance and restore hydrologic and sediment processes while providing for traditional agriculture and forestry activities, the ideal approach, at a minimum, needs to be able to incorporate or rank projects based on biological data (fish abundance and migratory behavior), type of process restored, and socioeconomic factors. These are likely a combination of qualitative and quantitative data that may vary in quality and level of detail, and even this limited list of criteria may require additional data collection. Criteria used in some prioritization approaches require extensive data collection or modeling that can only be done by individuals with specialized training or computer software, and thus may be costly or time consuming to implement.

Key factors to consider when selecting both the prioritization approach and criteria are data availability, resources or staff needed, cost of additional data collection, whether data are available for all project types or can be collected for all project types, and time needed to complete the prioritization process. In addition, the transparency and repeatability of the approach are very important, as restoration programs typically affect many different stakeholders.

6.4.1 Common prioritization strategies

Despite massive investments in restoration, there is no universally accepted approach for prioritizing restoration and habitat protection (Johnson *et al.* 2003). Before large-scale, publicly funded restoration efforts, projects were often chosen based on professional opinion or preference of the local biologist or restoration proponent. This has a number of drawbacks and usually does not provide a defensible approach for modern restoration programs or actions where numerous stakeholders, landowners, and funding agencies are involved. Perhaps more importantly, publicly funded restoration efforts are increasingly under scrutiny to show that restoration dollars are being spent effectively and wisely. In an effort to make more efficient use of funds and resources, a variety of more rigorous approaches for prioritizing actions have been developed in the last few decades. These range from relatively transparent methods based on project type or scoring systems to much more quantitative and complex statistical models (Beechie *et al.* 2008). Because many factors influence which prioritization strategy might be the most appropriate, we discuss the pros and cons of common strategies below. We categorize them here for ease of discussion and presentation, but these strategies or approaches are a continuum and many of the more complex approaches incorporate aspects of the others. The list of common restoration actions outlined in Table 6.2 is used to help demonstrate the differences between prioritization approaches. We provide additional detail on the last approach (multi-criteria decision analysis or MCDA; Section 6.4.1.7) because it is the most widely used, can incorporate information from the other approaches, and is one of the most flexible and transparent approaches.

6.4.1.1 Prioritizing restoration actions by project type

Projects are often selected based on the effectiveness of the restoration technique (e.g. barrier removal, sediment reduction, wood placement), the kind of habitat they restore (e.g. spawning, rearing, floodplain), or by a logical sequence or 'building block' approach that first restores processes, then physical form (e.g. Petersen *et al.* 1992; Roni *et al.* 2002). For example one approach, initially developed for recovery of salmon and watersheds in the western North America and later refined for application in other countries, uses a combination of project effectiveness metrics (longevity, probability of success, change in habitat, potential increase in fish abundance) and whether the technique restores watershed processes (Roni *et al.* 2002, 2008; Figure 6.3). Projects that have a high probability of success, restore watershed processes, and lead to long-term changes in habitat are given higher priority under this system. Those that focus more on creating or enhancing habitat are given lower priority, except in those cases where land-use constraints prevent higher-priority restoration or where rapid changes are needed to protect and improve habitat for endangered fishes. A similar approach is to rank projects based on the habitat type they restore rather than the technique used. For example, actions that restore floodplain side-channel or delta habitats might be given precedence over actions designed to improve habitat in small streams. Some of the criteria used in approaches based on restoration type or habitat type are also used as criteria in other prioritization approaches, particularly those that use many different criteria.

The advantages of prioritizing by project type are that it is relatively simple to follow as long as there is agreement on the ranking of the different project types and it does not necessarily require extensive data collection. The disadvantage of the approach is that it is a 'one size fits all' method which does not consider the site- or watershed-specific factors that are limiting habitat conditions or biotic production: it therefore does not consider which types of restoration are needed to achieve local restoration goals. Prioritization approaches based on restoration type or effectiveness are therefore best used as an initial screen and should be modified by using or collecting additional data or included as criteria in other approaches such as MCDA (Roni *et al.* 2002).

6.4.1.2 Refugia

An approach commonly used in conservation biology for both protection and restoration of habitat for terrestrial

Table 6.2 Common approaches or categories of approaches for prioritizing restoration (modified from Beechie *et al.* 2008).

Approach	Description (examples)	Pros and Cons
Project type and effectiveness	Ranks projects based on restoration effectiveness or type	Pros: good interim approach in cases where limited data on physical habitat conditions in a watershed are available; based on published reviews or restoration effectiveness Cons: starting point that should be modified as additional data are collected or used as one criterion in other approaches; not useful in ranking watersheds for restoration
Refugia	Focuses on protecting intact habitats with relatively healthy or existing populations (refugia), then typically proceeds outward from refugia so that restored habitats are located near an established source of colonists	Pros: good approach for single species or ecosystem type; typically used to rank watersheds, sub-watersheds or areas for protection or restoration Cons: can be more challenging to identify or implement for multiple species with different habitat requirements; better for ranking watersheds or areas than individual restoration projects
Species or habitat	Rank projects based on habitat area restored or projected increase in fish production	Pros: based largely on empirical fish and habitat data; ranking simply based on habitat area or increase in fish production can be relatively straightforward Cons: data on increase in fish production for different restoration techniques often do not exist; can be difficult to estimate habitat area affected by some projects
Capacity or life-cycle models	Uses life-cycle or habitat capacity models to estimate magnitude of project benefits	Pros: based on empirical data on specific species life stages and can be expanded to incorporate multiple species that may broadly represent ecosystem condition Cons: requires detailed habitat quality and quantity and fish abundance or survival data; life-cycle modeling can be complex and time consuming and can be difficult to apply at a project level
Cost, cost-effectiveness or cost-benefit analysis	Uses the cost, the cost-per-unit of benefit, or the economic benefit to rank and compare projects	Pros: provides a common currency to compare projects Cons: requires data on cost, effectiveness, and economic benefits that may not be readily available; different project types may have different response metrics that are not readily comparable; estimating economic benefits may be difficult
Computer models and conservation planning software	Uses statistical models or conservation planning software to predict potential restoration outcomes and determine high-priority habitats or projects; can incorporate a variety of physical or biological data, qualitative information, and professional opinion	Pros: can incorporate a variety of complex data types Cons: limited transparency; often only those who create and run the model or work extensively with the software understand its inner working; can be costly and time consuming to run although this varies greatly by the model or software used
Scoring and MCDA	A 'score sheet' approach in which multiple criteria (e.g. cost, size, effectiveness) are used to score projects, habitats, or watersheds for restoration	Pros: a simple, straightforward, and transparent system that can incorporate a variety of metrics and information from other prioritization approaches; easily modified to include new data; well-documented approach used in a variety of fields Cons: scoring and weighting system used can affect project rankings

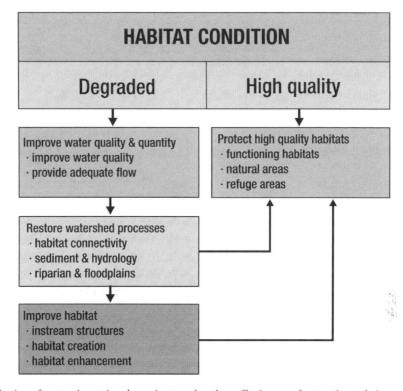

Figure 6.3 Prioritization of restoration actions by project type based on effectiveness of restoration technique and whether it restores underlying watershed processes (connectivity, transport water, sediment, or organic matter) or improves habitat (from Roni *et al.* 2008).

and aquatic species is the identification of intact refuge areas. This typically involves identification of important regions, ecosystems, watersheds, or habitats for the species of interest to determine priorities for protection and restoration (Beechie & Bolton 1999). This may also include protection of refuge or core areas and restoration of nearby areas to allow expansion and recovery or migration corridors for fauna. The *regufia* approach is largely based on island biogeography theory, with the assumption that these refugia will serve as key source areas for maintaining and rebuilding populations and ecosystems (Diamond 1976; Sarkar *et al.* 2006; Beechie *et al.* 2008) and on models and field studies that show that focusing restoration actions near healthy populations or colonists can lead to more rapid species recovery (Detenbeck *et al.* 1992; Huxel & Hastings 1999; Davey & Kelly 2007). The systems of parks in Africa, North America and elsewhere provide some of the earliest and most visible example of refugia for big game animals and unique ecosystems. Efforts for recovery and protection of endangered fishes have focused on identifying and

prioritizing healthy watersheds for protection and degraded watersheds for restoration (Beechie *et al.* 1996; Williams *et al.* 2007; Noss *et al.* 2009; Figure 6.4).

Refugia are identified and prioritized most typically with either GIS or computer models or through MCDA (discussed in Section 6.4.1.7). In the last 25 years numerous computer models and conservation biology software have been developed for identifying refugia for a variety of terrestrial and aquatic species (Sarkar *et al.* 2006; Moilanen *et al.* 2008; Lynch & Taylor 2010; Zafra-Calvo *et al.* 2010). In contrast, simple multimetric scoring approaches (MCDA), which often utilize GIS data on land cover and habitat condition to assist with scoring, have been widely used for identifying refuge and restoration areas for fishes (Nehlsen 1997; Williams *et al.* 2007; Noss *et al.* 2009; see also Section 6.4.1.7).

The strength of the refugia approach is that it focuses on protecting relatively healthy watersheds and populations, which is more cost effective in the short term than trying to restore degraded watersheds or unhealthy populations, and reduces the likelihood of local extirpation of the

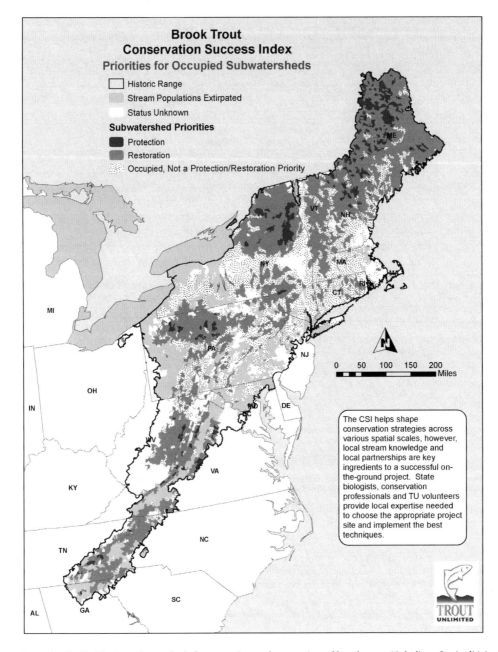

**Brook Trout
Conservation Success Index**

Priorities for Occupied Subwatersheds

Historic Range
Stream Populations Extirpated
Status Unknown

Subwatershed Priorities

Protection
Restoration
Occupied, Not a Protection/Restoration Priority

The CSI helps shape conservation strategies across various spatial scales, however, local stream knowledge and local partnerships are key ingredients to a successful on-the-ground project. State biologists, conservation professionals and TU volunteers provide local expertise needed to choose the appropriate project site and implement the best techniques.

Figure 6.4 Example of prioritization of watersheds for protection and restoration of brook trout (*Salvelinus fontinalis*) in the Eastern USA using Trout Unlimited's Conservation Success Index (Williams *et al.* 2007), a type of multi-criteria decision analysis. Image courtesy of Trout Unlimited.

species of interest (McGurrin & Forsgren 1997; Beechie *et al.* 2008). In addition, it typically focuses on protecting whole watersheds or sub-watersheds, is particularly well suited for prioritizing which watersheds or areas of watersheds to restore, and both scoring systems and computer models can be modified to incorporate connectivity or linkage between protected and restored areas (Moilanen *et al.* 2008). The disadvantages are that it is difficult to apply at a site or project level as it only addresses whether the location of the project is appropriate; if the refugia identified are small or fragmented, they tend to be more susceptible to disturbance (Sedell *et al.* 1990). It is therefore best suited for prioritizing reaches or watersheds rather than prioritizing individual projects within refugia, unless included as one factor in a multi-criteria approach.

6.4.1.3 Habitat area and increase in fish or other biota

Early efforts at conservation and restoration, and even ongoing efforts at recovering endangered species, have prioritized restoration based on individual species and their specific habitat needs. These approaches can be broken into two general categories: simple approaches for ranking restoration projects based on increases in biota (e.g. fish or other important taxa) or habitat area, and single- or multi-species capacity or life-cycle models that help rank habitat types for restoration. We discuss the capacity and life-cycle models in the following subsection.

Restoration actions are frequently prioritized partly by habitat area or length restored, by the predicted increase in fish or biota production, or by a combination of length and increase in biota (Roni *et al.* 2003; Beechie *et al.* 2008; Pini-Prato 2008). This involves simply ranking projects based on their estimated increase in fish numbers or densities, or habitat length or area. For example, Tables 6.3 and 6.4 provide common restoration projects and how their rankings would differ based on total stream area restored, total increase in fish production, or fish per kilometer restored. This approach requires reasonable estimates of habitat area and increases in fish or the biota of interest for all project types, which unfortunately is often not easily measured and not available for most species. Estimates of increases in fish or biota production from restoration techniques are relatively rare and typically focus on only a few types of projects (Roni *et al.* 2005). For example, the fish response to reconnecting isolated habitats, creating habitats, or placing wood or other structures in streams has been quantified for a

number of salmonid species (O'Grady 1995; Roni *et al.* 2008; Whiteway *et al.* 2010). In contrast, measurements of fish or aquatic biota responses to road removal, land protection, conservation easements (paying landowner to leave land in a natural state), riparian replanting, and other more process-based restoration techniques are not readily available and are difficult to quantify with even the best monitoring programs (Beechie *et al.* 2005; Lucchetti *et al.* 2005; Pollock *et al.* 2005). Even a metric as simple as total area or length of stream restored can be difficult to calculate and compare for projects as diverse as road removal, riparian planting, or placement of instream structures (e.g. what length of stream is affected by removal of 5 km of logging roads?). A consistent metric therefore may not be available to compare and rank all project types. To address this, groups sometimes prioritize only those projects for which data are available or make estimates of potential increases in fish abundance based on data available for other project types. Using some measure of increases in biota or habitat for each project is a common criterion used in many other prioritization approaches.

6.4.1.4 Capacity and life-cycle models for prioritizing habitats

Some of the earliest aquatic restoration efforts were focused on restoration of specific habitat types for Atlantic salmon (*Salmo salar*), brook trout (*Salvelinus fontinalis*), brown trout (*S. trutta*) or other salmonid species (White 2002; Chapter 1). These efforts evolved into focusing on identifying habitats that limit production for specific species in an effort to guide restoration, such as those developed for coho salmon (*O. kisutch*) and steelhead (*O. mykiss*) (Reeves *et al.* 1989; Beechie *et al.* 1994; Pollock *et al.* 2004). These methods are primarily based on estimates of habitat availability and habitat-specific density and survival for the species of interest. They have been used to identify the most important habitats for protection and restoration based simply on habitat types that limit population size, and also to evaluate losses of habitat by comparing historical and current habitat estimates. Many of the recent capacity models are similar to that developed by Reeves *et al.* (1989), which require season and life-stage specific estimates of habitat area and fish densities to calculate total production at each life stage and, ultimately, determining which life stage and habitat are bottlenecks limiting fish production.

Life-cycle models are more complex versions of simple limiting factors analyses based on habitat capacity (density). They can be used to identify a list of priority

Table 6.3 Hypothetical examples of common restoration projects and data commonly used to prioritize restoration actions within a large watershed. All numbers are hypothetical and for demonstration purpose only. Restoration type: Conn. = connectivity, Imp. = habitat improvement, Process = restores processes (riparian, sediment, etc.).

Project description	Water-shed	Stream length (km)	Cost (US$)	Increase in fish	Cost/ fish (US$)	Fish/ km	No. rare species	Restoration type
Replacement of culvert blocking fish passage	Deer Creek	3.0	250,000	1,530	163	510	1	Conn.
Large wood and instream structures	Clear Creek	2.0	60,000	610	98	305	2	Imp.
Replant 3 ha of riparian area	Clear Creek	1.0	48,000	110	436	110	2	Process
Fencing to exclude sheep grazing	Cold Creek	3.1	45,000	326	138	105	1	Process
Increase instream flows	Dry Creek	11.0	500,000	5,500	91	500	4	Flow
Reconnect side channel	Big River	2.9	200,000	875	229	302	2	Conn.
Remove 5 km of forest road	Big River	7.0	65,000	630	103	90	2	Process
Setback lower Bear River levee 500 m	Bear River	1.1	1,600,000	440	3,636	400	4	Conn.
Remove mill dam on upper Bear River	Bear River	10.0	2,500,000	5,050	495	505	3	Conn.
Remeandering of channelized stream	Bull Run	3.3	800,000	990	808	300	0	Imp.

habitats to assist in ranking projects (e.g. Nickelson & Lawson 1998; Karieva *et al.* 2000; Greene & Beechie 2004; Figure 6.5), or to assist in ranking of marine habitats for protection and restoration of rockfish (*Sabestes* spp.) or other aquatic and terrestrial fauna (Levin & Stunz 2005; Mangel *et al.* 2006). Life-cycle models typically use life-stage or age-specific survival or capacity estimates to determine life stages and habitats that are limiting production and thus, if addressed, would increase productivity and abundance of the species of interest. For example, life-cycle modeling by Karieva *et al.* (2000) using a Leslie Matrix approach suggested that restoration efforts for Chinook salmon (*O. tshawytscha*) on the Columbia River basin should focus on restoration of tributary habitat and estuarine habitat. This has led to large efforts

to restore these habitats throughout the 220,000 km² portion of the 673,396 km² basin presently accessible to threatened or endangered salmon populations.

The benefits of capacity and life-cycle models are that they use empirical data to estimate limitations or bottlenecks in survival (Beechie *et al.* 2008). The shortcomings are that they often require very detailed information on habitat and habitat use or life-stage-specific survival estimates and, more importantly, they typically focus on one important or endangered species. If another species becomes the focus of restoration, then the analyses need to be rerun and the focus of habitat restoration would change from one species to the next (Beechie & Bolton 1999). The need to consider multiple species has led to analysis for species complexes or species diversity, and

Table 6.4 Ranking of projects from Table 6.3 based on amount of habitat restored, cost, increase in fish numbers, cost benefit, number or rare species and the project type. Numbers in each column represent rankings for each project for each criterion (columns) with a 1 being the highest priority and 10 the lowest. Project types (last column) were given a ranking of 1 if the primary goal was to restore flow, 2 if restored connectivity, 3 if restored other processes, and 4 if it improved habitat. Note in this example two projects with an equal score are given the same rank.

Restoration project description	Stream length (km)	Cost (US$)	Estimated annual increase in fish #s	Cost/ fish	Fish/km	No. rare species present	Project type
Replacement of culvert blocking fish passage	6	6	3	5	1	4	2
Large wood and instream structures	8	3	7	2	5	3	4
Replant 3 ha of riparian area	10	2	9	7	8	3	3
Fencing to exclude sheep grazing	5	1	10	4	9	4	3
Increase instream flows	1	7	1	1	3	1	1
Reconnect side channel	7	5	5	6	6	3	2
Remove 5 km of forest road	3	4	6	3	10	3	3
Setback lower Bear River levee 500 m	9	9	8	10	4	1	3
Remove mill dam on upper Bear River	2	10	2	8	2	2	2
Remeandering of channelized stream	4	8	1	9	7	5	4

to the call for focusing on restoring ecosystems and processes that will create and maintain the natural mosaic of habitats rather than focusing on individual species (Beechie & Bolton 1999; Roni *et al.* 2002; Mangel *et al.* 2006).

6.4.1.5 Costs, cost-effectiveness, and cost-benefit analysis

Limited economic resources mean that restoration actions should be prioritized at least in part by economic costs and benefits. This is usually done by estimating total cost or cost-effectiveness or through a cost-benefit analysis (Plummer 2005). Estimating the cost of a restoration project can be straightforward for projects that have a detailed budget; however, this becomes more complex if some costs are not yet known (i.e. cost of land) or donations such as land, materials, or volunteer labor, a common part of many restoration projects, are considered. It is also important to include when the costs will be incurred and whether periodic maintenance of the project will be needed. Estimating benefits of a project is much more difficult than estimating costs, as costs are based on goods and services sold in the market while benefits include many intangibles not easily quantified in monetary terms.

One of the simplest approaches is to rank projects by total cost, but this ignores differences among projects in total area treated, benefits to resources, and when costs are incurred. Ideally, projects would be ranked either by cost-effectiveness or cost-benefit analysis. Ranking projects based on cost-effectiveness requires information on both cost and the measure of project effectiveness. Common measures of effectiveness include fish or biota abundance, diversity, increase in habitat area or length, or improvements in other ecosystem services such as water quality, drinking water, or recreation. Most prioritization of projects based on cost-effectiveness has occurred on similar project types such as wetlands or instream structures (Johnson & Lynch 1992; O'Grady 1995; Cederholm *et al.* 1997; Plummer 2005), although some studies have compared different project types (Bilsby *et al.* 1998). Comparing across project types requires a common metric of effectiveness that is available for all project types which, as mentioned previously, can be challenging to estimate. The metric chosen can have a considerable effect on priorities. For example, the

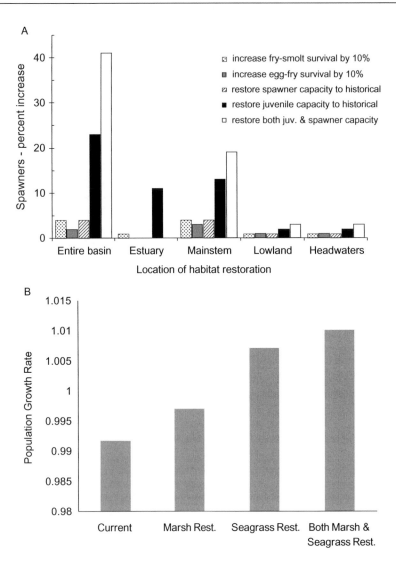

Figure 6.5 Identification of priority freshwater and estuarine habitats or restoration actions identified using life-cycle models for (A) Pacific salmon (*Oncorhynchus* spp.) (modified after Scheurell *et al.* 2006); and (B) red drum (*Sciaenops ocellatus*) in the Gulf of Mexico (modified after Levin & Stunz 2005).

rankings of the 10 projects in Table 6.3 changes substantially if they are prioritized on total stream length, cost, or cost per fish (Table 6.4). These analyses assume, of course, that the costs for restoring a given length of habitat are fixed or that each additional kilometer restored will cost the same. This is often not the case, as there may be some efficiency gained by restoring additional lengths of stream.

Restoration actions can also be prioritized based on their economic benefit or net economic benefit, commonly called a cost-benefit or (more accurately) a benefit-cost analysis (Thurston *et al.* 2009). These types of analysis are common in wetland and environmental protection and mitigation, but have been used less frequently for prioritizing restoration actions. This is partly because it requires an estimate of the benefits of a project in economic or monetary terms. This has been most frequently estimated for fish in terms of value of commercial harvest or sport-caught fish, but also as the value of a day spent fishing or the impact to the economy through jobs created or money spent on recreation (e.g. purchase of food, gas, lodging, fishing equipment) (Kennedy & Crozier 1991; Hickley & Tompkins 1998; Pitcher & Hollingworth 2002). Less tangible 'ecosystem services' such as the value of instream flows, value of improved water quality, recreation opportunity, and other services have also been estimated (Rietbergen-McCracken & Abaza 2000; Brouwer & Pearce 2007; Thurston *et al.* 2009).

Any effort to compare costs or benefits received in different years requires adjusting costs or benefits to a base year or calculating a present value based on a discount rate that adjusts for the time value of money (real discount rate) or the time value of money plus expected inflation (nominal discount rate) (Plummer 2005). Similarly, projecting costs of projects that occur in the future should be adjusted for inflation (Box 6.1).

Box 6.1 Adjusting restoration costs for the time value of money

The discount rate is the rate at which future dollars are discounted relative to current dollars. The importance of discount rates in valuing restoration projects can be demonstrated with a simple example of three projects that each cost $100,000 but incur costs in different years. Project 1 costs $100,000 today, project 2 will cost $100,000 but not be possible for three years, and project 3 will cost $25,000 a year starting today and for the next four years. Using the formula

$$\text{Present value} = \frac{V}{(1+r)t} \qquad (1)$$

where V is the value in dollars or other currency, r is the discount rate and t is the time in years, we see that present value of project 2 using a discount rate of 5% is $86,384. Project 3 would require calculating the present value of a stream of equal costs, which can be done with the formula

$$\text{Present Value} = V_0 \frac{1-(1+r)^{-n}}{r} \qquad (2)$$

Using this formula, the present value of project 3 would be $88,649. The same approach could be used to calculate the benefits of a project. For example, if the economic benefit of the project over the next 10 years is $1000 a year, the present value of those benefits at a discount rate of 5% would be $7722 rather than $10,000. We could also use this approach to project the future costs of a project if we know its present value and have some estimate of inflation. Detailed explanation of discounting and cost-benefit analysis for watershed restoration can be found in Plummer (2005), Thurston *et al.* (2009), or in a variety of economics texts.

Using cost and benefit to prioritize restoration is a way of efficiently allocating economic resources, but it has its limitations. It is very difficult to place economic value on non-market or intangible goods such as the value of the existence of a species, value of a healthy environment, or how people feel when they spend time in a natural area. Estimates of the value of these intangibles are site or project specific, can be controversial and have not been calculated for the vast majority of species or ecosystems (Heinzerling & Ackerman 2002). Economic analysis also assumes all projects are independent, which is often not the case. For example, a project that restores rearing habitat for juvenile fishes may be of little use if its success depends on a project that restores spawning habitat (Plummer 2005). It is important to note that economic analyses such as those described here are based on peoples' preferences and not on what might be best for an ecosystem. Prioritizing projects based solely on cost, cost-effectiveness or cost-benefit analysis can lead to an entirely different set of priorities than other approaches. Therefore, while these cost-based approaches can be useful, they should not be the only factor considered and are most useful when coupled with other metrics or approaches when ranking watersheds or projects for restoration.

6.4.1.6 Conservation planning software and computer models

As mentioned previously, a variety of computer models and software have been developed for identifying conservation areas and priority watersheds for the protection and recovery of rare and important flora and fauna throughout the world (Sarkar *et al.* 2006). The majority of these have focused on terrestrial species or marine reserves, but have more recently been modified and applied to conservation of freshwater fishes (Moilanen *et al.* 2008) and, only recently, specifically to restoration planning (Fullerton *et al.* 2010). These computer modeling efforts range from simple GIS exercises overlaying land cover, land use and species distribution layers to identify priority areas for protection and restoration – such as for grizzly bears (*Ursus arctos horribilis*) in Canada or salmon in the United States (Noss *et al.* 2009) – to more complex analysis that incorporate GIS data layers, professional opinion, life history models, and other biological and sociological information – such as those for endangered fishes (Villa *et al.* 2002; Abellán *et al.* 2005; Zafra-Calvo *et al.* 2010).

Software packages such as Marxan, ConsNet, Natureserve Vista, C-Plan, Zonation, and others have been developed to assist with conservation planning and many are available as shareware on the Internet. Most focus on physical and biological metrics, but incorporating socioeconomic factors such as the willingness of landowners to participate in conservation efforts can have a dramatic impact on prioritizing areas for protection and restoration (Guerrero *et al.* 2010). Conservation planning software is helpful for ranking watersheds for protection and restoration (Moilanen *et al.* 2008), but less effective for restoration projects. The more complex software packages and computer models, while often very useful, lack transparency, can be difficult to explain to constituents, and are sometimes beyond the capabilities of many watershed restoration practitioners.

Models designed specifically to prioritize restoration projects have often focused on one type of restoration such as barrier removal (Steel *et al.* 2004; Pini-Prato 2008), or examined trade-offs of implementing different restoration scenarios in a basin (Steel *et al.* 2009; Fullerton *et al.* 2010). Models predicting the effects of land use on habitat conditions and fish distribution can be used to develop decision trees for selecting restoration projects (Poppe *et al.* 2008). With the exception of these efforts to rate barrier removal or other specific techniques, decision-support models are typically basin or region specific and need to be developed individually for each watershed (i.e. Baker *et al.* 2004; Fullerton *et al.* 2010). As with many other methods for prioritizing watersheds and projects for restoration, simple and complex models are often combined for use with scoring systems (Filipe *et al.* 2004). The benefits of these modeling approaches are that they can incorporate many different types of information, but the more complex models and software may not be easily used or interpreted by all users.

6.4.1.7 Scoring and multi-criteria decision analysis

One of the most common approaches, and one that can incorporate many criteria including the previously described approaches, is a scoring system that uses many different criteria to determine project priorities. The ranking of potential decisions based on multiple criteria has a long history in engineering and management and has more recently been applied to environmental conservation and restoration (Belton & Stewart 2002; Pohekar & Ramachandran 2004). These procedures, collectively known as multi-criteria decision analysis (MCDA), range from simple scoring and weighting systems to complex computer models and software (Malczewski 1999; Belton & Stewart 2002; Pohekar & Ramachandran 2004). Indeed,

some of the computer models and conservation planning software include MCDA or are used to assist with MCDA.

There are numerous publications, texts and journals dedicated to MCDA. It has been applied to a variety of watershed management problems including: prioritization of watersheds and sub-watersheds for protection and restoration (Gellis *et al.* 2001); endangered fish species protection (Williams *et al.* 2007; Lynch & Taylor 2010); river corridor and riparian protection and management (Harris & Olson 1997; Hermans *et al.* 2007); forest road removal (Allison *et al.* 2004); native plant community protection and restoration (Palik *et al.* 2000); and selection of estuarine, nearshore, and marine areas for protection (Villa *et al.* 2002; Johnson *et al.* 2003; Diefenderfer *et al.* 2009). MCDA is increasingly used to prioritize restoration actions for endangered fishes throughout a region or watershed (Roni *et al.* 2003; Beechie *et al.* 2008). For example, Trout Unlimited has developed a simple MCDA to rank watersheds for protection of endangered trout (*Oncorhynchus* and *Salvelinus* spp.) throughout the United States (Williams *et al.* 2007; Figure 6.4).

Related to MCDA are a number of multimetric indices of habitat quality such as the index of biotic integrity (IBI; Karr & Chu 1999), multilevel concept for fish-based assessment (MuLFLA; Schmutz *et al.* 2000), or the fluvial audit approach developed to rate river reach quality (Sear *et al.* 2009). These methods, developed to assess stream health, are often used to assist with management decisions including prioritization of watershed or stream reaches for protection and restoration (see also Chapter 3).

The most common types of MCDA for prioritizing watersheds or actual restoration projects are relatively straightforward and transparent scoring systems. For example, scoring approaches that use experts to rate projects or areas for restoration are based largely on quantitative data, and are relatively easy to explain and understand. In practice, most use a combination of GIS or field data to provide quantitative information on a variety of factors and professional opinion to assist with scoring of the criteria selected. Common factors for prioritizing watersheds include land use, area or proportion of intact habitat or degraded habitat, road density, barriers to migration, water quality, invasive species, riparian or habitat condition, species present, biodiversity, presence or abundance of rare or endangered species, habitat condition index based on computer models, and many others (Nehlsen 1997; Gellis *et al.* 2001; Williams *et al.* 2007). Common criteria for prioritizing individual projects include areas or length restored; potential increase in fish or species of interest; the number of rare or endangered species that will benefit from the project; the cost, cost-benefit, and cost-effectiveness (cost/fish); and whether it addresses a factor limiting biotic production, restores a sensitive habitat, restores a key process (restoration type), or is a refuge or priority watershed (Beechie *et al.* 2008). Data from the previously described approaches may also be included as criteria.

Regardless of how simple or complex the utilized MCDA is, there are several key steps that must be followed: selection of which MCDA to use; selection of criteria for scoring; selection of scoring and weighting; data collection or analysis to assist with scoring (if applicable); scoring and ranking; and the final list of project priorities (Figure 6.1). Similar to any other prioritization method, the goals and objectives of the approach need to be clearly defined, and the technical team must decide on the criteria for restoration based on the objectives and values of the working group. It is critical that the group clearly define the criteria to use in the prioritization process and document their decision-making process

including why they chose or excluded certain criteria. Failing to document these factors can lead to conflict later when the final scores are calculated, or when the data are presented to a larger audience.

How the criteria will be scored, whether each criterion will receive equal weighting and how the final scores will be calculated is also important. Methods for scoring projects include simple rating (e.g. 0 to 3, 1 to 5, 1 to 10); fixed point scoring (100 points distributed among all projects); ordinal ranking (rank from least to most important); graphical weighting; paired comparisons; and normalization of weights (Yoe 2002). The most straightforward and common approach is a simple rating system that uses a consistent range of scores for each criterion. To ensure that scores are assigned consistently, each score should be well defined. For example, for prioritizing watersheds for protection of brook trout (*Salvelinus fontinalis*), Trout Unlimited used a 1 to 5 scoring system for each criteria, and clearly described what constituted a 1, 2, 3, 4 and 5 (Williams *et al.* 2007; Table 6.5).

Table 6.5 Examples of 5 of 20 criteria and scoring used to prioritize watershed protection and restoration of endangered trout species in the United States (Williams *et al.* 2007). Note that, similar to many multi-criteria decision analysis, some criteria are quantifiable (% historic habitat, life history diversity), others are qualitative (climate change), and others partially quantifiable but require a qualitative judgment for scoring (water quality, flow regime).

Indicator	Definition	General scoring rules
Water quality	Measured by presence of water-quality-limited stream segments: number of mines and point sources of pollution	5: high quality (no WQ limited segments) 4: high quality (minor pollution sources) 3: moderate to high quality 2: moderate quality with significant sources of pollution 1: poor quality
Flow regime	Measured by seasonal fluctuations and total flows compared to historic	5: flow regime unaltered 4: flow c. 90% of historic 3: flows c. 75% of historic 2: flows c. 50% of historic 1: flows highly modified, <50% of historic
Percent of historic stream habitat occupied by species of interest	Percent of historic stream habitat (km) currently occupied versus historic	5: > 50% occupied 4: 35–49% 3: 20–34% 2: 10–19% 1: <10%
Life history diversity	Number or life history forms present compared to presumed historic condition	5: all life history forms present 3: two or more life history forms present, but at least one absent 1: one life history form present and others absent
Climate change	Resistance to climate change impacts as a function of watershed connectivity, habitat conditions, and elevation gradients	5: high condition, connectivity 4: moderate condition moderate connectivity 3: moderate conditions but low connectivity 2: low conditions, low connectivity 1: very low conditions

Table 6.6 Simplified prioritization of projects from tables 6.3 and 6.4 using a multi-criteria decision analysis with four criteria (cost/fish, fish/km, number of rare species, and project type) with all criteria given equal weight and a four-point scoring system (0, 1, 2, 3). (1) Cost/fish: <$100/fish: 3; $100–499/fish: 2; $500–999/fish: 1; >$999/fish: 0. (2) Fish/km: >500 fish/km: 3; 300–499 fish/km: 2; 100–299 fish/km: 1; <100 fish/km: 0. (3) Rare species: 3 or 4 species: 3; 2 species: 2; 1 species: 1; 0 species: 0. (4) Project type: restores flows or connectivity: 3; restores processes: 2; improves habitat: 1. Total score is simple sum for each criterion for each project. The last column provides an example of how rankings might change if one criterion (fish/km) was double weighted (score for fish/km × 2).

Restoration project	Cost/fish	Fish/km	No. rare species present	Project type	Total score (equal weight)	Total score (double weight for fish/km)
Removal of culvert blocking fish passage	2	3	1	3	9	12
Large wood and instream structures	3	2	2	1	8	10
Replant 3 ha of riparian area	2	1	2	2	7	8
Fencing to exclude sheep grazing	2	1	1	2	6	7
Increase instream flows	3	3	3	3	12	15
Reconnect side channel	2	2	2	3	9	11
Remove 5 km of forest road	2	0	2	2	6	6
Setback lower Bear River levee 500 m	0	2	3	3	8	10
Remove mill dam on upper Bear River	2	3	3	3	11	14
Remeandering of channelized stream	1	2	0	1	4	6

The team also needs to determine whether some criteria should be given greater weight than others as, in many cases, some criteria are more important than others. The scores are then multiplied by the predetermined weight. Table 6.6 demonstrates how project ranking might vary using four simple criteria and a four-point scoring system both with and without weighting. We must be careful not to inadvertently weight criteria by using a different scoring system for different criteria. For example, if a system of 0 or 1 is used to rate presence or absence of a species, but 1 to 5 is used to score project cost, cost will be given greater weight in determining final score for projects. If different score ranges are necessary for each criterion, they may need to be normalized to a common scale. Including multiple similar criteria that are in essence different measures of the same factor (e.g. habitat area and habitat length) can unintentionally bias scoring systems towards that factor if each factor does not have the same number of criteria. Similarly, the technical team must decide how to treat projects for which no score can be given for some criteria. Leaving some criteria blank will result in lower scores for projects with incomplete information. Either the team must decide on a default

score, such as an average score, or the projects should be ranked with and without the criterion (or those criteria) to demonstrate how the priorities might change and the importance of that criterion (or those criteria).

The final step is to summarize the scores for each project or watershed by either summing or averaging the scores. In cases where professional judgment is used and projects are scored by multiple team members, the scores must be averaged or summed to provide a final ranking for each projects. While there are more complex ways of weighing and calculating scores, it is important to consider that the more complicated the method, the more difficult it will be for a broad audience to understand.

6.4.2 Selecting a prioritization approach

Project ranks can vary widely among the different approaches (Table 6.4) and there is no single best approach for prioritizing watersheds, habitats, or restoration projects. Which of the above approaches is appropriate depends largely on the goals of the group, availability of data, and time and resources available. MCDA approaches are often recommended because of their transparency and the ability to incorporate a variety of

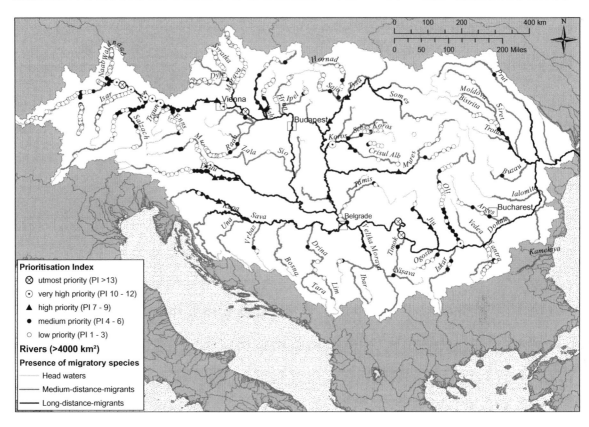

Figure 6.6 Prioritization of barriers to connectivity within the DRB using the Prioritization Index (PI) within habitat of long-distance and medium-distance migrants (see Box 6.2 for description of PI and methods).

qualitative and quantitative data, as well as professional opinion. While ranking can be used directly to set priorities (i.e. project with highest ranking is given top priority), it is often used to categorize projects or watersheds as high, medium, or low priority and determine sequencing of projects (Figure 6.4). For example, hundreds of barrier removal projects in the Danube River were categorized into high, medium, and low priorities based on their rankings (Figure 6.6; Box 6.2). It should also be noted that these approaches are usually used to determine the sequencing of restoration actions rather than to completely eliminate projects.

6.5 Completing analyses and examining rankings

Once a prioritization approach and criteria have been defined, the next steps are to collect the data, run an analysis or calculate scores, and determine project or watershed rankings. Obviously, the amount of time and resources needed will vary greatly by approach. Some approaches such as modeling may require both considerable data collection and computational time, while a basic MCDA approach may only require scoring projects based on simple information on project cost, area, and rough estimates of ecological benefit (Tables 6.3 and 6.5). The output of the procedure should be double-checked and reviewed to make sure there are no errors and that scores are accurate. If scores are correct but do not meet expectations, new criteria should be added with caution. There can be a temptation to add criteria or change scores for specific criteria so that a project receives the desired score which, if allowed, could lead to bias and undermine the purpose of the prioritization process. In some cases projects will have identical scores; this is not a problem if projects are simply grouped as high, medium, or low based on a range of scores. However, if the team wants to

Box 6.2 Case Study: Restoration of connectivity in the Danube Basin

Restoring habitat connectivity by removing migration barriers or installation of fish passes is key to improving the ecological health of riverine fish communities. Failing to address these is considered a major risk for failing to meet the environmental objectives under the European Union's Water Framework Directive (WFD). Consequently, restoring continuity is one of the main restoration actions within River Basin Management Plans of the EU member states (EEB 2010). This is particularly evident in large catchments like the more the 800,000 km² Danube River Basin (DRB), where most fish are migratory and numerous man-made barriers exist. More than 900 interruptions of river and habitat connectivity (migration barriers) are located in the DRB and its major tributaries (catchment areas >4000 km²), 56 of which are located in the mainstem Danube (ICPDR 2009). This has had particularly negative impacts on long-distance migrants (LDM) such as sturgeon (*Acipenser* spp.) and shad (*Alosa* spp.). These species once migrated from the Black Sea upstream several thousand kilometers to upper areas of the Danube and tributaries but are now blocked at the Iron Gate hydropower stations (km 863). Man-made barriers have also impacted medium-distance migrants (MDM or potadromous fishes) such as *Abramis brama*, *Abramis sapa*, *Acipenser ruthenus*, *Aspius aspius*, *Barbus barbus*, *Chondrostoma nasus*, *Hucho hucho*, and *Lota lota* which historically migrated up to 200 km (Waidbacher & Haidvogl 1998).

The overall goal of continuity restoration in the DRB is to allow fish migration throughout the entire basin. Due to the high number of barriers and limited resources, a prioritization index (PI) was developed to rank barriers for restoration using a modified scoring system with multiple criteria and weights. The specific objective of the PI was to provide a stepwise and ecologically efficient approach for implementation of restoration measures at a basin-wide scale. The goals and procedure for the PI were developed by Schmutz & Trautwein (2009) in collaboration with the DRB countries, the International Commission for the Protection of the Danube River (ICPDR), and experts from DRB countries. DRB countries provided data and the ICPDR was responsible for administration and integration of results in Danube River Basin Management Plan (DRBMP).

In order to guarantee an ecologically effective restoration process, five criteria that focus on the specific requirements of migratory fish species were selected (Table 6.7). The location of barriers within the river continuum is of major importance in order to provide access to spawning habitats. The availability of non-fragmented habitat is essential to guarantee minimum population sizes and sufficient genetic exchange between subpopulations (e.g. Fagan *et al.* 2002; Traill *et al.* 2007). The initial criteria for restoring connectivity are therefore based on the migratory behavior and habitat of LDM and MDM species in the DRB (criterion *migratory habitat*), which was determined from previous studies (Hensel & Holčík 1997; Waidbacher & Haidvogl 1998; Schmutz &

Table 6.7 Prioritization criteria and weighting factors for restoring continuity in the Danube River Basin

	Criteria	Scores
1.	Migratory habitat	
	• Long-distance migrants within Danube	4
	• Long-distance migrant habitat within Danube tributaries	2
	• Medium-distance migrants habitat	1
	• Short-distance migrants (head waters)	0
2.	First obstacle in river segment upstream of river mouth	
	• Yes – in Danube	2
	• Yes	1
	• No	0
3.	Distance from mouth (river segment)	
	• First river segment upstream of mouth	3
	• Second river segment upstream of mouth	2
	• Third river segment upstream of mouth	1
	• River segments upstream of third river segment	0
4.	Length of reconnected habitat (values in bracket are valid for Danube)	
	• >50 km (>100 km)	2
	• 20–50 km (40–100 km)	1
	• <20 km (<40 km)	0
5.	Protected site (Natura 2000)	
	• Yes	1
	• No	0

Trautwein 2009). Weirs representing barriers to fish migration were scored from 0 to 4 with those that affected long-distance migrants given a 4 and those that affected only short-distance migrants (SDM) given a 0.

The PI also needed to reflect the increasing migratory requirements from the headwaters down to the Danube. The Danube itself is the only pathway for long-distance migrants (LDM) from the Black Sea to the upstream spawning areas in the Danube and its tributaries. Hence, if the migration barrier was in the mainstem Danube it was given the highest priority in restoring river connectivity. In rivers with multiple barriers, obstacles located furthest downstream are given higher priority than those further upstream (criterion *first obstacle*). Moreover, the highest priority was given to most downstream barriers in the Danube River itself, as restoration of barriers further upstream will become effective for MDM and LDM only after restoration of downstream-located barriers. Consequently, the further an obstacle is located from the river mouth, the lower priority it received (criterion *distance from mouth*; Table 6.7).

Barriers were weighted by the length of the reconnected river segment (criterion *length of reconnected habitat*) to give higher weight to longer and less fragmented river segments (river stretch from tributary to tributary). For this criterion, different river length classes were defined for the Danube and the tributaries to account for different river sizes (Table 6.7). The final criterion was related to the environmental protection status. Obstacles within protected areas of the European Natura 2000 network receive higher priority because it is more likely that those river segments will be maintained in good condition or will be restored than unprotected river segments (criterion *protected site*).

The Prioritization Index (PI) for each barrier is calculated using the formula:

$$PI = migratory\ habitat \times (1 + first\ obstacle\ upstream + distance\ from\ mouth + length\ of\ reconnected\ habitat + protected\ site).$$

This is a modified scoring system where the PI weights the first and most important criteria (migratory habitat) by the cumulated weight of the four other criteria. This procedure guarantees that barriers within the Danube and affecting LDM receive higher scores than in the tributaries and within MDM habitats. This is to ensure that the index prioritizes projects based on the main goal of free migration routes. The maximum possible PI value is 36 and the minimum 0 (only in head waters). For graphical representation the PI was grouped into 5 priority classes: utmost (PI >13), very high (10–12), high (7–9), medium (4–6) and low (1–3).

These results reveal clear ecological priorities for restoring connectivity within the DRB. Barriers in the lower Danube are the highest priority with values ≥20 (Figure 6.6). In the upper Danube, the PI ranges over values 8–16 as long as the Danube is classified as LDM habitat. Within the LDM habitat, the obstacles in Bavaria generally receive higher values compared to Austria because longer habitats are reconnected and most obstacles are within protected Natura 2000 sites. Within the tributaries, the obstacles near the river mouth generally have higher PI values than obstacles located further upstream. Out of the 671 prioritized barriers, 29 (3%) are of the utmost priority for restoration, 99 (10%) of medium priority and 543 (58%) of low ecological priority. More than a quarter of the barriers are not currently priorities for ecological restoration (PI = 0) on the basin-wide scale as they are located in headwaters or artificial canals. Of particular importance are the Iron Gate Dams I & II in the lower Danube, which have resulted in sharp declines in most Danube sturgeon species (now endangered) with significant regional economic impacts on the productivity of fisheries (Hensel & Holčík 1997). As part of the DRBM Plan, the first step is a feasibility study to modify the Iron Gate Dams to allow free fish migration, particularly for sturgeon species.

The PI provides a guideline whereby the final decision where and when to restore connectivity depends on the technical feasibility to build fish passes or remove barriers, the costs involved, and also the relevance for national restoration and conservation programs. In this case, the Danube countries have identified 108 fish migration aids that will be constructed in the DRBD by 2015. In addition, more than 600 measures to restore river continuity interruptions will be undertaken in the second (2021) and third (2027) WFD cycle and some migration barriers will not be restored at all due to technical problems or disproportionate costs (ICPDR 2009).

distinguish between two similarly scored projects, some additional discussion may be needed to determine how to rank projects with similar scores.

A sensitivity analysis can be performed for models and complex MCDA to see the relative sensitivity of different criteria and their weights in the final total score. In general, this requires systematically setting the value of the criterion of interest at different levels, fixing the scores of the other criteria and examining the influence that variation in the criteria of interest has on the final project scores. This can be repeated for all criteria to understand which are most sensitive. For example, Levin & Stunz (2005) conducted a sensitivity analysis of their life-cycle model to show that most of the variability in population growth rate of red drum (*Sciaenops ocellatus*) is explained by larval and juvenile survival rates. This was then used to demonstrate which habitat restoration techniques might produce the largest increase in survival (Figure 6.5b).

Despite its rank, a variety of factors can influence whether a project is implemented including cost, permitting, willingness of landowners, and the planning horizon needed to implement and complete a project. That is, even the highest-ranking project or projects may not be feasible due to land-use constraints or if funds are not available. Nevertheless, these projects may ultimately be necessary to achieve restoration goals. For example, achieving fish recovery objectives for the Elwha River in Washington State requires removal of two large dams, but these removals have taken many years to realize because of legal and environmental issues and the number of studies that were required before demolition. In contrast, a low rank for a project does not necessarily mean it should never be implemented. In practice, lower-ranking projects are sometimes implemented first because they are either easy to implement or less expensive, or until factors preventing implementation of higher-priority projects are overcome. However, lower-priority projects often have lower environmental and biological benefits, and restoration goals may not be achieved even with a large number of lower priority projects.

Documentation of the approach and decision-making process used to develop the priorities is critical. Restoration projects within a basin or restoration program are frequently reprioritized annually as projects are completed, new funding becomes available, or new restoration and protection opportunities are identified. It is therefore very important that all these steps are documented so that funding partners, those reviewing the projects, and those proposing the projects can easily understand the process and the process can be consistently repeated periodically. This is most typically done by providing a brief write-up and report describing the process and results.

6.6 Summary

A suite of different methods and tools exists for prioritizing restoration actions within a watershed or across watersheds, or for prioritizing projects throughout a watershed or region. Regardless of which approach is chosen, several logical steps can be followed to ensure that the approach is transparent, repeatable, and achieves its objectives. These include identifying the objectives and scale, determining who will prioritize actions, selecting a prioritization approach and criteria, collecting the data to be used, and analyzing and ranking projects or watersheds. The objectives and the scale should have been partly outlined when the goals for restoration were defined in the assessment process. Well-defined goals with specific objectives are helpful in determining the prioritization approach and in guiding the selection of criteria. The individuals or technical team who will decide on the prioritization approach and select the criteria need to be identified either before or after definition of the goals. Common approaches to prioritizing watersheds or restoration actions include project type, habitat type or biota, life-cycle models, cost or cost-effectiveness, conservation planning software, or multi-criteria scoring systems which can incorporate a variety of approaches and criteria. Common criteria used in these approaches include biological (e.g. fish density, fish or macroinvertebrate diversity, presence or absence of key species), physical (e.g. instream habitat area, type or condition, riparian condition, fine sediment), economic (e.g. cost, cost-effectiveness), sociopolitical (e.g. ownership, protected

areas), and others that are difficult to categorize (e.g. restoration type or implementation time). Which approach is most effective is largely dependent on the goals of the restoration and the funds, data, and time available. However, given that most restoration actions are funded from public sources, the MCDA or scoring system approach is often recommended because it is typically transparent and can incorporate a variety of data and values. Complex multi-criteria models can also benefit from a sensitivity analysis to see which criteria have the largest effect on rankings. Because restoration programs often involve and may affect many diverse groups, the process for prioritization should be well documented and made available to all interested parties either as a written report or online. Finally, prioritization should be repeated periodically as new projects are identified, projects are completed, and new information becomes available.

6.7 References

Abellán, P., Sanchez-Fernandez, D., Velasco, J. & Millan, A. (2005) Conservation of freshwater biodiversity: a comparison of different selection methods. *Conservation Biology* **14**, 3457–3474.

Allison, C., Sidle, R.C. & Tait, D. (2004) Application of a decision analysis to forest road deactivation in unstable terrain. *Environmental Management* **33**, 173–185.

Baker, J.P., Hulse, D.W., Gregory, S.V. *et al.* (2004) Alternative futures for the Willamette River Basin, Oregon. *Ecological Applications* **14**, 313–324.

Beechie, T. & Bolton, S. (1999) An approach to restoring salmonid habitat-forming processes in Pacific Northwest watersheds. *Fisheries* **24**(4), 6–15.

Beechie, T., Beamer, E. & Wasserman, L. (1994) Estimating coho salmon rearing habitat and smolt production losses in a large river basin, and implications for habitat restoration. *North American Journal of Fisheries Management* **14**, 797–811.

Beechie, T., Beamer, E., Collins, B. & Benda, L. (1996) Restoration of habitat-forming processes in Pacific Northwest watersheds: a locally adaptable approach to salmonid habitat restoration. In: Peterson, D.L. & Klimas, C.V. (eds) *The Role of Restoration in Ecosystem Management*. Society for Ecological Restoration, Madison, Wisconsin, pp. 48–67.

Beechie, T.J., Veldhuisen, C.N., Beamer, E.M. *et al.* (2005) Monitoring treatments to reduce sediment and hydrologic effects from roads. In: Roni, P (ed.) *Monitoring stream and watershed restoration*, American Fisheries Society, Bethesda, Maryland, pp. 35–65.

Beechie, T., Pess, G., Roni, P. *et al.* (2008) Setting river restoration priorities: a review of approaches and a general protocol for identifying and prioritizing actions. *North American Journal of Fisheries Management* **28**, 891–905.

Belton, V. & Stewart, T.J. (2002) *Multiple Criteria Decision Analysis – an Integrated Approach*. Klewar Academic Publishers, Norwell, Massachusetts.

Bilsby, M.A., Cragg-Hine, D. & Hendry, K. (1998) Cost benefit analysis: its role in recreational fisheries development and management. In: Hickley, P. & Tompkins, H. (eds) *Recreational Fisheries: Social, Economic and Management Aspects*. Fishing News Books, Oxford, pp. 279–286.

Brouwer, R. & Pearce, D. (eds) (2007) *Cost-benefit Analysis and Water Resources Management*. Edward Elgar Publishing, Camberley, UK.

Cederholm, C.J., Bilby, R.E., Bisson, P.A. *et al.* (1997) Response of juvenile coho salmon and steelhead to placement of large woody debris in a coastal Washington stream. *North American Journal of Fisheries Management* **17**, 947–963.

Davey, A.J. & Kelly, D.J. (2007) Fish community responses to drying disturbances in an intermittent stream: a landscape perspective. *Freshwater Biology* **52**, 1719–1733.

Detenbeck, N.E., DeVore, P.W., Niemi, G.J. & Lima, A. (1992) Recovery of temperate-stream fish communities from disturbance: A review of case studies and synthesis of theory. *Environmental Management* **16**, 33–53.

Diamond, J.M. (1976) Island biogeography and conservation: strategy and limitations. *Science* **193**, 1027–1029.

Diefenderfer, H., Sobocisnki, K., Thom, R. *et al.* (2009). Multiscale analysis of restoration priorities for marine shoreline planning. *Environmental Management* **44**, 712–731.

EEB (European Environmental Bureau) (2010) *10 years of the Water Framework Directive: a toothless tiger? A snapshot assessment of the WU environmental ambitions. European Environmental Bureau.* Available from: http://www.eeb.org/?LinkServID=B1E256EB-DBC1-AA1C-DBA46F91C9118E7D&showMeta=0.

Fagan, W.F., Unmack, P.J., Burgess, C. & Minckley, W.L. (2002) Rarity, fragmentation, and extinction risk in desert fishes. *Ecology* **83**, 3250–3256.

Filipe, A.F., Marques, T.A., Seabra, S. *et al.* (2004) Selection of priority areas for fish conservation in Guadiana River Basin, Iberian Peninsula. *Conservation Biology* **18**, 189–200.

Fullerton, A.H., Steel, E.A., Lange, I. & Cara, Y. (2010) Effects of spatial pattern and economic uncertainties on freshwater habitat restoration planning: a simulation exercise. *Restoration Ecology* **18**, 354–369.

Gellis, A.C., Cheama, A. & Sheldon, M.L. (2001) Developing a geomorphic approach for ranking watersheds for rehabilitation, Zuni Indian Reservation, New Mexico. *Geomorphology* **37**, 105–134.

Greene, C.M. & Beechie, T.J. (2004) Habitat-specific population dynamics of ocean-type Chinook salmon (*Oncorhynchus tshawytscha*) in Puget Sound. *Canadian Journal of Fisheries and Aquatic Sciences* **61**, 590–602.

Guerrero, A.M., Knight, A.T., Grantham, H.S. *et al.* (2010) Predicting willingness-to-sell and its utility for assessing conservation opportunity for expanding protected area networks. *Conservation Letters* **3**, 332–339.

Harris, R., & Olson, C. (1997) Two-stage system for prioritizing riparian restoration at the stream reach and community scales. *Restoration Ecology* **5**(4S), 3442.

Heinzerling, L. & Ackerman, F. (2002) *Pricing the Priceless: Cost Benefit Analysis of Environmental Protection*. Georgetown Environmental Law and Policy Institute, Georgetown University, Washington DC.

Hensel, K. & Holčík, J. (1997) Past and current status of sturgeons in the Upper and Middle Danube River. *Environmental Biology of Fishes* **48**, 185–200.

Hermans, D., Erickson, J., Noordewier, T. *et al.* (2007) Collaborative environmental planning in river management: an application of multicriteria decision analysis in the White River Watershed in Vermont. *Journal of Environmental Management* **84**, 534–546.

Hickley, P. & Tompkins, H. (eds) (1998) *Recreational Fisheries: Social, Economic and Management Aspects*. Fishing News Books, Oxford, UK.

Huxel, G.R. & Hastings, A. (1999) Habitat loss, fragmentation, and restoration. *Restoration Ecology* **7**, 309–315.

ICPDR (International Commission for the Protection of the Danube River) (2009) *Danube River Basin District Management Plan*. ICPDR, Vienna.

Jay, A. (1971) *The Corporation Man: Who he is, what he does, why his ancient tribal impulses dominate the life of the modern corporation*. Jonathan Cape Ltd., London.

Johnson, D.L. & Lynch, W.E. Jr. (1992) Panfish use of and angler success at evergreen tree, brush, and stake-bed structures. *North American Journal of Fisheries Management* **12**, 222–229.

Johnson, G.E., Ebberts, B.D., Thom, R.M. *et al.* (2003) *An Ecosystem-Based Approach to Habitat Restoration*

Projects with Emphasis on Salmonids in the Columbia River estuary. Report no. PNNL-14412, Battelle Pacific Northwest National Laboratory, Sequim, Washington.

Karieva, P., Marvier, M. & McClure, M. (2000) Recovery and management options for Snake River spring/summer Chinook salmon in the Columbia River Basin. *Science* **290**, 977–979.

Karr, J.R. & Chu, E.W. (1999*) Restoring Life in Running Waters: Better Biological Monitoring*. Island Press, Washington DC.

Kennedy, G.J.A. & Crozier, W.W. (1991) Strategies for the rehabilitation of salmon rivers – post-project appraisal. In: Mills, D.H. (ed.) *Strategies for the Rehabilitation of Salmon Rivers*. London, Linnaean Society of London, pp. 46–63.

Levin, P.A. & Stunz, G.W. (2005) Habitat triage for exploited fishes: can we identify essential "Essential Fish Habitat?" *Estuarine, Coastal and Shelf Science* **64**, 70–78.

Lucchetti, G., Richer, K.O. & Schaefer, R.E. (2005) Monitoring acquisitions and conservation easements. In: Roni, P. (ed.) *Monitoring Stream and Watershed Restoration*. American Fisheries Society, Bethesda, Maryland, pp. 276–311.

Lynch, A.J. & Taylor, W.W. (2010) Evaluating a science-based decision support tool used to prioritize brook char conservation project proposals in the eastern United States. *Hydrobiologia* **650**, 233–241.

Malczewski, J. (1999) *GIS and Multicriteria Decision Analysis*. John Wiley and Sons, New York.

Mangel, M., Levin, P. & Patil, A. (2006) Using life history and persistence criteria to prioritize habitats for management and conservation. *Ecological Applications* **16**, 797–806.

McGurrin, J. & Forsgren, J. (1997) What works, what doesn't, and why? In: Williams, J.E., Wood, D.A. & Dombeck, M.P. (eds) *Watershed Restoration: Principles and Practices*. American Fisheries Society, Bethesda, Maryland, pp. 459–471.

Mears, P. & Voehl, F. (1997) *Team building*. CRC Press, Boca Rotan, Florida.

Moilanen, A., Leathwick, J. & Elith, J. (2008) A method for spatial freshwater conservation. *Freshwater Biology* **53**, 577–592.

Nehlsen, W.A. (1997) Prioritzation of watersheds in Oregon for salmon restoration. *Restoration Ecology* **5**, 25–33.

Nickelson, T.E. & Lawson, P.W. (1998) Population viability of coho salmon, Oncorhynchus kisutch, in Oregon coastal basins: application of a habitat-based life cycle model. *Canadian Journal of Fisheries and Aquatic Sciences* **55**(11), 2383–2392.

Noss, R., Nielsen, S. & Vance-Borland, K. (2009) Prioritizing ecosystems, species, and sites for restoration. In: Moilanen, A., Wilson, K.A. & Possingham, H. (eds) *Spatial Conservation Prioritization: Quantitative Methods and Computational Tools*. Oxford University Press, Oxford, pp. 158–171.

O'Grady, M. (1995) The enhancement of salmonid rivers in the Republic of Ireland. *Journal of the Chartered Institution of Waters and Environmental Management* **9**, 164–172.

Palik, B.J., Goebel, P.C., Kirkman, L.K., & West, L. (2000) Using landscape hierarchies to guide restoration of disturbed ecosystems. *Ecological Applications* **10**, 189–202.

Petersen, R.C., Petersen, L.B-M. & Lacoursiere, J. (1992) A building block model for stream restoration. In: Boon, P.J., Calow, P. & Petts, G.E. (eds) *River Conservation and Management*, John Wiley & Sons, New York, pp. 293–310.

Pini-Prato, E. (2008) River rehabilitation for fish: fish pass management in the Arno River Watershed. In: Gumiero, B., Rinaldi, M. & Fokkens, B. (eds) *IVth ECRR International Conference on River Restoration 2008*, Centro Italiano per la Riqualificazione Fluviale, Venice, Italy, pp. 257–266.

Pitcher, T.J. & Hollingworth, C.E. (2002) *Recreational Fisheries: Ecological, Economic and Social Evaluation*. Blackwell, London.

Plummer, M. (2005) The economic evaluation of stream and watershed restoration projects. In: Roni, P. (ed.) *Monitoring Stream and Watershed Restoration*. American Fisheries Society, Bethesda, Maryland, pp. 313–330.

Pohekar, S.D. & Ramachandran, M. (2004) Application of multi-criteria decision making to sustainable energy planning–A review. *Renewable and Sustainable Energy Reviews* **8**, 365–381.

Pollock, M.M., Pess, G.R., Beechie, T.J. & Montgomery, D.R. (2004) The importance of beaver ponds to coho production in the Stillaguamish River basin, Washington, USA. *North American Journal of Fisheries Management* **24**, 749–760.

Pollock, M.M., Beechie, T.J., Chan, S.S. & Bigley, R. (2005) Monitoring of restoration of riparian forests. In: Roni, P. (ed.) *Monitoring Stream and Watershed Restoration*. American Fisheries Society, Bethesda, Maryland, pp. 67–96.

Poppe, M., Schmutz, S., Muhar, S., Melcher, A., Hohensinner, S. & Trautwein, C. (2008) The prioritization of restoration measures in multiple-effected rivers: an Austrian approach. In: Gumiero, B., Rinaldi, M. & Fokkens, B. (eds) *IVth ECRR International Conference on River Restoration 2008*. Centro Italiano per la Riqualificazione Fluviale, Venice, Italy, pp. 251–256.

Preister, K. & Kent, J.A. (1997) Social ecology: a new pathway to watershed restoration. In: Williams, J.E., Wood, C.A. & Dombeck, M.P. (eds) *Watershed Restoration: Principles and Practices*. American Fisheries Society, Bethesda, Maryland, pp. 28–48.

Reeves, G.H., Everest, F.H. & Nickelson, T.E. (1989) *Identification of physical habitats limiting the production of coho salmon in western Oregon and Washington*. General Technical Report PNW-GTR-245, United States Department of Agriculture Forest Service Pacific Northwest Research Station, Portland, Oregon.

Rietbergen-McCracken, J. & Abaza, H. (2000) *Environmental Valuation: A Worldwide Compendium of Case Studies*. Earthscan Publications Ltd., London.

Roni, P., Beechie, T.J., Bilby, R.E., Leonetti, F.E., Pollock, M.M. & Pess, G.R. (2002) A review of stream restoration techniques and a hierarchical strategy for prioritizing restoration in Pacific Northwest watersheds. *North American Journal of Fisheries Management* **22**, 1–20.

Roni, P., Beechie, T. & Pess, G. (2003) Prioritizing restoration actions within watersheds. In: Beechie, T., Steel, E.A., Roni, P. & Quimby, E. (eds) *Ecosystem Recovery Planning for Listed Salmon: An Integrated Assessment Approach for Salmon Habitat*. NOAA Technical Memorandum NMFS-NWFSC-58, Seattle, Washington, pp. 60–73.

Roni, P., Hanson, K., Pess, G., Beechie, T., Pollock, M. & Bartley, D. (2005) *Habitat Rehabilitation for Inland Fisheries: Global Review of Effectiveness and Guidance for Restoration of Freshwater Ecosystems*. Fisheries Technical Paper 484. Food and Agriculture Organization of the United Nations, Rome, Italy.

Roni, P., Hanson, K. & Beechie, T. (2008) Global review of the physical and biological effectiveness of stream habitat rehabilitation techniques. *North American Journal of Fisheries Management* **28**, 856–890.

Roni, P., Pess, G., Beechie, T. & Morley, S. (2010) Estimating changes in coho salmon and steelhead abundance from watershed restoration: how much restoration is needed to measurably increase smolt production? *North American Journal of Fisheries Management* **30**, 1469–1484.

Sarkar, S., Pressey, R.L., Faith, D.P. *et al.* (2006) Biodiversity conservation planning tools: present status and challenges for the future. *Annual Review in Environmental Resources* **31**, 123–159.

Scheurell, M.D., Hillborn, R., Ruckelshaus, M.H. *et al.* (2006) The Shiraz model: a tool for incorporating anthropogenic effects and fish-habitat relationships in conservation planning. *Canadian Journal of Fisheries and Aquatic Sciences* **63**, 1596–1607.

Schmutz, S. & Trautwein, C. (2009) *Developing a methodology and carrying out an ecological prioritisation of continuum restoration in the Danube River Basin to be part of the Danube River Basin Management Plan*. Report prepared for the International Commission for the Protection of the Danube River (ICPDR), Vienna.

Schmutz, S., Kaufmann, M., Vogel, B., Jungwirth, M. & Muhar, S. (2000) A multi-level concept for fish-based, river-type-specific assessment of ecological integrity. *Hydrobiologia* **422**, 279–289.

Sear, D., Newson, M., Hill, C., Old, J. & Branson, J. (2009) A method for applying fluvial geomorphology in support of catchment-scale river restoration planning. *Aquatic Conservation: Marine and Freshwater Ecosystems* **19**, 506–519.

Sedell, J.R., Reeves, G.H., Hauer, F.R., Stanford, J.A. & Hawkins, C.P. (1990) Role of refugia in recovery from disturbances: modern fragmented and disconnected river systems. *Environmental Management* **14**(5), 711–724.

Steel, E.A., Feist, B.E., Jensen, D.E. *et al.* (2004) Landscape models to understand steelhead (*Oncorhynchus mykiss*) distribution and help prioritize barrier removals in the Willamette basin, Oregon, USA. *Canadian Journal of Fisheries and Aquatic Sciences* **61**, 999–1011.

Steel, E.A., Fullerton, A.H., Caras, Y. *et al.* (2009) A spatially explicit decision support system for managing wide-ranging species. *Ecology and Society* **13**(2), 50.

Thurston, H.W, Heberling, M.T. & Schrecongost, A. (2009) *Environmental Economics for Watershed Restoration*. CRC Press, Boca Rotan, Florida.

Traill, L.W., Bradshaw, J.A. & Brook, B.W. (2007) Minimum viable population size: a meta-analysis of 30 years of published estimates. *Biological Conservation* **139**, 159–166.

Turner, W. (1997) Achieving private-sector involvement and its implications for resource professionals. In: Williams, J.E., Wood, C.A. & Dombeck, M.P. (eds) *Watershed Restoration: Principles and Practices*. American Fisheries Society, Bethesda, Maryland, pp. 158–176.

Villa, F., Tunesi, L. & Agardy, T. (2002) Zoning marine protected areas through spatial multiple-criteria analysis: the case of the Asinara Island National Marine Reserve of Italy. *Conservation Biology* **16**, 515–526.

Waidbacher, H. & Haidvogl, G. 1998. Fish migration and fish passage facilities in the Danube: past and present. In: Jungwirth, M., Schmutz, S. & Weiss, S. (eds) *Fish Migration and Fish Bypasses*. Fishing News Books, Oxford, pp. 85–98.

White, R.J. (2002) Restoring streams for salmonids: where have we been? Where are we going? In: O'Grady, M. (ed.) *Proceedings of the 13th International Salmonid Habitat Enhancement Workshop*, Westport, County Mayo, Ireland, September 2002. Central Fisheries Board, Dublin, Ireland, pp.1–31.

Whiteway, S.L., Biron, P.M., Zimmermann, A. *et al.* (2010) Do in-stream restoration structures enhance salmonid abundance? A meta-analysis. *Canadian Journal of Fisheries and Aquatic Sciences* **67**, 831–841, doi: 10.1139/F10-021.

Williams, J.E., Haak, A.L., Gillespie, N.G. & Colyer, W.T. (2007) The conservation success index: synthesizing and communicating salmonid condition and management needs. *Fisheries* **32**(10), 477–492.

Yoe, C. (2002) *Trade-Off Analysis Planning and Procedures Guidebook*. IWR 02-R-2 Contract No. DACW72-00-D-0001, US Army Corps of Engineers, Institute for Water Resources, Alexandria, Virginia.

Zafra-Calvo, N., Cerro, R., Fuller, T., Lobo, J.M., Rodriguez, M.A. & Sarkar, S. (2010) Prioritizing areas for conservation and vegetation restoration in post-agricultural landscapes: a biosphere reserve plan for Bioko, Equatorial New Guinea. *Biological Conservation* **143**, 787–794.

7 Developing, Designing, and Implementing Restoration Projects

Peter Skidmore[1], Tim Beechie[2], George Pess[2], Janine Castro[3], Brian Cluer[4], Colin Thorne[5], Conor Shea[6], and Rickie Chen[7]

[1]Skidmore Restoration Consulting LLC, Montana, USA
[2]Northwest Fisheries Science Center, National Oceanic and Atmospheric Administration, Washington, USA
[3]US Fish and Wildlife Service, Oregon, USA
[4]Southwest Region, National Oceanic and Atmospheric Administration, California, USA
[5]University of Nottingham, Nottingham, UK
[6]US Fish and Wildlife Service, California, USA
[7]University of Washington, Washington, USA

7.1 Introduction

Restoration planning and implementation are grounded in an understanding of causes of degradation, the scale at which the problem must be addressed, and constraints on restoration opportunities (Chapters 2 and 3). The watershed assessments described in Chapter 3 illustrate how assessments of watershed- and reach-scale processes are used to identify the general types and scale at which restoration is needed. For example, impacts to sediment supply or hydrologic regime may cause reach-scale channel and habitat degradation, but restoration strategies may require watershed-scale actions to address the root causes of degradation and to restore impacted habitat. By contrast, reach-scale causes of degradation such as riparian modification or channel armoring can be addressed locally. In either case, addressing the cause of degradation requires development of restoration projects at specific sites and a logical sequence for planning, design, implementation, and monitoring those projects. Additional site-level analyses are therefore required to design each individual restoration project. In this chapter we describe and illustrate

a project development process that includes several key steps: identify the cause of the problem to be addressed, assess the project context, define project goals and objectives, conduct site-level investigations, evaluate alternative solutions, design the project to meet objectives and accommodate watershed context, and integrate implementation and monitoring plans into the final design.

A comprehensive approach to project development and implementation can substantially improve restoration practice and facilitate adaptive management in restoration programs. Projects typically proceed through three main phases (Skidmore *et al.* 2011):

• *planning*, which establishes the purpose and need for restoration, puts the project in a watershed context, and articulates the specific intentions of a project;
• *design*, which describes the details of the project and how it will be implemented and the project objectives accomplished;
• *implementation and monitoring*, which includes the actions taken to complete the project, checking to see that the project was implemented as designed, and evaluating whether the project had the desired habitat and biological effects.

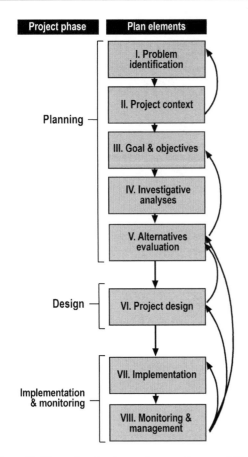

Figure 7.1 The project development process is sequential but may involve iterations as each step in the process brings additional information and perspective. Analyses conducted later in the development process may verify or invalidate assumptions asserted earlier in the process, and therefore warrant revisiting earlier steps in the process (based on Skidmore *et al.* 2011).

Within these phases of project development are eight distinct steps (Figure 7.1) (Skidmore *et al.* 2011).

The first step is problem identification, which should link the problem to its underlying root cause and describe the scale at which the problem should be addressed (this usually comes from the watershed assessment, Chapter 3). The second step is an assessment that establishes the planning, ecological, geomorphic, and socioeconomic context for the project. Much of this context information is also derived from the watershed assessment and human dimensions of restoration (Chapters 3 and 4). However, additional analyses are needed to help determine project context, goals, and design. The third step is defining the project goal and specific and measurable objectives,

which should clearly describe the purpose of the project and its anticipated outcomes. The fourth step is investigative analysis, which assesses physical, chemical, and biological features of a site to inform the alternatives evaluation and project design. The fifth step is alternatives evaluation, which identifies and compares alternative solutions, management strategies, and design concepts with each other and with a no-action alternative. Project design (Step 6) follows selection of a preferred alternative and includes a suite of project element designs that support objectives and communicate design details to all parties. Implementation (Step 7) includes managing impacts and risks associated with construction activities and documenting all project elements. Finally, measuring project compliance and performance is critical to adaptive management (Step 8), both for managing deviations from anticipated outcomes and for making adjustments to the watershed restoration program (see Chapter 8 for detailed discussion of monitoring).

The steps are sequential in that each is informed by the previous step; skipping any of these steps can significantly jeopardize project integrity. However, later steps may also inform previous steps and so the process is often iterative. For instance, analysis of alternatives may reveal that stated objectives are impractical or contradictory, requiring that objectives be revisited. The design process may also invalidate assumptions applied in evaluating alternatives and compel reconsideration of project details or alternatives. Hence, the final project design should be based on the integration of all the design steps to ensure consistency of information among the steps and to ensure that a project is planned, designed, and implemented in a way that is most likely to succeed.

Current restoration practices often shortcut the project development process or fail to monitor outcomes, resulting in projects that do not meet objectives or expectations (Pretty *et al.* 2003; Bernhardt *et al.* 2005; Stewart *et al.* 2009). Two of the most common pitfalls are lack of clearly defined project goals and objectives, and lack of a clear analysis process that identifies causes of ecosystem degradation and links this to design. A lack of clearly defined project goals and objectives propagates through the entire project development process, often resulting in projects that fail to address the identified problem. Moreover, where expectations have not been adequately defined with measurable objectives, there is no clear hypothesis that can be tested through monitoring and project success is difficult to evaluate (Bernhardt *et al.* 2007; Skinner *et al.* 2008). The second common pitfall is skipping key design steps that link the analysis of root causes to project design, thereby failing to fully consider the project context. This often leads to repeated use of a limited number of standard or 'typical' design drawings, even though the projects vary widely in geographic setting and stream types (Montgomery & Buffington 1997). Such designs are often inadequate or inappropriate because they lead to implementation of projects that include elements that may be unnecessary, cause unanticipated problems, or otherwise limit project success.

7.2 Identify the problem

Designing a restoration project requires identification of the underlying causes of river ecosystem degradation, as well as an understanding of scale at which those causes occur (Chapters 2 and 3). Accurate problem identification may require the expertise of a number of scientific disciplines, and it will improve the project planning process and ultimately contribute to more sustainable and successful projects (Chapter 3). Unfortunately, common approaches to restoration project development often focus on treating observed symptoms (structure) rather than identifying and resolving underlying causes (processes or functions). For example, water quality assessment often points to eroding banks as sources of excess sediment (Montana Department of Environmental Quality 2009), leading to engineering solutions such as armoring or reconstructing banks using rock structures or bioengineering techniques. Such actions treat the symptom instead of addressing the underlying cause of bank failure, which may be (1) a site-scale issue such as vegetation removal, bank trampling, or local channel manipulation; (2) a reach-scale issue such as historic floodplain logging or channel straightening; or (3) a watershed-scale issue, such as sediment accumulation or change in flood duration due to land-use changes upstream. Treating symptoms may alleviate short-term concerns and allow for rapid project development, but may also result in project benefits that are unsustainable. For example, a fish biologist may conclude that a limiting factor (symptom) for fisheries is a lack of rearing habitat due to increased sediment, and that a project objective should be to create more rearing habitat. Rearing habitat, however, is a structural component of a stream that is formed by active geomorphic processes. While artificial bank stabilization and creation of new rearing habitat may meet project objectives in the near-term, it may soon be lost in the absence of restored stream processes that create and maintain habitat.

As discussed in Chapter 3, the process of problem identification should recognize that root causes of degraded stream condition or structure may occur at much different spatial and temporal scales than the symptoms (altered habitat or biota or unstable reach conditions). Evaluation of problems and their causes should therefore take place in the context of a multidisciplinary watershed assessment, which typically:
• determines the status of watershed controls including hydrology and sediment regimes;
• identifies changes that have occurred within the watershed and stream corridor that may explain changes in stream conditions at varying scales; and
• translates data collected into information that contributes to selection of alternatives and design at an appropriate scale.
Common problems leading to a determination that restoration is necessary include habitat constraints, altered channel conditions, or water quality degradation. Habitat constraints, sometimes referred to as limiting factors, are commonly described for a single species of interest and can be further reduced to life-stage specific constraints (Beechie et al. 1994). The identification of habitat constraints is critical to guiding restoration priorities, but it must be coupled with analysis of processes that sustain habitat (Skidmore et al. 2011). Change in channel structure and condition is often the result of manipulation of the stream, but may also be a response to changes in watershed controls such as sediment supply and hydrologic regime (Brierley & Fryirs 2008). Common channel changes can be simply categorized as aggradation, degradation, width adjustment, or planform change (Leopold et al. 1964). Change in channel structure or condition is commonly associated with disruption of the sediment regime and resulting imbalance between sediment supply and local sediment transport capacity (Kondolf 1997). However, land-use histories may complicate the identification of such causal relationships, especially legacy impacts from activities such as wetland drainage, channelization, mining, beaver trapping, log jam removal, logging, and other land uses (Wohl 2004).

Water quality impairments may include changes in temperature, pH, dissolved oxygen, and excess inputs of nutrients, heavy metals, water-borne pathogens, and fine-grained sediment (Miller & Miller 2007). These impairments may affect any level of the food chain, having either direct or indirect effects on higher trophic levels. In a restoration context, even successful restoration of physical processes may be inadequate to achieve restoration goals if water quality is a dominant biological constraint. Water quality problems can be caused by either point or non-point sources. Point-source pollution is easier to identify and generally easier to treat than non-point-source pollution. Solutions to non-point-source water quality problems usually require broad-scale actions and treatments, such as land-use regulation or other watershed-scale treatment to address sources and delivery of fine sediments. Metals, pesticides, nutrients, and organics can all bind to fine sediments and can be distributed throughout a stream ecosystem including across flood-plains and in-channel deposits, even if they originate from a point source such as a mine (Marcus *et al.* 2001).

7.3 Assess project context

All restoration project elements and actions should consider the project context, where context refers to factors that influence restoration opportunities and potential outcomes. Here we focus on three key components of context: existing watershed plans, the project's setting within the watershed, and the socioeconomic context. Where watershed restoration plans have been developed, a proposed project should usually be identified as a priority action in the plan. Watershed assessments and planning provide context that includes recognition of geomorphic, ecologic, social, regulatory, and economic constraints, as well as the overarching restoration goal (Koehn *et al.* 2001; Moss 2004; Beechie *et al.* 2010). Where watershed assessment and planning have not been conducted, it is critical to consider the extent to which assessment of key habitat-forming processes have informed project development. A lack of process assessments reduces confidence in the project design, whereas detailed analysis of driving processes supports the design of projects with higher likelihood of success. Notably, while restoration projects are often conducted independently of broader watershed plans (Bernhardt *et al.* 2005), an encouraging trend toward restoration and recovery planning at a watershed scale is emerging (Steel *et al.* 2008; www.moriverrecovery.org; www.edenriverstrust.org.uk; www.chesapeakebay.net/restrtn.htm).

Stream restoration projects are often constrained by social and economic obstacles (Miller & Hobbs 2007), including regulatory constraints (Chapter 4). Individual landowners and communities often value protection of private property and traditional economic values (e.g. water supply, land development, power generation or recreation) above other ecological services provided by

rivers, and these values may present obstacles to achieving restoration objectives. Even where proposed project objectives have general stakeholder and community support, funding limitations may impose limits on what is otherwise possible. Socioeconomic constraints typically become more prominent as the scope and spatial extent of restoration increases. Perhaps as a function of these socioeconomic constraints, most stream restoration to date has focused on isolated, reach-scale efforts with narrowly defined objectives (Bernhardt *et al.* 2005; Beechie *et al.* 2008). Consideration of socioeconomic and political context, in addition to physical or ecological assessments conducted at a watershed scale, can greatly inform project development by identifying important stakeholders, additional funding opportunities and unlikely partnership opportunities. For example, while agriculture and conservation values were traditionally at odds with each other, they are now more frequently finding common ground in the face of rapid development or other pressures on use of stream water.

The ecological and geomorphic setting of the project determines what is physically and biologically possible in a restoration project. Geomorphic context includes processes that govern creation and maintenance of channel structure and associated habitat, including human actions and land uses that affect hydrologic and sediment regimes, impose physical constraints on dynamic processes, or change the character of channel boundaries. Changes in watershed controls (hydrologic regime and sediment regime) that significantly influence channel character and processes at a project site may not be apparent at a reach scale. For example, reach-scale analysis may indicate that a channel is incised, but only a watershed-scale analysis will be sufficient to relate this incision to changes in hydrologic or sediment regimes that are the ultimate cause of incision and to inform selection of an appropriate solution. Social and cultural expectations may favor stream types thought to be 'natural' or 'healthy' (e.g. meandering rivers with pools and riffles; Wohl 2004), yet these stream types may be unsustainable given existing hydrologic and sediment regimes and inherent channel boundary characteristics. For example, creation of habitat features in channels crossing alluvial fans, conversion of braided channels to single-thread channels, or stimulating perennial flow in intermittent or semi-arid streams should not be expected to be sustainable in most cases (Abbe *et al.* 2002).

The use of reference reaches to help understand geomorphic and biological restoration objectives is common in North America (Wheaton *et al.* 2004), but less common in Europe and other parts of the world where un-impacted reference reaches are rare (Statzner *et al.* 2005; Comiti *et al.* 2009). However, many of the dominant factors influencing channel processes cannot be adequately evaluated – or even identified – at a reference reach scale. For example, local sediment transport analyses have become common and indeed may be essential for project design, but it is only through consideration of sediment transfer at a watershed scale that these analyses can be appropriately applied (Wilcock *et al.* 2009). Geomorphic and historic investigations intended to determine reference conditions may be confounded by the complexities of legacy impacts such as deforestation, logging, dam construction, beaver removal (North America and Europe), channelization and draining of wetlands, and other human impacts (e.g. Massong & Montgomery 2000; Wohl 2005; Pollock *et al.* 2007; Walter & Merritts 2008). For instance, apparent channel stability following systematic disruptions to channel processes and conditions associated with legacy impacts may reflect a different suite of processes and conditions and ecological potential than those that existed historically. In such instances, we must also address the question of what point in time to use as baseline for restoration. Similarly, many basins are experiencing a steady change in hydrologic regime associated with climate change (Mote 2003), which requires recognition that future conditions may not be adequately represented by past conditions. This is especially true for stream temperature and hydrologic analyses, as it may not be ecologically possible to restore certain fisheries where water temperature and the seasonality of flows are critical to life history transitions (Scheuerell *et al.* 2009).

Ecological context refers to the suite of ecological processes influencing a project site, as well as the constraints put on these processes (Miller & Hobbs 2007). Common elements of ecological context are riparian and aquatic species compositions, riparian disturbance, successional processes, and food web structure. The presence or introduction of invasive species may also provide relevant ecological context. The effects of land-use impacts, such as urban development, agricultural practices, harvesting of mature riparian forest, or native species extirpation may be considered irreversible in practical management timeframes (Brooks & Brierley 2004; Brooks 2006). However, the structure and condition of urban or agricultural streams can still be substantially improved in terms of habitat value or ecological function (Riley 1998; Talmage *et al.* 2002) or otherwise enhanced to accommodate irreversible changes in drivers and processes.

7.4 Define project goals and objectives

From start to finish, a river restoration project should be evaluated relative to its goals and objectives. Project-level goals differ from the watershed-scale goals discussed in Chapter 3 in that they are specific to anticipated outcomes of an individual project, rather than a statement of the ecological aims of a watershed restoration plan. A goal in this context provides a guiding image (or leitbild) for a restoration project, which is a simple statement of a desired outcome that informs stakeholders, project funders, project designers, and those implementing and evaluating a project (Kern 1992). To provide this guiding image, a goal statement will define the desired outcome, efficiently express the intent of a project, and provide a reference for evaluating all project actions. Ideally, goal statements describe intended outcomes without being prescriptive about the means to those outcomes. Prescriptive goal statements (i.e. those that prescribe stabilization, reconstruction, or reconfiguration) are more likely to constrain the options for achieving goals or to misrepresent the intended goal of improved ecological conditions. Examples of typical prescriptive project goal statements include 'restore historic planform alignment', 'stabilize stream banks or streambed,' and 'return channel bed elevation to historic elevation and grade.' While goals will necessarily vary among projects, restoration project goals should typically include the following intentions (adapted from Palmer *et al.* 2005).

1. *Improve processes that sustain structure and conditions that support a natural ecosystem.* Structure and condition refer primarily to the health and state of habitat and water quality, and are reasonable and appropriate desired outcomes for restoration. Improved structure and condition are common if not universal intentions for restoration projects. However, the structure and condition of a stream are the outcome of dynamic processes; without these processes in place, structure and condition are at best temporary. Increasingly, restoration project goal statements acknowledge the process foundation for desired conditions, and explicitly state that restoration of process is paramount.

2. *Emphasize resilience.* Resilience refers to the capacity to bounce back from disturbances. Natural systems, unconstrained and un-impacted by human actions, are typically resilient to natural disturbances. In fact, disturbance may be a fundamental component of maintaining ecosystems (Pickett *et al.* 1989). Resilience is therefore an encompassing concept that should be integral to restoration goals and guide restoration projects to create self-sustaining systems that require minimal future action to maintain intended outcomes.

Project objectives define specific outcomes of a project and refine the intent of a project as stated in the goal. While the terms 'goals' and 'objectives' are often used interchangeably, they represent distinctly different concepts (Chapter 3). Goals are statements of vision that define project intent, whereas objectives are statements of specific and measurable outcomes. Project objectives should provide sufficient detail to define the actions necessary to achieve the goal within a specified timeframe. A useful framework for establishing objectives is the SMART acronym (Box 7.1) (Doran 1981). Objectives that meet SMART criteria provide an outcome target for the design process, and establish a basis for post-project appraisals and monitoring. Where restoration design involves numerous analyses from varying disciplines,

each analysis contributes to a design solution (Skidmore *et al.* 2011). The design process may then implement predictive, quantitative analyses to determine the probability of achieving project objectives. Similarly, verification that project outcomes meet project objectives requires that monitoring is specifically conducted to measure outcomes.

Where objectives are not fully SMART, they are often refined further using performance standards, targets, or other measurable and time-bound descriptors. There are many different and equally adequate means to articulate SMART objectives, and many different names for these various categorizations. However, the basic principle that a project development process requires articulation of SMART outcomes holds true regardless of the way these outcomes are defined or broken down. To conduct a design process that is linked to objectives and to be able to adequately measure project success relative to these objectives, an outcome must ultimately be described through measurable parameters and target values for those parameters within an established timeframe (Table 7.1).

An additional benefit of SMART objectives, or any suite of descriptive indices applied to define objectives in measurable parameters, is that they can also serve as the foundation for monitoring and post-project appraisal (Bernhardt *et al.* 2007; Skinner *et al.* 2008). When objectives are articulated as measurable outcomes and defined with target values for those outcomes, monitoring can evaluate success relative to stated project objectives and a

---

Box 7.1 SMART objectives

- Specific: clear, concise statements that specify what you want to achieve
- Measurable: articulated using parameters that can be measured before and after project implementation
- Achievable: geomorphically and ecologically possible and socially acceptable
- Relevant: clearly related to the identified problem and supportive of the project goal
- Time-bound: bound by a specified time frame

Table 7.1 Measurable objectives include specific parameters to be measured, target values for those parameters, and a timeframe when targets are expected to be reached.

Goal	Objective	Parameter	Target	Timeframe
Increase recruitment of target species	Restore access to existing spawning habitat	Number of stream miles with unrestricted passage	Greater than 95% of available spawning habitat is unrestricted by passage barriers during spawning season	Within 2 years
	Restore access to historic extent of rearing habitat	Hectares of rearing habitat with continuous connection to main channel	75% of estimated historic extent of rearing habitat	Within 5 years
Restore degraded saltmarsh to healthy state	Increase abundance of native vegetation	Percent cover of native species	Greater than 40% cover of native plant species	Within 2 years
	Improve ability to provide habitat for native fish	Population size of native fish species	10% increase in population using saltmarsh	In 3 out of 5 following years

monitoring protocol can be developed to specifically test each objective. If articulated objectives are translated into hypotheses or statements of projected outcomes, then monitoring becomes an exercise in testing hypotheses that restoration actions will lead to intended outcomes within predicted timeframes.

The integrity and value of goals and objectives can be evaluated by asking 'Do the goals and objectives address the problem, cause, and context?' A restoration project goal serves to express an intended outcome that addresses the identified problems with consideration and accommodation of socioeconomic, ecological, and physical contexts. Outcomes articulated as objectives may be either structural or functional, and in both cases should include an action, a measurable target, and timing to reach the target. A structural objective focuses on the distribution, abundance, and physical condition of some element of the ecosystem (e.g. a particular organism or component of the environment) and is measured as a point-in-time value. A functional objective focuses on the processes that sustain an organism or environmental component, and is measured as change or rate over time. Restoration efforts and monitoring of outcomes is generally considered more robust when both structural and functional objectives are represented (Sandin & Solimini 2009). Examples of structural objectives include:

1. replace blocking culvert with a fish passage structure maintaining a bed slope less than 3%, maximum velocity of 1.5 m/s at the 90% exceedence discharge, and minimum depth of 30 cm at base flow;

2. restore native riparian vegetation to include at least 4 native woody species within 3 years; or

3. create a minimum of 2 acres per stream kilometer of off-channel rearing habitat in the first year of restoration actions.

Examples of functional objectives include:

1. establish fish passage to at least 7 km of essential fish habitat within 1 year,

2. re-establish channel bank erosion rates such that no more than 10% of banks are actively eroding over a 5-year period within 30 years; or

3. restore overbank flow to an average duration of 5% in 3 out of 5 years within 10 years.

7.5 Investigative analysis

Investigative analyses provide a foundation for characterizing site conditions prior to restoration actions, evaluating probable outcomes of alternatives, and design-

ing the final project (Skidmore *et al.* 2011). The extent of analysis required for alternatives evaluation and project design varies greatly depending on project type. For example, instream projects often require a greater number of analyses and more integration of those analyses than upland restoration actions. Investigative analyses conducted to inform alternatives evaluation are typically also necessary for eventual design analysis (Copeland *et al.* 2001; Miller & Skidmore 2003). Basic investigations that inform alternatives analysis and ultimately design for instream and riparian or floodplain projects include survey and mapping, hydrologic investigation, hydraulic modeling, sediment transport analysis, geomorphic investigation, and geotechnical analysis (Brown & Pasternack 2009; Skidmore *et al.* 2011). Baseline investigations and monitoring that also serve to inform alternatives development and design include vegetation and weed assessment, instream and floodplain habitat assessment, water quality assessment and various biological assessments such as macroinvertebrate or bird surveys (refer to Chapter 3 for more information on assessment).

7.5.1 Investigative analyses for in-channel restoration projects

In-channel restoration projects generally require more detailed investigative analyses than less complicated watershed-scale or riparian projects. In this section we describe detailed investigative analyses that are commonly used for in-channel restoration projects: (1) survey and mapping; (2) hydrologic investigation; (3) hydraulic modeling; (4) sediment transport analysis; (5) geomorphic investigation; (6) geotechnical assessment; and (7) uncertainty and risk. Most projects do not require all of these analyses, but each in-channel project requires some combination of these analyses depending on the goals and objectives of the project.

7.5.1.1 Maps and surveys

Project maps show a project site in its watershed and valley context and serve as the basis for illustrating project options and expectations through the project development process. Project maps are used to portray project alternatives relative to existing conditions and are used for permitting, design, and monitoring. Project maps can be developed from a combination of existing map data sources such as topographic, remote sensing images, or existing geographic information system (GIS) data, as well as from project-specific surveys. Investigative analyses commonly include watershed-scale maps (Figure 7.2), project-scale base maps (Figure 7.3) and project survey

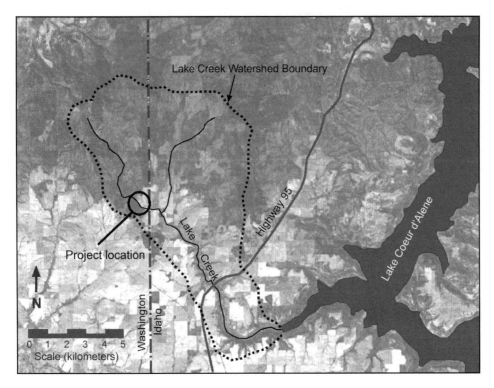

Figure 7.2 Location map for a channel remeandering project on West Fork Lake Creek, Washington State, USA. Site maps, cross-sections, and longitudinal profiles for this project are shown in Figures 7.3–7.5. Background image copyright 2012 Google.

data that detail channel and floodplain topography and dimensions. Table 7.2 lists data sources and applications of common project survey and map information that are relevant to all restoration efforts requiring some degree of communication, such as for permitting. Most restoration projects will benefit from map illustration, although the level of survey detail necessary or appropriate will vary significantly among project types and project communication requirements. For example, detailed map illustration or survey data may not be necessary for restoration efforts based primarily on floodplain revegetation. However, revegetation efforts may be more successful if the planning process is informed by assessments and modeled floodplain inundation and moisture, which may require a detailed topographic survey.

Maps are typically developed in either a computer-aided design (CAD) format or a GIS format. The integration of CAD and GIS enables sophisticated and robust analysis and portrayal of project elements and alternatives. Topographic surveys are used to generate a digital portrayal floodplain surface topography and rough channel dimensions and serve as the basis for

numerous other investigative analyses. They may be derived from a land-based manual survey of site topography or from remote sensing technologies, such as a Light Detection and Ranging (LiDAR) survey. However, for most instream projects, a manual survey of the channel is necessary to derive adequate-resolution survey data. These data can then be integrated with LiDAR data. In addition, a topographic survey may be necessary for many hydraulic model applications and analyses, and will enable the evaluation of inundation at varying flows. As an alternative to detailed topographic survey, surveyed channel and floodplain cross-sections may be adequate for all but the most sophisticated hydraulic modeling. Topographic surveys typically extend laterally to or beyond the floodplain margins and upstream and downstream of the project a sufficient distance to capture relevant topographic features, such as side channels and berms.

When projects include some degree of channel manipulation or benefit from any form of hydraulic modeling, topographic data may be supplemented with detailed channel profile and cross-section data. A longitudinal profile is a survey of the channel bed and water surface,

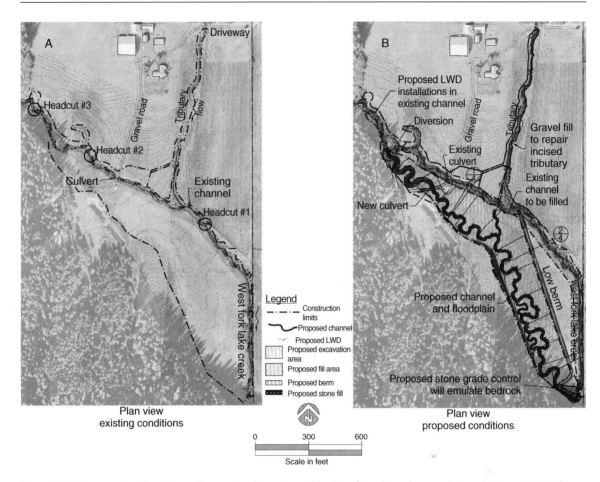

Figure 7.3 Site maps of existing stream alignment and new channel location for a channel remeandering project on West Fork Lake Creek, Washington State, USA showing (A) the pre-project channel location and (B) the proposed remeandered channel location. Illustration courtesy of Inter-Fluve, Inc.

and a number of inflection points along the floodplain (Harrelson *et al.* 1994; Figure 7.4). Channel cross-sections are lateral surveys across the channel and floodplain, typically including a minimum of top of bank, toe of bank slope, and deepest point in the channel (Harrelson *et al.* 1994; Figure 7.5). While survey data are commonly gathered by surveyors or engineers, it is beneficial to have a hydrologist or geologist in consultation to ensure that relevant topographic features are adequately characterized.

7.5.1.2 Hydrologic investigation
Hydrology has been characterized as the master variable affecting stream ecology (Schlosser 1985; Doyle *et al.* 2005), because it exerts a strong influence on ecological communities and life history strategies for instream and riparian species (Poff *et al.* 1997; Richter *et al.* 2003). For

instream projects or any project conducted in riparian or floodplain environments, hydrologic investigations characterize the flow regime or volume and timing of flow in a channel and on its floodplain typically through statistical parameters such as magnitude, frequency, duration, and rate of change of stream flow (Poff & Ward 1989). The flow regime is strongly influenced by dominant climatic character, elevation, geography, and land use of the contributing watershed. While each river system is characterized by a unique hydrographic signature, regional hydrograph signatures can be consistent in character among associated watersheds. Land uses that change the infiltration and evapotranspiration character of the watershed can have profound impacts on the hydrologic regime and associated ecological community (Booth & Jackson 1997; see also Chapters 2 and 3).

Table 7.2 Project maps are derived from varying data sources and illustrate spatial information about the project setting and project plans. Maps are also illustrated in Figures 7.2 and 7.3.

Map type	Map content	Data sources	Relevance and application
Watershed map	Watershed boundary and features; land use; project location	Topographic maps, aerial photos, public GIS databases	Illustrates relevant features and land use that influence hydrology and sediment supply and stream condition (Beechie *et al.* 2002)
Project base map	Project area and vicinity; infrastructure, structures and utilities; existing topography; property boundaries; floodplain limits	In addition to those used for watershed map: LiDAR, topographic survey, vegetation and channel surveys (planform, section and longitudinal)	Serves as platform for engineering drawings at varying scales; illustrates site and project constraints and opportunities; establishes baseline conditions; provides topographic data for hydraulic and geomorphic modeling and other engineering analyses
Project plan map	Project alternatives; project features and elements; implementation features	In addition to those used for above, draft and final drawings of proposed or alternative future conditions	Illustrates proposed floodplain and channel site conditions and features under restoration relative to existing conditions; illustrates implementation features such as site access, staging, storage; illustrates monitoring and maintenance features
Topographic survey	Digital topographic surface, channel longitudinal profile, channel and floodplain cross-sections	Land-based survey data, LiDAR, other remote-sensing-derived data	Source data for project base maps; hydraulic, geomorphic, and sediment transport analyses; revegetation plan maps; project plan maps; and project implementation analyses, such as for cut-and-fill calculation

Hydrologic analyses characterize the seasonal and inter-annual variation in stream flow including overbank flood flows, and provide perspective on how these parameters have changed over time as a result of watershed land use as well as how they may change in the future under predicted or alternative land-use scenarios and in the face of climate change. An understanding of the hydrologic setting, including land-use impacts to the hydrologic regime and potential future scenarios, is paramount to understanding causes of identified problems and probable future stresses for instream and riparian environments (Beechie *et al.* 2010). For example, analyses may reveal changes in the frequency (probability) or duration (percent of time) of channel-forming flows, which may be a leading cause of observed degradation or may indicate the likelihood of future channel instability (Allan 1995). Similarly, changes in the frequency of floodplain inundation documented through hydrologic analysis can explain changes in riparian and floodplain community composition or degradation.

Hydrologic investigations provide the foundation for both understanding the probable causes of observed problems and for developing restoration alternatives and subsequent designs for instream projects or any project within a riparian or floodplain environment. Hydrologic analyses are used to characterize three general discharge bands: low flows, channel-forming flows, and flood flows (Bragg *et al.* 2005). These three discharge categories can serve as the basis for performing design-related analyses. Often referred to as base flow, low flow is the highest duration flow and is strongly associated with habitat quality and availability. Low-flow periods are associated with limiting factors stemming from stresses such as water withdrawal, water temperature, pollution, and predation (Poff *et al.* 1997).

Channel-forming flows are higher flows that drive stream processes, including erosion or sediment transport and deposition. Determining channel-forming flows is therefore a critical component of channel design projects. These flows may be represented by one of three related discharges: effective discharge (Q_{eff}), bankfull discharge (Q_{bf}), or a return interval such as the 2-year flow (Q_2) (Andrews 1980; Emmett & Wolman 2001; Shields *et al.* 2003). Effective discharge is the discharge

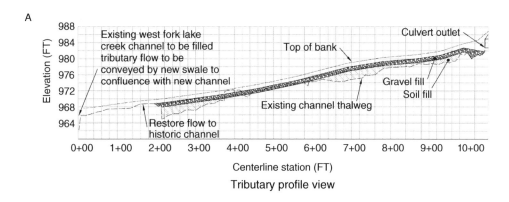

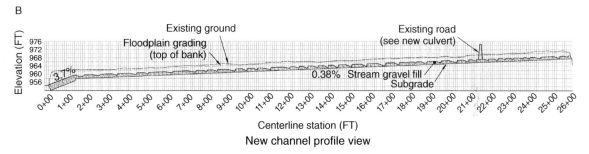

Figure 7.4 Typical longitudinal profiles for the channel remeandering project on West Fork Lake Creek, Washington State, USA. (A) Upper profile is for the tributary extension at the north end of the project site, and (B) lower panel shows the long profile for the remeandered reach. Site plan is shown in Figure 7.2. Figures courtesy of Inter-Fluve, Inc.

range that transports the most sediment given its magnitude and frequency of occurrence; these flows accomplish the most geomorphic work relative to other flow ranges (Wolman & Miller 1960; Doyle *et al.* 2005, 2007). Bankfull discharge is the discharge at which stream flow just begins to exceed its banks and access the floodplain (Leopold *et al.* 1964). It is often considered 'the stream level that corresponds to the discharge at which channel maintenance is most effective' (Dunne & Leopold 1978). This is typically considered the 1.5 to 2.0 year flood (Dunne & Leopold 1978), although the flood recurrence level may vary considerably based upon ecoregion (Castro & Jackson 2001).

While Q_{bf} and Q_2 are commonly employed as the basis for channel design, Q_{eff} is regarded as a more robust representation of channel-forming flow because it is specific to each stream and it combines a flow-frequency distribution and sediment rating curve to estimate sediment load as a function of discharge (Biedenharn *et al.* 2000). By contrast, Q_{bf} or Q_2 are derived primarily from empirical studies in alluvial streams in humid regions and perennial streams in semi-arid environments (Biedenharn *et al.*

2000; Soar & Thorne 2001). The application of either Q_{bf} or Q_2 to restoration design may be inappropriate for non-alluvial or non-equilibrium streams or streams with ephemeral flow regimes, abundant large wood or confinement by bedrock (Skidmore *et al.* 2011). This emphasizes the need for hydrologic analyses to be founded on a strong understanding of basin characteristics, watershed processes, and training in hydrologic investigations.

Flood flows are those that overtop channel banks, inundating some or all of the floodplain (Leopold *et al.* 1964). Flood flows deposit sediment, nutrients, seeds, and plant propagules on floodplains contributing to riparian vitality, replenishing shallow aquifers, and recruiting large wood and organic matter to the channel (Gomez *et al.* 1999; Postel & Ritcher 2003). Hence, flood flows form the basis for the design of floodplain features (Tompkins & Kondolf 2007), and are also the basis for design of channel features intersecting infrastructure or other property concerns.

A restoration design, whether for concept development or detailed final design, will typically account for low-flow discharge, a 'bankfull' dimension or effective

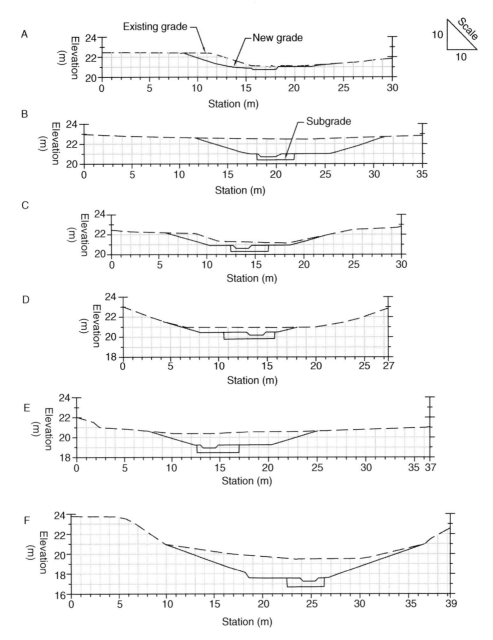

Figure 7.5 Typical cross-sections showing existing and proposed channel and floodplain cross-section dimensions for the Sawmill Pond reach of the Eel River, Massachusetts, USA. 'Existing grade' is the current sediment surface in the former mill pond, 'new grade' is the proposed riparian and floodplain grade after excavation of sediments, and 'subgrade' is the proposed new channel base of alluvial fill elevation. A–F are cross-sections arranged from upstream to downstream. Figure courtesy of Inter-Fluve, Inc.

discharge, and flood or overbank discharges. Hydrologic analyses are used to develop a specific target value, or range of values, for each of these categories (Skidmore *et al.* 2011). For example, where low-flow habitat has been identified as a constraint to resident fish, modeling of in-channel habitat or channel enhancements will require a specific low-flow discharge value, such as the mean daily flow during typical dry seasons. Similarly, where significant channel modifications are proposed, the computed effective discharge can be used to design channel cross-section dimensions, grade, and substrate using various hydraulic relations or models described in the next section. Riparian reforestation design will also be informed by inundation depths and frequencies of floodplain flow, derived from flow statistics. These values can then be applied to varying levels of design, from simple cross-section hydraulic calculations to complex multi-dimensional hydraulic modeling. Where stream restoration is intended to address the physical response of a channel to changes in hydrologic regime, the aim of those projects is usually to address combined ecological responses to changes in hydrologic regime and physical channel change. However, where limiting factors or other identified problems are most influenced by changes in flow regime (e.g. flow regulation or abstraction), addressing these problems through restoration of flow may be the most effective means to achieving restoration goals (Beechie *et al.* 2010).

7.5.1.3 Hydraulic modeling

Whereas hydrologic investigation is used to characterize the timing and volume of flow, hydraulic modeling is used to characterize or predict the forces of moving water that act on channel boundaries, including obstacles within the channel (Shields *et al.* 2003). Hydraulic characteristics of flow within the channel at varying scales are largely responsible for determining both micro- and macro-scale channel form and associated habitat. Additionally, the habitat preferences of many aquatic organisms include the hydraulic character of physical habitat (Poff *et al.* 1997). Forces associated with moving water are largely dependent upon velocity and are also complex, three-dimensional, and influenced not only by channel morphology and boundary conditions in the project reach but also by conditions in reaches immediately upstream and downstream (Shields *et al.* 2003). Details of the velocity distribution are difficult to measure, model, or predict. Nonetheless, analytical tools and models are readily applied in investigations of existing conditions, restoration alternatives, and channel design

development. Models may simulate flow in one, two, or three dimensions, and selection of an appropriate model depends on questions being asked and design specifications (Skidmore *et al.* 2011). Regardless of the modeling approach, hydraulic analysis has become a fundamental component of investigative analysis for evaluating alternatives. Models and analyses used to evaluate concept-level alternatives can be further developed to refine project designs for the selected alternative.

Hydraulic analyses allow a design team and project reviewers to evaluate site- and reach-scale flow conditions, sediment dynamics, the extent of inundation at varying flows, and effects of structural elements such as large wood, rock weirs, or bank stabilization (Doyle *et al.* 2007). In general, hydraulic analyses are used for four types of problems in restoration design. First, they are used to model water surface profiles through the project area and upstream and downstream, for the purpose of predicting inundation and flooding under different project designs (Figure 7.6; Abbe *et al.* 2002). Second, they are used to estimate forces acting on the streambed for sediment transport analyses and to design project elements such as substrate composition and mobility (Shields *et al.* 2003). Third, hydraulic analyses are used to estimate forces acting on stream banks (e.g. critical shear stress) and to develop streambank design alternatives and details, such as selection of erosion control fabric used in constructing stream banks (Shields *et al.* 2003). Lastly, hydraulic analyses can be used to estimate velocity distributions for evaluating and predicting microhabitat conditions under various alternatives (Harvey & Railsback 2009).

The level of analysis should be appropriate to the problem being investigated, the details necessary for design, and the data and expertise available to build and run models. Models available are either hydraulic models, which assume steady flow and require discrete steps to evaluate hydrologic conditions, or hydrodynamic models, which account for unsteady and gradually varied flow (ASCE 1996). While more sophisticated modeling may generate more detailed information and allow for glossier illustrations, it may instill a false sense of precision and may not be justified given project design objectives. In addition, a model may generate more detailed output than is valid given uncertainty or resolution of input data. Once an appropriate model is selected, numerous design scenarios and restoration alternatives can be readily evaluated (Skidmore *et al.* 2011). Table 7.3 provides a summary of the applications, limitations, and outputs of different levels of hydraulic analysis. The level of input

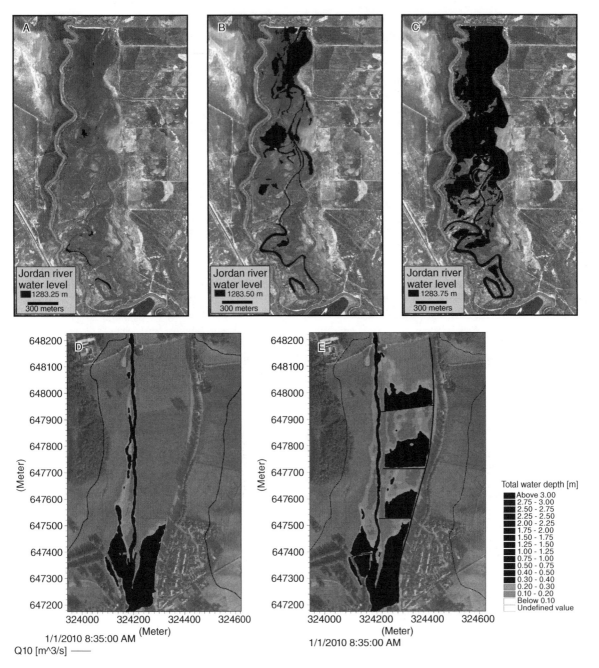

Figure 7.6 Examples of using the 1D HEC-RAS (Hydrological Engineering Center – River Analysis System) model to predict inundation of wetlands at (A) low; (B) moderate; and (C) high flows as part of a playa wetland restoration project on the Jordan River, Utah, USA and the 2D MIKE21 model to predict flow depths, velocities, floodplain inundation patterns, and flood storage under existing and design scenarios for the Darnhall reach of the Edellston Water, UK. The 2D examples show (D) existing and (E) design configurations with the addition of cross-floodplain berms to increase flood storage in the design scenario. Figures (A–C) adapted from SWCA *et al.* (2008); Figures (D) and (E) courtesy of cbec, Inc. (See Colour Plate 12)

Table 7.3 Comparison of hydraulic modeling approaches. The complexity and sophistication of modeling generally increases with the number of dimensions analyzed.

Number of dimensions	Outputs	Applications	Limitations
1D, at-a-section	• Mean downstream velocity across a cross-section	• Relatively uniform and single-thread channels • For basic evaluation of existing or proposed conditions • To estimate incipient motion • Where considerable uncertainty in actual hydraulic conditions is acceptable	• Cannot account for up- or downstream influence • Not appropriate for use in multi-thread channel or overbank flow • Not appropriate for evaluation of structures or where infrastructure may be at risk • Not appropriate where tides exist or where upstream or downstream conditions may impact flow at a section
1D model	• Mean downstream velocity across a cross-section • Water surface profile among multiple cross-sections • Inundation extent	• Where influence of upstream or downstream conditions is of concern, such as where backwatering is expected • Computation of inundation for given flows • For determination of influence of local controls on hydraulics • For estimation of extent of flooding associated with given flows	• Not appropriate for evaluating multi-thread channels, side channels, or complex cross-sections • Not appropriate where tidal influence exists
2D model	• Depth-averaged velocity and direction of flow • Width-averaged velocity, and direction of flow	• Where tidal influence is of concern • To evaluate influence of multiple channels • For evaluation of floodplain flow • Allows for greater detail in presentation of model output with color representation of variable velocity • To supplement 1D models where secondary currents are common such as through bends, channel junctions, and around structures	• 2D modeling may be cost-prohibitive • Requires detailed survey data or design detail to run model
3D model	• Flow fields consisting of downstream, horizontal and vertical components	• Design or evaluation of specific elements to evaluate complex local distribution of velocity, shear stress, and fluid drag	• Limited to applications where detailed and sophisticated design and analysis warrants significant expense • Requires detailed survey data or design detail to run model

data, effort, expense, and expertise required generally increases in a non-linear fashion with the number of dimensions in output.

7.5.1.4 Sediment transport analysis

The input and movement of sediment through a river network is second only to hydrology in its influence on the character, behavior, and potential of a stream system. Sediment transport analyses are fundamental to identifying reach-scale problems associated with either excess or insufficient sediment supply due to upstream land use or channel impacts (Montgomery & Buffington 1997). Sediment transport analyses are particularly applicable to the design of in-channel restoration projects, where they are used to evaluate the sediment transport performance of various project alternatives and to evaluate design iterations of a selected alternative.

Even simple sediment transport analyses can greatly enhance understanding of the implications of various design alternatives at varying flows. Basic sediment transport analysis can be conducted for a single design flow, usually for channel-forming flow, or for the full range of flows observed or anticipated. Sediment transport analyses are often used to support design iterations or alternatives where changes in channel dimensions among iterations result in changes in sediment transport. For example, design iterations may consider a range of channel dimensions and their impact on the capacity of a channel to transport the available bed sediment. Channel dimensions that do not result in sufficient force on the streambed to transport the sediment delivered to a given channel reach are likely to aggrade. In contrast, channel dimensions that exaggerate or concentrate forces necessary to transport sediment may lead to channel bed incision.

Sediment transport analyses can be categorized as either incipient motion analysis or sediment discharge analysis (Dunne & Leopold 1978). Incipient motion analyses predict the maximum size of a particle that can be transported for a given flow and are dependent upon channel dimensions, roughness, and slope (Dunne & Leopold 1978). This calculation is useful for determining the erosion potential of existing or constructed channels, and for determining appropriate substrate size for constructed channels or gravel augmentation downstream of dams. Incipient motion analysis may be adequate for some design applications where bed material load is low and the risk associated with a possible imbalance between sediment supply and transport capacity is tolerable (Newbury & Gaboury 1993). For example, this is often

sufficient for bedrock-dominated systems where grade controls exist or where increased risk is accepted due to a limited project budget. When greater certainty is needed, incipient motion analysis can be confirmed or corrected based on field observations of the largest particles moved during floods.

In contrast to incipient motion analysis, sediment discharge analysis is usually necessary where sediment load is substantial and maintaining the sediment balance is an essential project objective, or where projects influence hydraulic conditions through an extended reach or multiple reaches. Sediment discharge analyses allow the characterization of reach-scale sediment balance by calculating the rate and size distribution of bed material transported, either by the range of observed flows or for design flows. Output of sediment discharge calculations is integrated over a range of flows for a given period of time to determine the volume of sediment transported into and through a project reach. Where sediment balance is not zero over a specified period of time and flows, the difference between incoming load and transport capacity through a reach indicates the potential for scour or fill (Figure 7.7). Models that estimate sediment discharge can also be used to estimate sediment discharge and grain-size distributions under various flows, which is especially useful for evaluating gravel augmentation scenarios downstream of dams. For example, in the Trinity River, California the model iSURF was used to estimate grain-size distributions that would result from various levels of gravel augmentation (Figure 7.7; Gaeuman 2008).

Sediment transport analyses require input of size distributions of bed and substrate materials within the project reach as well as in upstream supply reaches. Sampling methods depend on the character of bed materials. For coarse-bedded streams, sampling is commonly and simply conducted using pebble counts. Bulk sampling and sieve analysis may be necessary to quantify finer components of bed substrate (Church *et al.* 1987). Sediment transport analyses also require hydraulic inputs including channel and flow characteristics. While transport analysis can be conducted using only measured gradations and hydraulic inputs, it is preferable to measure or calculate actual sediment loads to calibrate calculated rates. Uncalibrated sediment transport calculations are prone to large errors and, even for calibrated models, sediment transport analyses may produce significant errors (Gomez & Church 1989; Wilcock *et al* 2009). Considerable judgment and professional experience is necessary to select and apply appropriate analysis equations and methods.

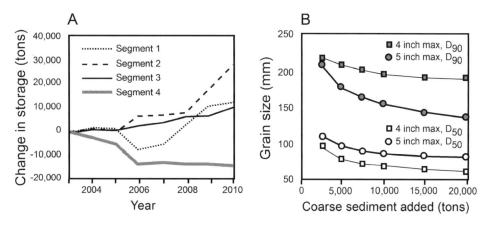

Figure 7.7 (A) Examples of reach-scale sediment discharge analysis and budgeting to evaluate potential effects of gravel augmentation downstream of Lewiston Dam in the Trinity River, California, and (B) modeling of equilibrium grain-size distributions with varying levels of gravel augmentation. Segments 1–4 in (A) are from upstream to downstream, and the analysis of proposed sediment augmentation in segments 1–3 indicated that sediment storage should increase in the upper three segments. In B, equilibrium D_{50} (median grain size) and D_{90} (90th percentile) estimates are based on iSurf modeling of different quantities of gravel addition just downstream of Lewiston Dam (x axis) and two different sediment input size ranges (sand to 4-inch gravels and sand to 5-inch gravels). The analysis showed that augmentation of at least 10,000 tons per of particles <5 inches per year would achieve a suitable gravel size for salmon and steelhead spawning. Figure (A) adapted from Gaeuman & Krause (2011); Figure (B) adapted from Gaeuman (2008).

7.5.1.5 Geomorphic investigation

Geomorphic investigations are generally concerned with understanding stream processes and boundary conditions that influence channel character, and predicting the potential channel responses to changes in these processes or boundary conditions (Dietrich *et al.* 2003). The ultimate function of most geomorphic analyses is to evaluate whether a stream channel is stable or changing and whether the current state of a stream is the result of anthropogenic disturbance or is simply the natural character of the stream given its geomorphic setting (Montgomery & MacDonald 2002). Specifically, geomorphic analyses can be used to characterize equilibrium dynamics of existing, historic, or other reference conditions, explain departures from those conditions, and help identify causes of change or instability (Simon & Castro 2003). Additionally, geomorphic analyses are used to explain or predict channel response to observed or anticipated changes in hydrology, sediment, or boundary conditions (Abbe *et al.* 2002). These analyses can also be used to determine whether causes are local and direct, thereby implying a reach-scale remedy or, of a broader scale and indirect cause, implying a remedy that may require addressing hydrologic or sediment variables. The perspective gained through geomorphic investigations therefore informs the selection of restoration strategies and whether they should be watershed- or reach-scale,

active or passive, focus on in-channel or floodplain processes, or emphasize upland management or channel corridor management (Beechie *et al.* 2010).

A fundamental objective of many geomorphic investigations, critical to developing appropriate management or restoration strategies, is the characterization of equilibrium conditions in a dynamic channel and the explanation of departures from equilibrium conditions (Shields *et al.* 2003). For example, erosion, deposition, and associated dynamic processes, such as lateral migration within an alluvial stream channel, are natural and essential to sustaining an aquatic ecosystem. Lateral migration of a channel may occur in the context of an equilibrium channel where migration rates do not vary significantly over time, or may occur primarily as a result of channel incision or aggradation (disequilibrium). Where historic or reference conditions or regimes are available and characterized as in equilibrium, geomorphic analyses can be used to evaluate channel boundary conditions and to identify erosion and deposition rates that may serve as restoration targets. For example, if rates of lateral migration increase through time but bed elevation and width-to-depth ratios remain relatively constant over several decades (Leopold 1973), the basic relationship between channel inputs (hydrologic regime and sediment supply) and channel processes is likely in equilibrium; increases in lateral migration may however be related to

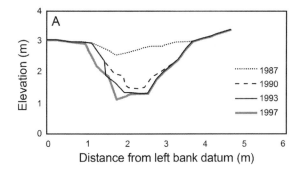

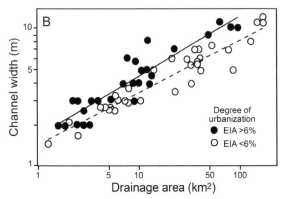

Figure 7.8 (A) Repeat cross-section surveys of small urban streams indicate effects of increased urban runoff, and (B) regional surveys of numerous cross-sections can be used to develop regional regressions for forested and urbanized streams. Adapted from Booth & Jackson (1997) and Booth & Henshaw (2001).

changes in boundary conditions, such as loss of riparian and floodplain vegetation. Where natural channels are not in equilibrium, or where historically equilibrium channels are trending toward a different channel state, geomorphic analyses can be used to characterize the trajectory of change as a means to predict ultimate outcomes or to predict restoration trajectories following restoration actions.

Dramatic changes or persistent trends in bed elevation, particularly when coupled with changes in width-to-depth ratios, are typically an indication of unstable or disequilibrium channel conditions (Figure 7.8; Lisle 1982; Booth & Henshaw 2001). Geomorphic analyses are used to evaluate these trends over decades and, in the context of geologic setting, to determine whether the observed condition is the result of anthropogenic actions. Understanding the historical channel character and change through time may reveal that changes in equilib-

rium conditions coincide with, and are likely the result of, significant change in land-use practices at the watershed scale. For example, urbanization may result in channel incision due to increased runoff, or an upstream dam may cut off sediment supply and lead to channel incision and lateral erosion associated with evolution of incised channels (Booth & Jackson 1997). Alternatively, investigations may indicate that disequilibrium conditions have persisted for decades and that the geologic setting can explain an imbalance between sediment supply and sediment transport. For example, streams that cross alluvial fans and flashy streams in semi-arid regions may be inherently unstable. These streams may perpetually alternate between aggrading, incising, or avulsing states in response to natural sediment pulses or large floods over periods of many years to decades. Restoration efforts to enhance or stabilize such streams often prove futile because natural channel dynamics overwhelm constructed features.

Geomorphic analysis may also include channel classification, which is occasionally applied by project sponsors (Simon *et al.* 2007) although it is not an essential step in either geomorphic investigation or channel design (Chapter 3; FISRWG 1998). Classification may be useful for communicating information about observed conditions and for considering the response potential of the stream to disturbance (Montgomery & Buffington 1998). However, most classifications do not account for or convey information about temporal changes in forms and patterns (Kondolf 1995). Channel classification therefore does not provide a sufficient basis for identifying dominant processes or predicting likely responses to disturbance or restoration.

7.5.1.6 Geotechnical assessment

The term 'geotechnical' refers to properties of earth and soil materials as they relate to the stability of slopes, such as stream banks. In a stream restoration context, geotechnical assessment is primarily used to understand mechanisms behind bank erosion, which is a common impetus for channel restoration and stabilization projects intended to address loss of property or protection of infrastructure (Thorne *et al.* 1996). Geotechnical assessment may also be used to understand mechanisms behind slope failure, such as those associated with road cuts or forestry practices or for steep and failing slopes adjacent to stream channels. While bank erosion is a natural process within streams in equilibrium, and indeed is fundamental to many processes that create and maintain habitat, excessive bank erosion can be a sign of channel instability and can contribute to water quality degradation (Simon & Downs 1995). Evaluating rates of bank erosion and channel migration are parameters typically investigated in geomorphic analyses, but understanding the causes and mechanisms of failure for bank erosion may require geotechnical analyses. An understanding of the mechanism of failure can help in determining the cause of bank erosion, whether it is due to human disturbance and at what scale, or whether it is occurring within the context of natural channel equilibrium. Similarly, the design of stream banks for stabilization or restoration often requires geotechnical analyses.

Bank retreat typically results from a combination of fluvial erosion and geotechnical failure. Fluvial erosion, or entrainment of bank materials by stream flow, is usually concentrated in the lower portion of the bank profile due to a higher frequency and longer duration of flow, higher shear stress, and lack of vegetation or roots along the lower bank (Brooks & Brierley 2002). In larger rivers, the channel depth often exceeds the rooting depth and the river can more easily erode its banks and naturally develop dynamic and complex channel patterns (Beechie *et al.* 2006). Similarly, bank erosion in degraded or incised channels often leads to bank collapse or geotechnical failure. Bank collapse occurs when factors that cause failure, such as gravitational force and internal saturation of the bank, exceed those that hold it together, including roots and soil cohesion (Simon & Downs 1995). Land-use impacts that may increase the potential for bank failure include soil compaction near the bank, loss of bank vegetation and associated roots, intense drying that leads to cracking, excessive saturation of the bank, and rapid drawdown of the water surface that generates high positive pore pressure, particularly in poorly drained soils (Sidle *et al.* 2006).

Restoration projects that include channel reconstruction should pay specific attention to streambank design. The design of reconstructed stream banks should consider both fluvial erosion forces along the bank and geotechnical properties of the bank materials. Constructed banks that are higher than the anticipated rooting depth will require special attention to geotechnical factors, and may require that internal stabilizing features be engineered. The primary design consideration for constructed banks is whether they are intended to be deformable over time (subject to erosion) or non-deformable (permanently stabilized). The challenge inherent in constructing deformable banks is protecting them from erosion while vegetation that eventually provides root binding of soil becomes established. Common approaches include the use of biodegradable stabilizing fabric to hold banks together or diversion of high flows for a number of years.

7.5.1.7 Uncertainty and risk

The final design for any in-channel restoration project will contain both uncertainties and risks. Uncertainty refers to unknowns that cannot be resolved in project design, whereas risk refers to the probability and likely consequences of project failure. Both uncertainty and risk may be reduced to some extent in the design phase, but doing so often requires considerably more time and a larger budget. Nevertheless, it is important for a project team to identify, articulate and acknowledge uncertainties and risks during the alternatives evaluation process so that stakeholders are fully informed about the selected alternative. Similarly, it is important to acknowledge that there is some degree of inherent unpredictability in restoration efforts, and that the exact outcome is rarely known or predicted. However, it is usually possible to predict a reasonably small range of potential outcomes to deal with permitting issues.

Uncertainty in restoration project planning can be separated into natural variability or knowledge uncertainty. Natural variability includes all sources of stochastic variation, whereas knowledge uncertainty refers to our inability to measure or model something (Hall 2003). It is possible to reduce knowledge uncertainty by improving the model form, decreasing measurement error, and acquiring local data rather than extrapolating from other locations. However, significant reductions in uncertainty are usually costly. By contrast, uncertainty resulting from natural variability cannot be reduced, although it can be characterized through probability distribution. For example, we cannot know what the peak flow will be a year from now, but with a sufficient record of historical flows we can determine the probability of each flow in the year ahead (Skidmore *et al.* 2011).

In developing projects, the implications of uncertainties can be acknowledged and evaluated by following four basic steps. First, define and quantify the sources of uncertainty in input data, analyses, and models (Anderson *et al.* 2003). Second, estimate uncertainties in analysis and model outputs (Steel *et al.* 2009). Uncertainty in inputs will propagate through analytical output and into design. Third, consider the potential consequences of uncertainties (Anderson *et al.* 2003). The range of possible outcomes, or unforeseen outcomes, can be explored quantitatively or qualitatively to express the impact of natural variability or knowledge uncertainty (Anderson *et al.* 2003). Lastly, communicate uncertainties to stakeholders. Uncertainty in outcomes is a critical component of evaluating risk and relaying this to stakeholders. Scenario modeling is a valuable approach to incorporating uncertainty and evaluating alternatives, particularly in light of the range of potential futures (Limbrick *et al.* 2000). For example, hydraulic modeling of project outcomes under a range of possible future flow conditions can facilitate both the evaluation of uncertainty and the selection of a more resilient design.

Risk can be defined as the combination of the chance of a particular event (probability) and the impact that the event would cause if it occurred (consequence) (Sayers & Meadowcraft 2002). In the context of river restoration, quantifiable risks include cost or liability to the property or project owner, or potential displacement or death of biota. The desire to reduce risk must be balanced with project objectives, the degree of uncertainty, and cost. For example, the integration of engineering practice into restoration planning and design often leads to over-design and stabilization to reduce the risk, even where it compromises the natural function of the stream system. This may be exacerbated by high levels of uncertainty, where a lack of knowledge increases the perception of risk. Where more data or modeling can decrease uncertainty, this may lead to greater accuracy in describing probabilities, which may in turn facilitate a more accurate portrayal of risk. However, additional data and analyses may be costly, and the value of decreasing uncertainty or more accurately predicting risk must be balanced against the increased cost of doing so. Ultimately, evaluating a wide range of potential outcomes during the project design phase can help reduce the risk that a project will result in undesired outcomes.

7.5.2 Investigative analyses for other restoration actions

As with in-channel actions, other watershed restoration actions also benefit from investigative analyses to help identify alternatives and focus project design. Restoration needs such as sediment reduction, restoring riparian functions and increasing floodplain connectivity and dynamics are identified in the watershed assessment and restoration inventories discussed in Chapter 3. However, analyses needed to identify alternative solutions for those restoration needs are often conducted during project design. For example, a restoration objective might be characterized as 'reduction of landslide sediment from roads' or 're-establish riparian vegetation communities along specified reaches,' and further investigation at each

site is needed to determine the best course of action. In this case, such investigations might involve identification of road failure hazards and the types of road rehabilitation actions that might solve those problems, or investigations to identify appropriate species mixes and planting options for riparian restoration sites. In this section we briefly describe the types of investigative analyses that might be needed for design of site-specific actions that address reach- or watershed-scale processes.

Restoration actions that address watershed-scale processes such as hydrology, sediment supply, and nutrient problems are identified through analyses of hydrologic alteration, partial or full sediment budgets, and identification of non-point-source pollution areas (Chapter 3). These assessments identify key restoration needs, and provide information needed to set project goals and identify spatially explicit objectives. Once the key goals and objectives for watershed-scale actions have been identified, solving the problem then becomes a matter of designing and implementing projects at multiple sites. These site-level designs require additional investigations to identify project alternatives and to develop a project design for the preferred alternative. For example, hydrologic models such as those described earlier in this chapter might be used to help determine whether stormwater detention ponds will sufficiently absorb peak flows from a newly developed area, or perhaps that peak flows cannot be realistically ameliorated in a particular setting. Similarly, models of nutrient or pollutant sources and transport might be used to evaluate the likely effectiveness of a restoration action at a particular site on the entire nutrient or pollutant budget.

For sediment reduction actions, the required site-level investigative analyses depend on the erosion mechanism and land use. For example, sediment reduction in the forested mountains of the Pacific Northwest often focuses on reducing landslides from forest roads to achieve sediment reduction objectives. Depending on site conditions and road construction techniques, reducing coarse sediment delivery from road-related landslides may require removing roads and reshaping hillslopes, removing side-cast material on steep slopes, or reducing the risk of failure at stream crossings by increasing culvert sizes, installing bridges, or constructing fords. Each road segment and stream crossing must therefore be investigated to identify appropriate solutions, although in most cases there is no formal analysis of various options. Nevertheless, there are common elements to the investigation including estimating stream discharge at crossings, characterizing channel slope and size, measuring slopes of side-cast and the hillslope itself, and examining road construction material. Solutions are generally chosen based on effectiveness at reducing landslides, short- and long-term cost, and future use of the road. For instance, a road with little likelihood of future use may be removed and re-contoured to match the original topography, which involves considerable short-term expense but no future maintenance. By contrast, a heavy-use road might be reconstructed with reduced side-cast material and bridges over larger streams, which maintains use of the road but reduces the major landslide hazards.

In agricultural areas where effective solutions typically include changes in land-use practices rather than specific restoration actions, there may be little formal analysis of alternatives. Rather, landowners may choose to reduce erosion from fields by changing tilling practices (e.g. switching to no-till seeding, which maintains ground cover and roots year-round), or to reduce sediment delivery to streams by developing vegetated buffer strips that filter fine sediment from overland flow before it reaches a stream. Investigative analyses might therefore include assessment of pathways by which water and sediment is routed to streams, measuring slopes of fields or pastures to help estimate erosion potential, looking for evidence of rilling or gullying, and examining riparian buffer conditions. Based on such investigations as well as assessments of cost and impact to agricultural production, each landowner elects to alter land-use practices based on their own interests in stream restoration and the economics of altering their land-use practices.

Investigative analyses for most reach-scale projects will consist of those already described in the previous section for in-channel restoration actions. However, some restoration actions – such as restoring stream flows or riparian vegetation – may require additional or different evaluations. For example, control of grazing impacts on streams may include geomorphic investigation or geotechnical assessment (Sections 7.5.1.5 and 7.5.1.6), but it may also include simple surveys of locations of livestock access and degradation of banks. This easily acquired additional information is important for identifying appropriate actions for reducing livestock impacts on streams. For example, fencing may be a simple solution to localized bank erosion problems, while in other cases fencing may solve the erosion problem but may also necessitate an alternative water source for livestock. Investigations for some project designs might therefore include examination of impacts to agricultural or other interests in addition to investigations of reach-scale processes and alternative solutions.

Riparian restoration projects may also require specific investigations to identify and design appropriate restoration options. As discussed briefly in Chapter 5, key elements of investigation for riparian revegetation projects might include measuring depth to the water table, characterizing soil type, and identifying species that can survive well in the local environment (e.g. Hall *et al.* 2011). In most cases, some understanding of native plant species distributions can inform the selection of suitable species (e.g. Harris 1999), although species may also be selected based on other functional characteristics where restoring native communities is not a realistic option. Understanding these factors can help to identify appropriate species as well as the likely need for short-term irrigation of plants, protection from browsing animals, or specialized planting techniques to increase survival (e.g. Hall *et al.* 2011).

Addressing water withdrawals on smaller streams and rivers not blocked by dams (e.g. throughout the western United States) often involves two types of problems. The first problem is water loss from the stream during summer low flows, and the second is entrainment of fish into diversions or pumps and subsequent mortality. The most common investigation for water losses is an assessment of instream flow needs for key biota (usually fish). Such instream flow analyses usually examine useable habitat areas for fish at specific life stages based on depth and velocity distributions at varying stream flows (often called habitat suitability criteria). These investigations quantify available habitat area for species and life stages at various flows for the purpose of identifying how much stream flow should be retained to support stream fishes, usually using a model such as the Physical Habitat Simulation System (PHABSIM; Waddell 2001). The second problem is direct loss of fish at streamflow diversions and pumps. The most common investigation to address this problem is an inventory of screening devices at each diversion or pump to ensure that fish are not drawn into the irrigation or water supply system (e.g. Washington Department of Fish and Wildlife 2009).

7.6 Evaluate alternatives

The investigative analyses commonly lead to several possible restoration alternatives that can meet project goals and objectives. A comprehensive evaluation of these alternatives ensures that the selected restoration approach meets project goals and objectives, provides value relative to no action, and poses no unnecessary or unavoidable risk to natural resources (Bean & Rowland 1997; Parnell 2000). While the logic of considering alternatives prior to embarking on a restoration design seems obvious, it is common in the river restoration industry to embark on restoration design without evaluating alternatives. This is evidenced by the common use of a limited suite of design approaches despite widely varying geographic variables, project contexts, watershed conditions, project objectives, and stream types. Where rote structural designs are proposed, it is impossible to determine whether the design approach has been developed through a logical process or whether a preconceived and potentially inappropriate design solution has been applied without due consideration of the problem, project context, and goals and objectives. Thoughtful and comprehensive evaluation of alternatives can ensure that connections between causes of identified problems and project designs are established (Alexander *et al.* 2006).

Many provincial, state, or federal regulatory frameworks do not require an alternatives analysis for stream restoration projects (Brookes 1990; Bean & Rowland 1997). Remarkably, funding entities are often similarly reticent to require an evaluation of alternatives (Skidmore *et al.* 2011). The implications of embarking on design without evaluation of alternatives are significant: there may be another alternative that poses less risk, can be done less expensively, or will have a higher probability of achieving desired outcomes. There may also be a passive alternative that presents a more sustainable solution, even if achieving outcomes requires longer timeframes.

A systematic evaluation of alternatives will compare probability of success, risk, cost, and timeframe for each project objective, including a no-action alternative (Anderson *et al.* 2003; Skidmore *et al.* 2011). To evaluate the probability of success, the alternatives evaluation often develops conceptual level designs. Indeed, it is difficult to evaluate alternatives or test basic feasibility and consistency with objectives without some degree of design. For example, to evaluate channel restoration alternatives where disrupted sediment causes channel instability and restored sediment supply is a project objective, modeling of sediment transport through a proposed channel configuration will be necessary. This of course necessitates some level of channel design for each alternative. Site surveys and investigations of hydrology, hydraulics, sediment transport, and geomorphology will be needed to identify causes of problems, evaluate constraints relative to alternatives, and to conduct a sediment transport analysis to support the feasibility of the selected alternative.

Alternatives evaluations commonly employ a matrix consisting of defined project objectives in rows and parameters for comparison in columns (i.e. success, risk, cost). Values for each parameter can be weighted to reflect stakeholder priorities. One of the most important evaluation parameters is the probability of achieving each objective, or degree to which the objective is likely to be met. Developing predicted values for each objective and alternative as well as the degree of uncertainty in predicted values helps quantify differences between the alternatives (Anderson *et al.* 2003). Identification of the risk of failure for each objective and the risk to resources for each alternative is also essential. To consider probable costs of varying alternatives, design concepts must be developed sufficiently to estimate unit costs for elements and treatments. For larger and more complicated projects, additional important considerations include long-term operations and management costs and the timeframe required for project completion and achieving objectives.

7.7 Project design

Once a restoration alternative has been selected, the final project design can begin. A sound technical design implies that investigative analyses have informed selection of project elements to meet objectives, an appropriate design approach has been applied, and documentation is sufficient for any third party to understand the justification and derivation of proposed designs. The project design phase develops discrete project elements and ensures that these elements work in concert to meet project objectives. Ultimately, the project design process develops the 'what' and 'how' of project implementation and should ensure that: (1) project elements collectively support project objectives; (2) design criteria are defined for all project elements; (3) project elements allow or promote stream processes that create and maintain the stream system; (4) the technical basis of design is appropriate and sound for each project element; and (5) design documentation is sufficient in scope and detail to explain and justify project details.

No universally accepted standards of practice exist to guide project design for the restoration of rivers, particularly where restoration strategies include significant modifications to the stream channel (Miller & Skidmore 2003; Palmer *et al.* 2005; Skidmore *et al.* 2011). Given the absence of accepted design standards or methods and the variation in restoration project goals and objectives, it is critical that a design team follow a logical sequence of investigation of causes of observed problems, evaluation of alternative remedies, and justification of designs relative to specific project objectives. The investigations described previously will inform the design process and provide a framework for modeling or testing outcomes. Following selection of a preferred alternative, the design process is characterized by the following sequence:

1. specify project elements that will meet project objectives;
2. establish design criteria for project elements that define expectations;
3. develop design details to meet criteria for each element; and
4. verify that elements are mutually supportive of project objectives.

In this section we first describe three basic design approaches used for in-channel restoration projects, and then describe each of the four design steps. General design considerations vary greatly by specific project types however, and a detailed discussion of these is beyond the scope of this chapter or volume. Chapter 5 provides some general considerations by project type as well as sources for additional information on detailed design considerations for various project types.

7.7.1 Design approaches

Many restoration strategies, such as upland land-use management or floodplain or streambank revegetation, can be accomplished with minimal or no engineering design. However, for projects that involve channel or floodplain modification, designs are necessary to guide the implementation or construction of the project. Designs illustrate what a project will consist of and how it will look when it is complete, but do not necessarily indicate how it will perform. While the relationship between form and process in stream systems has been researched for decades (Leopold & Maddock 1953), the application of those studies to channel restoration and design is an emerging and evolving science. The basic challenge in channel restoration design is to identify a channel form that is appropriate for local hydrologic and sediment regimes, suited to existing bed and bank materials, and is sized and graded to provide continuity of sediment transport.

There are three general design approaches: analog, empirical, and analytical (Shields 1996; FISRWG 1998; Watson *et al.* 1999; Fripp *et al.* 2001; Skidmore *et al.* 2001). An analog approach uses geomorphic templates from historic or other channel reaches as the basis for design of a restoration reach. This approach is also

referred to as the 'reference reach' approach. Analog approaches to design for river restoration are commonplace, partly because they are intuitive and simple and can be applied with a minimum of analysis. They are intuitive in the sense that if an adjacent or nearby stream appears to be in equilibrium, it seems logical that it would serve as a good template for restoration of a stream that is degraded or not in equilibrium. Analog approaches are generally only appropriate when the hydrologic and sediment regimes and boundary conditions of the analog are the same as for the project reach, where upstream and downstream reaches are in equilibrium, and where causes of problems and their solutions are relatively local (Hey 2006).

An empirical approach uses equations derived empirically from large datasets to relate channel dimensions to parameters defining flow regime, sediment regime, and boundary characteristics. Datasets from which empirical relations are derived may be local, regional, or global and determine the relevance to a restoration project area. Where datasets are concerned primarily with channel dimensions, the empirical approach may also be referred to as the 'hydraulic geometry' approach. Empirical equations represent average conditions, although more advanced forms of equations may also include range of variability which can facilitate design of variability within a channel. Application of an empirical design approach is generally subject to the same qualifications as for analog design (Skidmore *et al.* 2011). However, a risk associated with the application of analog and empirical approaches is that investigative analyses, particularly those that evaluate hydraulics and sediment transport and which would help to identify problem causes at watershed scales, are often bypassed or disconnected from the design process. This can lead to designs based on untested assumptions of channel stability and the erroneous assumption that restoration of form leads to restoration of other ecosystem processes and attributes (Pretty *et al.* 2003; Sandin & Solimini 2009).

The analytical approach uses process-based or theoretically derived equations for flow hydraulics and sediment transport to design a channel and floodplain geometry that supports sediment continuity of flow and sediment regimes. An approach relying primarily on modeled conditions is an analytical approach, sometimes referred to as 'predictive.' Analytical approaches offer an advantage over both analog and empirical approaches in that they are not dependent on the existence of equilibrium conditions; they are therefore particularly well suited to restoration or stabilization projects where existing channels are not stable. Analytical approaches are, however, dependent on sufficient quantity and quality of input data and may be limited by the scientific uncertainty of analytical models. Analytical approaches differ fundamentally from analog and empirical approaches in that they provide opportunity to simulate process-response mechanisms in stable and unstable channels, allowing design practitioners to investigate potential causes of observed problems as well as to design solutions that may not be represented in reference conditions. Analytical methods also provide a mechanism to test and compare probable outcomes of designs. Ultimately, many projects employ a hybrid approach, combining elements of analog, empirical, and analytical approaches as necessary.

7.7.2 Specify project elements that will meet project objectives

While projects vary tremendously in the level of design that is necessary, the design process for river restoration is typically a multidisciplinary and iterative process of developing and justifying details for discrete project elements and tying those elements together. Project elements are distinct project components that can be designed independently, but together constitute a holistic restoration design. Project elements may be features intended to remedy sources of problems, such as levee removal or floodplain revegetation. They may also be channel features including cross-section, bank, or planform alignments, or they may be features intended to provide specific in-channel value, such as log jams. The design of each element may be developed independently and by individuals with specific and relevant expertise but,

ultimately, the design process will have to piece the elements together and test their influences on each other. To ensure that project elements are appropriate, clear links should be identified between the findings of the investigative analyses (Section 7.5) and the purpose of each project element.

7.7.3 Establish design criteria for project elements that define expectations

Design criteria are specific, measureable attributes of project elements that clarify the purpose of each element and state how each element will perform to meet one or more project objectives (Miller & Skidmore 2003; Table 7.4). Design criteria can be categorized as either prescriptive or performance criteria. Prescriptive criteria describe specific required attributes of project elements, such as how it will be constructed or implemented. Performance criteria describe specific performance attributes of the design element, such as desired depths and velocities in a fish passage structure. While it may be easier to develop designs for prescriptive criteria and to measure their success, they do not necessarily result in project elements that meet objectives. Performance criteria are better suited to direct correlation to structural or functional project objectives, but it may be more difficult to establish measurable design attributes for them. While design criteria are rarely stated explicitly in design documentation, they are often implicit to the design process. Specifying criteria in the design documentation helps to:

• provide explicit design targets for each project element to clarify intent and anticipated outcomes;

• facilitate understanding among project stakeholders of expectations and inherent risks;

Table 7.4 Examples of design criteria for several hypothetical project elements. Design and implementation of restoration projects are guided by these design criteria, which are based on the project goal and objectives.

Goal	Objective	Project element	Design criterion
Restore reach-scale hydrologic processes and connectivity	Reconnect channel and floodplain at time of project implementation	Modify channel dimensions by raising bed with check dams	Channel capacity will be equal to effective discharge 50% of floodplain area will be inundated at the 5-year discharge
		Reintroduce beaver	Minimum of 1 breeding pair per km
	Increase inundated floodplain area to 80% of historic floodplain within 2 years	Setback levees	Flood risk to existing structures and infrastructure will not increase Levees will contain the 125-year flow

• provide acceptable or allowable tolerance ranges for performance of project elements,

• allow engineers and scientists to narrow performance expectations to within specific bounds, thereby limiting the liability associated with uncertain environmental conditions outside of those bounds; and

• establish measurable attributes that serve as the basis for project monitoring.

7.7.4 Develop design details to meet criteria for each element

Project plans include all of the design details for each project element, as well as drawings that illustrate the arrangement of elements within the project area. Design details convey a reasonable understanding of the intent of each element for stakeholders and permit agencies, and should be sufficient in detail for unambiguous implementation. Investigative analyses are developed to understand sources of problems and to select from among project alternatives, and they can be refined with the addition of project details to test whether proposed elements meet project objectives. For example, channel cross-section dimensions developed as a design detail for a channel reconstruction serve as new cross-sections for hydraulic analysis and sediment transport analysis, and allow comparison between existing and design conditions. Similarly, streambank design details can be evaluated using hydraulic analyses to test their stability under different flows relative to specific design criteria for channel bank stability.

7.7.5 Verify that elements address project objectives

Developing designs for specific project elements is an iterative process. Changes to details of one element may necessitate changes to other elements. In some cases, new information or perspective on constraints to specific design elements may warrant revising some or all design criteria. At the end of the design process, all project elements must mutually support specific project objectives and meet their individual design criteria. Verification will require final runs of the suite of investigative analyses and documentation of their output such that designs are justified, defensible, and documented.

7.7.6 Communicating project design

Effective communication of project designs to stakeholders, funders, permitting agencies, and implementers is critical to project success. Project proponents have the responsibility of ensuring that project goals, plans, and expected outcomes are effectively communicated. This is especially important where restoration efforts address problems that affect public interests or degrade ecosystem services, including water quantity and quality, wildlife, fisheries, and a host of social concerns including recreation, property, infrastructure, flood control, and water supply. Early involvement of stakeholders, particularly in establishing goals and objectives, is critical to effective and efficient project development, as the opportunity to influence the project diminishes and costs for making change increase as the project develops (Figure 7.9; see also Chapter 4).

The format and level of detail for communicating project goals, plans, and eventual outcomes should be tailored to the interests and demands of constituents for each specific project. While each project is potentially unique in its communication needs and requirements, these can be loosely organized around the interests of stakeholders, permitting agencies, and the implementation team. Stakeholders will have interest in every step of the project process, with an emphasis on project outcomes in particular, but may not require a significant level of detail about project design or implementation. Stakeholder communication is a two-way street, with project proponents soliciting input from stakeholders early in the project, and stakeholders having a responsibility to participate in project development.

Stream restoration projects are subject to review and permitting by a variety of regulatory bodies including federal, provincial, and state and local entities whose responsibility is to protect the public interest including water quality, transportation infrastructure, natural resources, and zoning ordinances. There is tremendous variation in permits required depending on project type, location, and resources affected, and project planners must consult with national, provincial, and local agencies to determine the appropriate requirements. Communication with permitting agencies is often initiated through a formal permit submittal process, with explicit reporting requirements that may include levels of project detail that can only be derived from relatively complete designs. While rules governing permitting do not typically provide for early involvement of permitting agency staff in project development, there are obvious benefits in early and continued communication with project proponents and permitting agencies.

Project implementation includes all actions necessary to execute a project, which may range from changes in upland management to complete reconstruction of stream channels. While implementation teams often

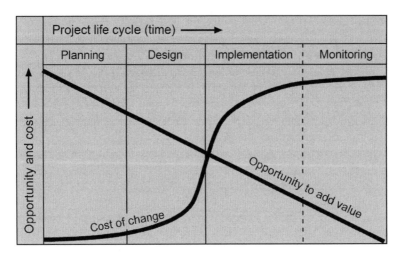

Figure 7.9 Opportunity and cost of change in project development. The opportunity for constituents to influence or make changes to a project plan generally decreases with time. The costs of making changes to a project increase as a project is developed; once design is advanced, the costs of making changes may become prohibitive.

include project proponents and the design team, particularly in providing oversight, additional parties are often contracted to construct part or all of a project, particularly where significant construction actions are involved. Contracting for project implementation requires clear communication of the project goals and detailed specification of all project elements to ensure that the implementation team executes a project as planned. Communication of project goals, plans, and outcomes is accomplished through design reports and project plans and specifications (Table 7.5).

7.7.6.1 Design reports
A design report is typically prepared for stakeholders and permitting agencies to explain how specific project elements address project goals and objectives, and to document the analytical process that supports the selection and integration of project elements. For the design team and project proponent, the process of articulating and justifying the design elements and their relation to project goals and objectives supports internal design review and contributes to a robust design process. For stakeholders, a design report will provide an explanation of how the project plan relates to project goals. Permitting agencies will look for sufficient detail in a design report to answer any questions relating to how and why designs were developed, as well as how the project may affect resources of interest or concern to agencies. A design report may also be the best protection against

liability in the event of project failure (Skidmore *et al.* 2011). The level of integrity of a design report can be a strong indication of the level of integrity brought to the design process. Design reports are developed incrementally to document concepts, evaluation of alternatives, and design methods. Iterations of design reports may be necessary to communicate progress and to provide sufficient information for review by constituents.

7.7.6.2 Plans and specifications
Plans or drawings are often developed iteratively along with the design, portraying design elements for review and ultimately serving as instructions to the implementation team. Incremental development of plans typically follows four phases: (1) concept plans; (2) preliminary plans; (3) draft final plans; and (4) final plans. These phases correspond to the design development process and communication requirements of constituents. Concept plans provide a sufficient level of detail for stakeholders and permitting agencies to compare alternatives, typically representing roughly 20–30% of design progress. Concept plans commonly include a map of all project elements and activities, identified constraints, and 'typical' drawings of anticipated project elements such as structures, bank treatments, or crossings. For the selected restoration alternative, preliminary plans are of sufficient detail to communicate exact intentions and locations of project elements. These will typically include a map of the project area, locations of all installations or reconfigu-

Table 7.5 Typical reporting frameworks for communicating project designs, plans, environmental assessments, and monitoring results to varying constituents (S: stakeholders; P: permitting agencies; I: implementation team; R: restoration community) through phases of project development.

Type of document or report	Constituency	Goals and objectives	Project elements	Criteria	Investigative analyses summary	Data inputs	Design methods	Alternatives and outcomes	Plans	Specifications	Cost estimate	Monitoring plan	As-built survey	Outcome report
1. Design report	S, P, R	x	x	x	x	x	x	x						
2. Plans & specifications	I, P, R	x							x	x	x			
3. Environmental assessment	S, P	x	x		x			x						
4. Monitoring results	S, P, I, R	x	x									x	x	x

rations, rough grading, and implementation logistics such as project sequencing, access, and storage and staging. They represent roughly 50–65% design completion, and are generally adequate for final permitting and detailed cost estimating. Draft final plans include all details necessary to communicate construction practices and instructions, and sufficient to develop final cost estimates; they typically reflect a 90% level of design. Details include all information required to plan and implement construction including survey notes, plans for regulatory compliance, access and staging, disturbance limits, and site remediation or revegetation plans. Final plans are prepared to serve as the basis for contract bid documents and are sufficient in detail for contractors to plan all aspects of implementation and to develop firm bids and timeframes for project execution. Final plans are accompanied by final specifications and a final design report, and serve as the basis for comparison with eventual as-built drawings.

Some projects may lend themselves to a 'design-build' approach, where responsibility for design and construction are contracted to a single entity or team and the design team is present to direct construction. Design-build contracts may require less detail in plans and specifications, where the contract documents specify deliverables and outcomes rather than prescribing how to execute the project. Design-build does not necessarily infer simplicity. Instead, it may appropriately be employed where project areas are rapidly changing or where the complexity and uncertainty of the environment precludes detailed

planning. This may be appropriate where significant and unknown constraints or complications are anticipated such as with contaminated sites, areas with presumed cultural resources, or uncertainty in subsurface conditions.

Any project in a stream ecosystem or intended to improve a stream ecosystem may affect ecosystem attributes such as water quality or native species, as well as the ecosystem functions (Sandin & Solimini 2009). An environmental assessment estimates or predicts impacts on these natural resources resulting from the project. Environmental assessments are often required for publically sponsored projects, where 'sponsoring' refers to any involvement in funding, permitting, or implementation. Environmental assessments may be relatively concise documents prepared for stakeholders and permitting agencies and generally include a description of the proposed project, its goals and objectives, alternatives considered including a no-action alternative, and justification for the selected alternative (Table 7.5).

7.8 Implementation

Project implementation is the installation or construction of project elements and the management of unexpected construction issues or environmental impacts. Project implementation should be a dominant consideration through the design process, resulting in a design that is practical to implement and acceptable to stakeholders

and permitting agencies. Additionally, considering implementation issues during the design process will help avoid or minimize undesirable or unavoidable effects of implementation. A project development team with a foundation in construction management will facilitate the consideration of implementation feasibility, practicality, and impact during the setting of project objectives, alternatives evaluation, and the project design, particularly where restoration actions include considerable construction. Project plans, design sheets, and design narratives are used to articulate specific implementation procedures, including maintenance requirements. While some restoration actions, such as grazing management or riparian revegetation, may be relatively benign in terms of implementation logistics, implementation issues that are included in implementation plans for more active restoration and instream activities include:
• site access and construction/implementation sequencing that minimizes in-channel work;
• dewatering and re-watering of the stream channel with consideration of anticipated storm flow where in-channel work is required;
• protection and salvage of aquatic organisms;
• site protection plans including sediment and erosion control, spill control and response, do not disturb areas, and minimizing equipment impacts;
• sourcing and staging of materials;
• safety;
• revegetation and weed control; and
• site protection and maintenance after completion.
All of these implementation considerations may influence the feasibility of alternatives. The iterative nature of project development and design is such that as design details develop implementation issues are considered, and designs are modified to reflect practical considerations inherent to implementation. Experience with construction management will provide a design team with perspective on the uncertain nature of project implementation including unanticipated site conditions, unpredictable weather, and qualifications and experience of contractors, and can greatly reduce the number of design iterations needed. In the absence of direct project implementation experience within the project design team, independent contractor review of plans as they develop can ensure that construction feasibility concerns are addressed during the design process.

Project implementation does not end when construction activities are complete, as large instream projects are rarely completed in one season or one hydrologic year. Short- and long-term operations may also be necessary to maintain a project's restoration trajectory. This is particularly important for floodplain and riparian elements, including revegetation. Revegetation, whether it is necessary to mitigate disturbance from construction actions or is the primary restoration action, often requires considerable operational maintenance such as irrigation or control of competing vegetation. Revegetation of banks and floodplains also requires the greatest amount of time to reach a restored condition. Indeed, for forested floodplains riparian maturity may take many decades to centuries. Many channel restoration projects fail less because of what was or was not done in the channel than what was or was not done on the floodplain, particularly during the first years after implementation.

7.9 Monitoring

Monitoring evaluates project-specific measures of success, facilitates adaptive management, and contributes to the broader purpose of learning from past projects (see also Chapter 8). Monitoring reports are summaries of project performance in light of project goals, objectives, and predicted outcomes (Kondolf 1998; Downs & Kondolf 2002). Constituents with an interest in monitoring reports may include: stakeholders who want to be assured that a project results in anticipated outcomes or informed when it is not performing as expected and needs corrective action; permitting agencies who want to be assured that a project performs as stipulated in negotiated permits; and implementation teams and the broader restoration community who want access to project lessons learned from monitoring (Table 7.5). Monitoring reports are strengthened by comprehensive as-built surveys that document actual implementation at the time of project execution. Without as-built surveys it is difficult to establish trends in outcomes except through the assumption that the project was completed as specified. Moreover, monitoring reports are most useful when monitoring plans and protocols closely match project goals and objectives. SMART objectives that specify measurable and time bound outcomes (see above) establish a foundation for measurement of specific parameters used to determine project success.

7.10 Case studies

To illustrate the design process and products, we describe three case studies that represent a range of complexities in project design. The first example is removal of the Number 1 Dam on the Chichiawan River in Taiwan which had a simple set of design criteria and constraints, leading to a relatively simple plan for removing the dam and monitoring the response of the last population of land-locked Formosan salmon (*Oncorhynchus masou formosanus*) in Asia to a restored migratory pathway. The second example illustrates development of a riparian restoration project along an incised stream in a semi-arid region of northwestern USA, focusing on the use of planting experiments to evaluate alternative solutions to re-establishing key riparian species. The third example is a complex stream and estuary restoration project in the Skagit River basin, USA. This project is characterized by a suite of design objectives and constraints, which results in a multi-faceted project to restore limited estuary functions.

7.10.1 Removal of the Number 1 Dam, Chichiawan River, Taiwan

Even conceptually simple restoration actions such as dam removal follow the sequential design steps outlined in this chapter and Figure 7.1. This process is illustrated by the removal of the Number 1 Dam in the Chichiawan River in Taiwan (Figure 7.10). The key problem to be addressed in this headwater basin was that dams blocked migratory pathways for the last remaining population of the land-locked Formosan salmon in Asia (Step 1, problem identification). From the watershed assessment, the broader project context (Step 2) indicated that several dams had been identified that blocked salmon migration, and the Number 1 Dam was the largest dam (22 m high) that blocked an important migration route between the lower portion of the Chichiawan River and its headwater

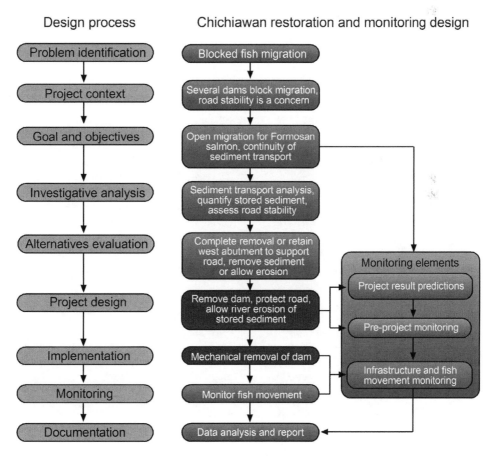

Figure 7.10 Schematic diagram of the design process for the Chichiawan #1 Dam removal, including steps to monitor project effectiveness.

tributaries. It was therefore identified as the highest priority barrier removal, although other dams must also eventually be removed to fully restore migration routes. This dam and the other dams were originally installed to trap sediment and prevent sediment from filling a larger reservoir downstream. However, the reservoir behind the Number 1 Dam has completely filled with sediment, negating its function as a sediment storage dam.

The primary goal of this project was defined simply: remove the dam to restore the main migration pathway for Formosan salmon in this portion of the Chichiawan River (Step 3). Specific objectives of the project were to completely remove the dam from the main river channel, and to allow the stored sediment to erode naturally and pass downstream. However, the removal project was also constrained by the presence of a main road just upslope of the dam's west abutment. The investigative analyses therefore included assessment of potential effects of releasing the stored sediment, as well as an assessment of potential effects on road stability (Step 4). The sediment transport analysis indicated that the amount of sediment stored behind the dam was less than the annual amount delivered to the downstream reservoir, and that dam removal and sediment release would have minimal effect on reservoir filling. The road stability analysis led to the conclusion that the road could be at risk if the west abutment was removed.

Because of sediment transport and road stability concerns, the design process considered two sediment management alternatives (Step 5): (1) removal of sediment prior to dam removal and (2) allowing river erosion of the reservoir sediments after dam removal. The design also considered two removal alternatives: (1) complete removal of the dam and (2) removing the dam from the main channel but leaving the west abutment to support the existing road. From these alternatives, the selected alternative was to remove the dam completely from the main channel but leave the west abutment, and to allow river erosion of stored sediments. The final project design was quite simple, including details on how much of the dam to remove and how to manage sediment during removal (Step 6). The dam removal was accomplished by mechanical breakup of concrete using a large jackhammer mounted on an excavator (Step 7; Figure 7.11).

Finally, the project incorporated an implementation and monitoring plan designed to measure the effects of the dam removal on movement of Formosan salmon (Step 8). The monitoring plan focuses on radio-telemetry

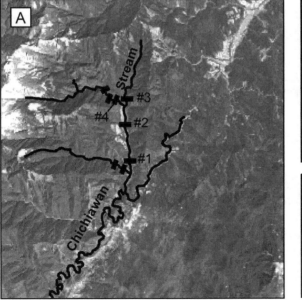

Figure 7.11 (A) Location of dams on the Chichiawan River, Taiwan and photos of the dam (B) prior to removal and (C) nearing completion of the removal (photos (B) and (C) Rickie Chen). Note that a portion of the dam on the left side of the photo was left intact and buried in rubble to support the hillslope and road just out of frame on the left. Background image in (A) copyright 2012 Google, 2012 Kingway Ltd., 2012 Cnes/Spot Image, 2012 Geoeye, and 2012 Digital Globe.

tracking of tagged salmon, which were expected to increase movement in the basin after dam removal. The monitoring design first tracked movement of 10 adult Formosan salmon before dam removal to establish baseline movement patterns, then tracked movements of an additional 10 salmon during the removal and 10 salmon after removal. The sampling during removal was intended to identify whether salmon were driven away from the dam by the deconstruction efforts or sediment erosion, while monitoring after removal and sediment erosion was intended to assess salmon migration in the restored channel.

7.10.2 Bridge Creek riparian restoration

Riparian restoration along Bridge Creek, Oregon, USA is one component of restoring salmon habitat in incised channels in the John Day River basin, which is a tributary to the Columbia River. The specific problem addressed by the riparian restoration effort is loss of the willow (*Salix* spp.) and black cottonwood (*Populous balsamiferia trichocarpa*) riparian vegetation (Step 1), which historically provided shade over the stream and reduced stream temperatures. The restoration sites in Bridge Creek are in the semi-arid, sagebrush-steppe region of the Columbia River basin, where riparian vegetation was historically a relatively narrow band of willow, cottonwood, and red osier dogwood (*Cornus sericea*) along the stream (Step 2). However, riparian restoration alone will not address the underlying problem of channel incision and lowering of the water table from 1 to 3 m, which was one of the likely causes of loss of riparian vegetation. Additional project components therefore addressed channel incision by encouraging beaver colonization and creation of beaver dams at key locations using beaver dam support structures which were intended to increase sediment and aquifer storage, raise the channel bed and water table, and increase beaver habitat (Pollock *et al.* 2007; Beechie *et al.* 2008). However, bed aggradation and raising the water table will likely take many decades, and the goal of the riparian restoration was to re-establish willow and cottonwood stands prior to recovery of the water table elevation without requiring irrigation (Step 3).

In order for riparian restoration to succeed in this semi-arid environment, the investigative analyses, alternatives evaluation, and project design (Steps 4–6) were conducted together in an experiment that evaluated the success of riparian planting (i.e. tree survival) as a function of planting depth and protection against browsing by deer, elk, and other wildlife (Figure 7.12). Two planting depths were tested: a standard planting depth which did not reach the water table, and rooting live stakes into the water table using a motorized auger to drill down 1–2 m. Browse protection options included: (1) wire fence cages; (2) plastic tree shelters; and (3) no protection. Each combination of rooting depth and browse protection was tested on willow, cottonwood, and dogwood stakes. The experiments revealed that 80% survival of willow and 70% survival of cottonwood were achieved when live stakes were planted deep enough to penetrate the water table and 1-m-tall plastic tree shelters were used to protect against browsing. However, there was still significant browsing that stunted willow and cottonwood growth, and additional experiments found that 6-foot shelters provided better browse protection. Survival was <20% for dogwood on all scenarios, so dogwood was omitted from the planting design.

The final planting design (Step 6) included planting live stakes of willow and cottonwood at a spacing of approximately 6 m, and the species mix was predominantly cottonwood because reaching the water table required stakes between 1 and 2 m tall. The zone within which plantings could reach the water table was delineated in GIS by mapping the area that was not more than 2 m above the water table, because 2 m was the maximum depth of the motorized augur (Figure 7.12). Implementation (Step 7) was straightforward, using the motorized augur to ensure that live stakes penetrated the water table when planted. The 6-foot tree shelters were used because they better protected the live stakes from browsing by wildlife. The monitoring plan was simply to quantify survival in the first few years after planting to ensure that the project design and objectives continued to be successful at other sites along the Bridge Creek restoration area.

7.10.3 Fisher Slough Restoration, Skagit River, Washington, USA

Fisher Slough is a freshwater estuarine system in the Skagit River Basin, western Washington, USA. The problem to be addressed through restoration was reduced floodplain connectivity and ineffective passage of juvenile salmon at the tidegate (Step 1). Estuarine habitat has been identified as a critical restoration priority for salmon recovery in the Skagit River system, and Fisher Slough historically provided important rearing habitat for several species of salmon including Chinook salmon (*Oncorhynchus tshawytscha*), coho salmon (*Oncorhynchus kisutch*), chum salmon (*Oncorhynchus keta*), and steelhead (*Oncorhynchus mykiss*). The area was also a migratory corridor to spawning habitat for adult coho and chum salmon, in smaller

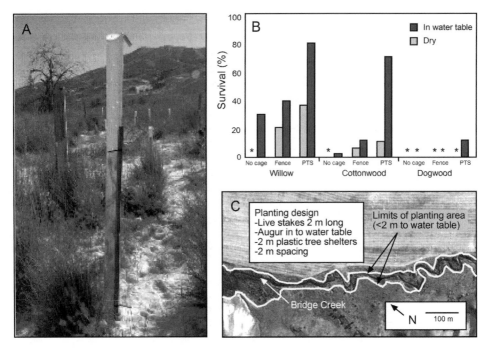

Figure 7.12 Design of a riparian planting project in Bridge Creek, Oregon, USA. The investigative analyses and alternatives evaluation revealed that live stakes planted within the water table had highest survival, and that (A) plastic tree shelters (PTS) provided the best browse protection (photo Jason Hall). (B) The selected alternative included planting live willow and cottonwood stakes deep enough to reach the water table and protecting them from browsing by wildlife using plastic tree shelters; asterisks indicate survival was zero (adapted from Hall *et al.* 2011). (C) Aerial view of project site showing the planting area within which 2-m-long live stakes could reach the water table, and the planting plan (photo Jason Hall).

tributaries (Step 2). A number of streams with a total watershed area of approximately 6000 hectares converge at Fisher Slough which has been channelized, confined by levees, and regulated by floodgates.

The goal of the Fisher Slough Restoration project is to 'maximize the area influenced by natural stream and tidal processes, allow for a broad range of ecosystem variability, restore estuarine rearing habitat for juvenile salmon to the maximum extent possible, and improve flood protection and storage capacity for Carpenter Creek and Fisher Slough' (TNC 2007). More specifically, the objectives of the Fisher Slough project were to improve tidal exchange and salmon passage through the tidegate, and to reconnect approximately 25 hectares of floodplain and marsh to the Fisher Slough channel (Step 3). The Fisher Slough Restoration project area was also constrained by levees, a downstream floodgate, and ditch and road crossings (Figure 7.13). Project objectives were separated into primary objectives that drove alternatives evaluation and project design, and secondary objectives to pursue unless they compromised primary objectives.

Primary objectives included creating a diverse array of native vegetative communities, establishing freshwater tidal Chinook salmon-rearing habitat, providing fish passage for adult coho and chum spawning access, and improving flood storage to protect agricultural uses of adjacent properties. Secondary objectives included improving habitat opportunity for cold-water fish other than Chinook, increasing habitat opportunity for migratory birds, identifying opportunities to address ongoing sediment transport and flooding issues in tributaries, and the ability to export lessons learned.

The primary investigative analyses (Step 4) included hydraulic modeling of tidal flux, flood scenarios, modeling sediment and geomorphic evolution of fluvial and estuarine habitat with numerical models, estimating habitat outcomes using functional response models, and predicting future vegetative communities using regional analogs. These analyses first informed the alternatives analysis, and then contributed to the final project design. Three alternative approaches to restoration were evaluated and compared to a no-action alternative,

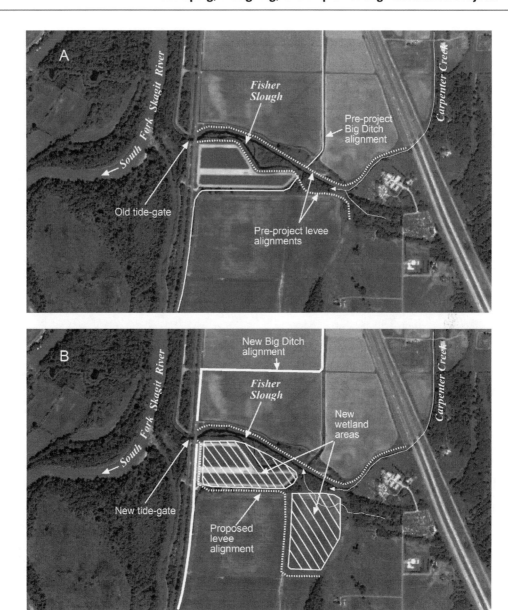

Figure 7.13 Overview of Fisher Slough site (A) prior to restoration and (B) showing proposed restored conditions and project elements. Adapted from TNC (2007); background image copyright 2012 Google.

using evaluation criteria based on the principal project objectives and cost (Step 5; Table 7.6). Alternatives varied primarily in approaches to reconfiguring ditch crossings in relation to the levee setback (Figure 7.13). For each criterion, a single normalized value was generated for each alternative (Table 7.6). The analysis revealed that, for many objectives, Alternatives 2 and 3 produced similar outcomes. However, in addition to producing the greatest habitat outcomes, which is a primary project objective, the long-term sustainability criterion was considerably greater for Alternative 3 than for other alternatives. Hence, the alternatives analysis not only assisted constituents in

Table 7.6 Project alternatives ranking for Fisher Slough Restoration, Skagit River, USA. The Fisher Slough Restoration expands and restores estuarine processes by setting back levees. Project goals include both freshwater estuary restoration and flood control. Project alternatives scores are the sum of 4 criteria, excluding project cost. Each criterion is the normalized value of a metric developed from a set of factors representing project objectives (TNC 2007). Alternatives varied primarily in how existing infrastructure was managed. Alternative 1 left existing ditches and crossings in place, thereby limiting levee setback opportunity; Alternative 2 extended existing crossings to expand levee setbacks; and Alternative 3 relocated and concentrated all infrastructure crossings at the downstream project boundary, improving maintenance access and opportunities for future additional levee setback.

Alternative	Project cost (2008) US$	Habitat output	Flood storage	Sedimentation capacity	Long-term sustainability	Total score
w/o project	———	0.10	0.17	0.12	0.20	0.59
Alternative 1	4,380,100	0.51	0.92	0.94	0.30	2.67
Alternative 2	5,267,300	0.94	1.00	1.00	0.40	3.34
Alternative 3	5,794,800	1.00	0.87	0.81	0.80	3.48

selecting a restoration approach, it also provided a thorough justification of proposed project elements by carrying initial investigative analyses through to predicting project outcomes.

The primary restoration technique was a levee setback to increase estuarine habitat area; additional restoration components included reconfiguring the floodgate to naturalize tidal flux within the project area, reconfiguring tidal and stream channels, and planting wetland, marsh, and floodplain vegetation. The final project design (Step 6) included three main components implemented in phases: (1) replacement of floodgates to allow more natural tidal flux in lower Fisher Slough; (2) realignment of a major drainage ditch (the former Fisher Creek); and (3) realigning the south levee of Fisher Slough to reopen tidal inundation of 25 hectares of estuary habitat. The implementation plan required replacement of the floodgate first in order to restore more natural tidal flux (Step 7). After that, the agricultural drainage ditch realignment was required to allow continued drainage function after the levee setback occurred, and finally the levee setback expanded and restored natural wetland and floodplain functions in lower Fisher Slough. The monitoring plan followed these phases, with initial monitoring focusing on evaluating effects of the new floodgate on habitats and salmon abundance (Step 8; Beamer *et al.* 2011). Continued monitoring after phases 2 and 3 will determine whether restored floodplain habitats are functioning as expected, and whether new floodgates combined with increased habitat capacity result in higher Chinook salmon production from the project location.

7.11 Summary

Understanding watershed processes, defining restoration goals, developing watershed assessments, and identifying and prioritizing restoration actions all inform the planning, design, and implementation of restoration projects. In this chapter we have outlined a series of steps and analyses for planning projects and identified three phases of the project-scale development process: planning, design, and implementation and monitoring (see Chapter 8 for additional monitoring information). Planning establishes the purpose and need for restoration, puts the project in a broader context, and articulates the specific intentions of a project. Design answers the question of how the project will be implemented by analyzing processes governing the project area's condition and describing specific actions that will be conducted. Implementation constitutes actions taken to conduct the project and includes monitoring, which verifies outcomes.

We also noted that it is critical to identify how the project restores watershed processes, and how it accommodates socioeconomic opportunities and constraints. Only then can there be an articulation of an overall project goal, project objectives, and measurable criteria. Project objectives need to address identified problems, be geomorphically, ecologically, and socially achievable, and specify measurable outcomes and time frames. An alternatives evaluation should also be conducted, and uncertainty and risk should be identified and incorporated into the alternatives analysis where possible. Project design selects the most appropriate alternative and specifies project elements that are in concert with project objectives. Design criteria for each project element guide the project design phase and typically help define project expectations. Project communication (project reports and public outreach) should occur early and throughout the planning, design, and implementation process. Lastly, project implementation should be documented and evaluated with compliance and effectiveness monitoring.

7.12 References

Abbe, T.B., Pess, G.R., Montgomery, D.R. & Fetherston, K. (2002) Integrating Engineered Log Jam Technology into Reach-Scale River Restoration. In: Montgomery, D.R., Bolton, S. & Booth, D.B. (eds) *Restoration of Puget Sound Rivers*. University of Washington Press, Seattle, Washington, pp. 443–482.

Alexander, C.A.D., Peters, C.N., Marmorek, D.R. & Higgins, P. (2006) A decision analysis of flow management experiments for Columbia River mountain whitefish (*Prosopium williamsoni*) management. *Canadian Journal of Fisheries and Aquatic Sciences* **63**, 1142–1156.

Allan, J.D. (1995) *Stream Ecology: Structure and Function of Running Waters*. Chapman & Hall, New York, New York.

Anderson, J.L., Hilborn, R.W., Lackey, R.T. & Ludwig, D. (2003) Watershed restoration – adaptive decision making in the face of uncertainty. In: Wissmar, R.C. & Bisson, P.A. (eds) *Strategies for Restoring River Ecosystems: Sources of Variability and Uncertainty in Natural and Managed Systems*. American Fisheries Society, Bethesda, Maryland, pp. 203–232.

Andrews, E.D. (1980) Effective and bankfull discharge of streams in the Yampa Basin, western Wyoming. *Journal of Hydrology* **46**, 311–330.

ASCE (1996) *River Hydraulics*. Technical Engineering and Design Guides as adapted from the U.S. Army Corps of Engineers, No. 18. American Society of Civil Engineers, ASCE Press, New York, New York.

Beamer, E., Henderson, R. & Wolf, K. (2011) *Juvenile Salmon, Estuarine, and Freshwater Fish Utilization of Habitat Associated with the Fisher Slough Restoration Project in 2010*. Report prepared for The Nature Conservancy, Seattle, Washington.

Bean, M.J. & Rowland, M.J. (1997) *The Evolution of National Wildlife Law*. Praeger, Westport, Connecticut.

Beechie, T., Beamer, E. & Wasserman, L. (1994) Estimating coho salmon rearing habitat and smolt production losses in a large river basin, and implications for restoration. *North American Journal of Fisheries Management* **14**, 797–811.

Beechie, T.J., Pess, G.R., Beamer, E.M., Lucchetti, G. & Bilby, R.E. (2002) Role of watershed assessments in recovery planning for salmon. In: Montgomery, D.R., Bolton, S. & Booth, D.B. (eds) *Restoration of Puget Sound Rivers*. University of Washington Press, Seattle, Washington, pp. 194–225.

Beechie, T.J., Liermann, M., Pollock, M.M., Baker, S. & Davies, J. (2006) Channel pattern and river-floodplain dynamics in forested mountain river systems. *Geomorphology* **78**(1–2), 124–141.

Beechie, T.J., Pess, G.R., Pollock, M.M., Ruckelshaus, M.R. & Roni, P. (2008) Restoring rivers in the 21st century: science challenges in a management context. In: Beamish R.J. & Rothschild, B.J. (eds) *The Future of Fisheries Science in North America*, Springer, Heidelberg, pp. 695–716.

Beechie, T.J., Sear, D.A., Olden, J.D., Pess, G.R., Buffington, J.M., Moir, H., Roni, P. & Pollock, M.M. (2010) Process-based principles for restoring river ecosystems. *BioScience* **60**, 209–222.

Bernhardt, E.S., Palmer, M.A., Allan, J.D. *et al.* (2005) Restoration of U.S. Rivers – a National Synthesis. *Science* **308**, 636–637.

Bernhardt, E.S., Sudduth, E.B., Palmer, M.A. *et al.* (2007) Restoring rivers one reach at a time: Results from a survey of U.S. River Restoration Practitioners. *Restoration Ecology* **15**, 482–493.

Biedenharn, D.S., Copeland, R.R., Thorne, C.R., Soar, P.J., Hey, R.D. & Watson, C.C. (2000) *Effective Discharge Calculation: A Practical Guide*. U.S. Army Corps of Engineers Technical Report No. ERDC/CHL TR-00-15. Washington, DC.

Brierley, G.J. & Fryirs, K.A. (eds) (2008) *Spatial Considerations in Aquatic Ecosystem Management, in Geomorphology and River Management: Applications of the River Styles Framework*, Blackwell Publishing, Malden, Massachusetts.

Booth, D.B. & Jackson, C.R. (1997) Urbanization of aquatic systems – degradation thresholds, stormwater detention, and the limits of mitigation. *Water Resources Bulletin* **33**, 1077–1090.

Booth, D.B. & Henshaw, P.C. (2001) Rates of channel erosion in small urban streams. In: Wigmosta, M.S. & Burges, S.J. (eds) *Land Use and Watersheds: Human influence on Hydrology and Geomorphology in Urban and Forest Areas*. Water Science and Application **3**, 17–38. American Geophysical Union, Washington, DC.

Bragg, O.M., Black, A.R., Duck, R.W., & Rowan, J.S. (2005) Approaching the physical biological interface in rivers: a review of methods for ecological evaluation of flow regimes. *Progress in Physical Geography* **29**, 506–531.

Brookes, A. (1990) Restoration and enhancement of engineered river channels: Some European experiences. *Regulated Rivers: Research & Management* **5**, 45–56. doi: 10.1002/rrr.3450050105.

Brooks, A.P. (2006) *Design Guideline for the Reintroduction of Wood into Australian Streams*. Land & Water Australia, Canberra, Australia.

Brooks, A.P. & Brierley, G.J. (2002) Mediated equilibrium: the influence of riparian vegetation and wood on the long-term evolution and behaviour of a near-pristine river. *Earth Surface Processes and Landforms* **27**, 343–367.

Brooks, A.P. & Brierley, G.J. (2004) Framing realistic river rehabilitation targets in light of altered sediment supply and transport relationships: lessons from East Gippsland, Australia. *Geomorphology* **58**, 107–123.

Brown, R.A. & Pasternack, G.B. (2009) Comparison of methods for analyzing salmon habitat rehabilitation designs for regulated rivers. *River Research and Applications* **25**, 745–772.

Castro, J.M. & Jackson, P.L. (2001) Bankfull discharge recurrence intervals and regional hydraulic geometery relationships: patterns in the Pacific Northwest, USA. *Journal of the American Water Resources Association* **37**, 1249–1262.

Church, M.A., McLean, D.G. & Wolcott, J.F. (1987) River bed gravels: sampling and analysis. In: Thorne, C.R., Bathurst, J.C. & Hey R.D. (eds) *Sediment Transport in Gravel-bed Rivers*, John Wiley & Sons, Chichester, UK, p. 43–88.

Comiti, F., Mao, L., Lenzi, M.A. & Siligardi, M. (2009) Artificial steps to stabilize mountain rivers: a post-project ecological assessment. *River Research and Applications* **25**, 639–659.

Copeland, R.R., McComas, D.N., Thorne, C.R., Soar, P.J., Jones, M.M. & Fripp J.B. (2001) *Hydraulic Design of Stream Restoration Projects*. U.S. Army Corps of Engineers Technical Report ERDC/CHL TR-01-28. Washington, DC.

Dietrich,W.E., Bellugi, D., Sklar, L.S., Stock, J.D., Heimsath, A.M. & Roering, J.J. (2003) Geomorphic transport laws for predicting landscape form and dynamics. In:

Iverson, R.M. & Wilcock, P. (eds) *Prediction in Geomorphology*. American Geophysical Union, Washington, DC, pp. 103–132.

Doran, G.T. (1981) There's a S.M.A.R.T. way to write management's goals and objectives. *Management Review* **70**(11, AMA FORUM), 35–36.

Downs, P.W. & Kondolf, G.M. (2002) Post-project appraisal in adaptive management of river channel restoration. *Environmental Management* **29**, 477–496.

Doyle, M.W., Stanley, E.H., Strayer, D.L., Jacobson, R.B. & Schmidt, J.C. (2005) Effective discharge analysis of ecological processes in streams. *Water Resources Research* **41**, W11411.

Doyle, M.W., Shields, F.D., Boyd, K.F., Skidmore, P.B. & Dominick, D.D. (2007) Channel-forming discharge selection in river restoration design. *Journal of Hydraulic Engineering* **133**(7), 831–837.

Dunne, T. & Leopold, L.B. (1978) *Water in Environmental Planning*. Freeman, San Francisco, California.

Emmett, W.W. & Wolman, M.G. (2001) Effective discharge and gravel-bed rivers. *Earth Surface Processes and Landforms* **26**, 1369–1380.

FISRWG (Federal Interagency Stream Restoration Working Group) (1998) *Stream Corridor Restoration: Principles, Processes, and Practices*. National Technical Information Service, US Department of Commerce, Springfield, Virginia.

Fripp, J., Copeland, R. & Jonas, M. (2001) An Overview of the USACE Stream Restoration Guidelines. In U.S. Army Corps of Engineers, *Proceedings of the Seventh Federal Interagency Sedimentation Conference, Reno, Nevada*.

Gaeuman, D. (2008) *Recommended Quantities and Gradation for Long-term Coarse Sediment Augmentation Downstream from Lewiston Dam*. Report TM-TRRP-2008-2, Trinity River Restoration Program, Weaverville, California.

Gaeuman, D. & Krause, A. (2011) *2010 Bed Material Sediment Budget Update, Trinity River, Lewiston Dam to Douglas City, California*. Report TM-TRRP-2011-2, Trinity River Restoration Program, Weaverville, California.

Gomez, B. & Church, M. (1989) An assessment of bed load sediment transport formulae for gravel bed rivers. *Water Resources Research* **25**, 1161–1186.

Gomez, B., Eden, D.N., Hicks, D.M., Trustrum, N.A., Peacock, D.H. & Wilmshurst, J.M. (1999) Contribution of floodplain sequestration to the sediment budget of the Waipaoa River, New Zealand. In: Mariott, S.B. & Alexander, J. (eds) *Floodplains: Interdisciplinary*

Approaches. Geological Society, London, Special Publication 163, pp. 69–88.

Hall, J.W. (2003) Handling uncertainty in the hydroinformatic process. *Hydroinformatic* **5**, 215–232.

Hall, J., Pollock, M. & Hoh, S. (2011) Methods for successful establishment of cottonwood and willow along an incised stream in semiarid eastern Oregon, USA. *Ecological Restoration* **29**, 261–269.

Harrelson, C.C., Rawlins, C.L. & Potyondy, J.P. (1994) *Stream Channel Reference Sites: An Illustrated Guide to Field Technique.* Gen. Tech. Rep. RM-245, U.S. Department of Agriculture, Washington, D.C.

Harris, R.R. (1999) Defining reference conditions for restoration of riparian plant communities: examples from California, USA. *Environmental Management* **24**(1), 55–63.

Harvey, B.C. & Railsback, S.F. (2009) Exploring the persistence of stream-dwelling trout populations under alternative real-world turbidity regimes with an individual-based-model. *Transactions of the American Fisheries Society* **138**, 348–360.

Hey, R.D. (2006) Fluvial geomorphological methodology for natural stable channel design. *Journal of the American Water Resources Association* **42**, 357–374.

Kern, K. (1992) Rehabilitation of streams in south-west Germany. In: Boon, P.J., Calow, P. & Petts, G.E. (eds) *River Conservation and Management.* John Wiley and Sons, New York, pp. 93–124.

Koehn, J.D., Brierley, G.J. Cant, B.L. & Lucas, A.M. (2001) River Restoration Framework. *Land and Water Australia Occasional Paper 01/01.* Canberra, Australia.

Kondolf, G.M. (1995) Geomorphological stream channel classification in aquatic habitat restoration: uses and limitations. *Aquatic Conservation: Marine and Freshwater Ecosystems* **5**, 127–141.

Kondolf, G.M. (1997) Hungry water: Effects of dams and gravel mining on river channels. *Environmental Management* **21**, 533–551.

Kondolf, G.M. (1998) Lessons learned from river restoration projects in California. *Aquatic Conservation* **8**, 39–52.

Leopold, L.B. (1973) River Channel Change with Time – An Example. *Geological Society of America Bulletin* **84**, 1845–1860.

Leopold, L.B. & Maddock, T. Jr. (1953) *The Hydraulic Geometry of Stream Channels and Some Physiographic Implications.* US Geological Survey Professional Paper 252. US Government Printing Office, Washington, DC.

Leopold, L.B., Wolman, M.G. & Miller, J.P. (1964) *Fluvial Processes in Geomorphology.* Freeman, San Francisco, California, 522 pp.

Limbrick, K.J., Whitehead, P.G., Butterfield, D. & Reynard, N. (2000) Assessing the potential impacts of various climate change scenarios on the hydrological regime of the River Kennet at Theale, Berkshire, south-central England, UK. *The Science of the Total Environment* **251/252**, 539–555.

Lisle, T.E. (1982) Effects of aggradation and degradation on riffle-pool morphology in natural gravel channels, northwestern California. *Water Resources Research* **18**, 1643–1651.

Marcus, W.A., Meyer, G.A. & Nimmo, D.R. (2001) Geomorphic control of persistent mine impacts in a Yellowstone Park stream and implications for the recovery of fluvial systems. *Geology* **29**(4), 355–358.

Massong, T.M. & Montgomery, D.R. (2000) Influence of sediment supply, lithology, and wood debris on the distribution of bedrock and alluvial channels. *Geological Society of America Bulletin* **112**, 591–599.

Miller, D.E. & Skidmore, P.B. (2003) Establishing a Standard of Practice for Natural Channel Design Using Design Criteria. In: Montgomery, D.R. Bolton, S.M. Booth, D.B. and Wall, L. (eds) *Restoration of Puget Sound Rivers.* University of Washington Press, Seattle, Washington.

Miller, J.R. & Hobbs, R.J. (2007) Habitat Restoration – Do We Know What We're Doing? *Restoration Ecology* **15**(3)**,** 382–390.

Miller, J.R. & Miller S.M.O. (2007) *Contaminated Rivers: A Geomorphological-Geochemical Approach to Site Assessment and Remediation.* Springer, Dordrecht, Netherlands.

Montana Department of Environmental Quality. (2009) *Lower Blackfoot Total Maximum Daily Loads and Water Quality Improvement Plan.* Report C03-TMDL-03. Montana DEQ, Helena, Montana.

Montgomery, D.R. & Buffington, J.M. (1997) Channel-reach morphology in mountain drainage basins. *Geological Society of America Bulletin* **109**, 596–611.

Montgomery, D.R. & Buffington, J.M. (1998) Channel processes, classification, and response. In: Naiman, R.J. & Bilby, R.E. (eds) *River Ecology and Management: Lessons from the Pacific Coastal Ecoregion.* Springer, New York, NY.

Montgomery, D.R. & MacDonald, L.H. (2002) Diagnostic Approach to Stream Channel Assessment and Monitoring. *Journal of the American Water Resources Association* **38**, 1–16.

Moss, T. (2004) The governance of land use in river basins: prospects for overcoming problems of institutional interplay with the EU Water Framework Directive. *Land Use Policy* **21**, 85–94.

Mote, P.W. (2003) Trends in snow water equivalent in the Pacific Northwest and their climatic causes. *Geophysical Research Letters* **30**, doi: 10.1029/2003GL0172588.

Newbury, R.W. & Gaboury, M.N. (1993) *Stream Analysis and Fish Habitat Design: A Field Manual.* Newbury Hydraulics, Gibsons, British Columbia.

Palmer, M.A., Bernhardt, E.S. Allan, J.D. *et al.* (2005) Standards for ecologically successful river restoration. *Journal of Applied Ecology* **42**, 208–217.

Parnell, I.J. (2000) *Use of Design Analysis to Design a Habitat Restoration Experiment.* Master's thesis. Simon Fraser University, Canada. Report No. 276.

Pickett, S.T.A., Kolasa, J., Armesto, J.J. & Collins, S.L. (1989) The ecological concept of disturbance and its expression at various hierarchical levels. *Oikos* **54**, 129–136.

Poff, N.L. & Ward, J.V. (1989) Implications of streamflow variability and predictions of lotic community structure: a regional analysis of streamflow patterns. *Canadian Journal of Fisheries and Aquatic Sciences* **46**, 1805–1818.

Poff, N.L., Allan, J.D., Bain, M.B. *et al.* (1997) The natural flow regime: a new paradigm for riverine conservation and restoration. *BioScience* **47**, 769–784.

Pollock, M.M., Beechie, T.J. & Jordon, C.E. (2007) Geomorphic changes upstream of beaver dams in Bridge Creek, an incised stream channel in the interior Columbia River basin, eastern Oregon. *Earth Surface Processes and Landforms* **32**, 1174–1185.

Postel, S. & Richter, B. (2003) *Rivers for Life: Managing Water for People and Nature.* Island Press, Washington, DC.

Pretty, J.L., Harrison, S.S., Shepherd, D.J., Smith, C., Hildrew, A.G. & Hey, R.D. (2003) River rehabitiltation and fish populations: assessing the benefit of instream structures. *Journal of Applied Ecology* **40**, 251–265.

Richter, B.D., Mathews, R., Harrison, D.L. & Wigington, R. (2003) Ecologically sustainable water management: managing river flows for ecological integrity. *Ecological Applications* **13**, 206–224.

Riley, A.L. (1998) *Restoring Streams in Cities.* Island Press, Washington, DC.

Sandin, L. & Solimini, A.G. (2009) Freshwater ecosystem structure–function relationships: from theory to application. *Freshwater Biology* **54**, 2017–2024.

Sayers P.B., Hall, J.W. & Meadowcroft, I.C. (2002) Towards risk-based flood hazard management in the UK. *Proceedings of the Institution of Civil Engineers: Civil Engineering* **150**(Special Issue 1), 36–42.

Scheuerell, M.D., Zabel, R.W. & Sandford, B.P. (2009) Relating juvenile migration timing and survival to adulthood in two species of threatened Pacific salmon (*Oncorhynchus* spp.). *Journal of Applied Ecology* **46**, 983–990.

Schlosser, I.J. (1985) Flow regime, juvenile abundance, and the assemblage structure of stream fishes. *Ecology* **66**, 1484–1490.

Shields, F.D., Jr. (1996) Hydraulic and hydrologic stability. In: Brookes, A. & Shields, F.D. Jr. (eds) *River Channel Restoration: Guiding Principles for Sustainable Projects.* John Wiley & Sons Ltd, Chichester, U.K.

Shields, F.D., Jr., Copeland, R.R., Klingeman, P.C., Doyle, M.W. & Simon, A. (2003) Design for stream restoration. *Journal of Hydraulic Engineering* **129**(8), 575–584.

Sidle, R.C., Ziegler, A.D., Negishi, J.N., Nik, A.R., Siew, R. & Turkelboom, F. (2006) Erosion processes in steep terrain – truths, myths, and uncertainties related to forest management in Southeast Asia. *Forest Ecology Management* **224**, 199–225.

Simon, A. & Downs, P.W. (1995) An interdisciplinary approach to evaluation of potential instability in alluvial channels. *Geomorphology* **12**, 215–232.

Simon, A. & Castro, J. (2003) Measurement and analysis of alluvial channel form. In: Kondolf G.M. & Piegay, H. (eds) *Tools in Fluvial Geomorphology.* John Wiley and Sons, Chichester, UK, pp. 291–322.

Simon, A., Doyle, M., Kondolf, M., Shields, F.D. Jr., Rhoads, B. & McPhillips, M. (2007) Critical evaluation of how the Rosgen classification and associated "Natural Channel Design" methods fail to integrate and quantify fluvial processes and channel response. *Journal of the American Water Resources Association* **43**(5), 1–15.

Skidmore, P.B, Shields, F.D., Doyle, M.W. & Miller, D.E. (2001) Categorization of approaches to natural channel design. In: *Proceedings of the ASCE Wetlands Engineering and River Restoration Conference*, Reno, Nevada.

Skidmore, P.B., Thorne , C.R., Cluer, B., Pess, G.R., Castro, J., Beechie, T.J. & Shea, C.C. (2011) *Science Base and Tools for Evaluating Stream Engineering, Management and Restoration Proposals.* US Dept. Commerce, NOAA Tech. Memo. NMFS-NWFSC-112.

Skinner, K., Shields, F.D. Jr. & Harrison, S. (2008) Measures of success: Uncertainty and defining the outcomes of river restoration schemes. In: Darby, S. & Sear, D. (eds) *River Restoration: Managing Uncertainty in Restoring Physical Habitat.* John Wiley & Sons, London.

Soar, P.J. & Thorne, C.R. (2001) *Channel Restoration Design for Meandering Rivers.* ERDC/CHL CR-01-1. U.S. Army Engineer Research and Development Center, Vicksburg, Mississippi.

Statzner, B., Bady, P., Doledec, S. & Scholl, F. (2005) Invertebrate traits for the biomonitoring of large European rivers: an initial assessment of trait patterns in least impacted river reaches. *Freshwater Biology* **50**, 2136–2161.

Steel, E.A., Fullerton, A., Caras, Y., Sheer, M.B., Olson, P., Jensen, D., Burke, J., Maher, M. & McElhany, P. (2008) A spatially explicit decision support system for watershed-scale management of salmon. *Ecology and Society* **13**(2), 50. Available at: http://www.ecologyand society.org/vol13/iss2/art50/.

Steel, E.A., Beechie, T.J., Rukelshaus, M.H., Fullerton, A.H., McElhany, P. & Roni, P. (2009) Mind the Gap: Uncertainty and model communication between managers and scientists. In: Knudsen, E.E. & Michael, J.H. Jr. (eds) *Pacific Salmon Environmental and Life History Models: Advancing Science for Sustainable Salmon in the Future.* American Fisheries Society, Symposium 71, Bethesda, Maryland, pp. 357–374.

Stewart, G.B., Bayliss, H.R., Showler, D.A., Sutherland, W.J. & Pullin, A.S. (2009) Effectiveness of engineered in-stream structure mitigation measures to increase salmonid abundance: a systematic review. *Ecological Applications* **19**(4), 931–941.

SWCA et al. (2008) *Legacy Nature Preserve Annual Monitoring and Site Management Report.* Final report submitted by SWCA Environmental Consultants, Salt Lake City, to the Utah Department of Transportation and HDR Inc., Salt Lake City, Utah.

Talmage, P.J., Perry, J.A. & Goldstein, R.M. (2002) Relation of instream habitat and physical conditions to fish communities of agricultural streams in the northern Midwest. *North American Journal of Fisheries Management* **22**, 825–833.

Thorne, C.R., Allen, R.G. & Simon, A. (1996) Geomorphological river channel reconnaissance for river analysis, engineering, and management. *Transactions of the Institute of British Geographer* **NS 21**, 469–483.

The Nature Conservancy (TNC) (2007) *Fisher Slough – Preferred Restoration Plan.* Skagit County, Washington. Final Report.

Tompkins, M.R. & Kondolf, G.M. (2007) Systematic post-project appraisals to maximize lessons learned from river restoration projects: case study of compound channel restoration projects in Northern California. *Restoration Ecology* **15**, 524–537.

Waddell, T.J. (ed.) (2001) *PHABSIM for Windows User's Manual and Exercises.* US Geological Survey Open-File Report 2001-340. US Geological Survey, Fort Collins, Colorado.

Walter, R.C. & Merritts, D.J. (2008) Natural streams and the legacy of water-powered mills. *Science* **319**, 299–304.

Washington Department of Fish and Wildlife (2009) *Fish Passage and Surface Water Diversion Screening Assessment and Prioritization Manual.* Washington Department of Fish and Wildlife, Olympia, Washington.

Watson, C.C., Biedenharn, D.S. & Scott, S.H. (1999) *Channel Rehabilitation: Processes, Design and Implementation.* Guidelines prepared for workshop presented by US Environmental Protection Agency.

Wheaton, J.M., Pasternack, G.B. & Merz, J.E. (2004) Spawning habitat rehabilitation – 1. Conceptual approach and methods. *International Journal of River Basin Management* **2**(1), 3–20.

Wilcock, P., Pitlick, J. & Cui, Y. (2009) Sediment transport primer: estimating bed-material transport in gravel-bed rivers. Gen. Tech. Rep. RMRS-GTR-226. Fort Collins, CO. US Department of Agriculture, Forest Service, Rocky Mountain Research Station. 78 p.

Wohl, E. (2004) *Disconnected Rivers – Linking Rivers to Landscapes.* Yale University Press, New Haven, Connecticut.

Wohl, E. (2005) Compromised rivers: Understanding historical human impacts on rivers in the context of restoration. *Ecology and Society* **10**(2), 2.

Wolman, M.G. & Miller, J.P. (1960) Magnitude and frequency of forces in geomorphic processes. *Journal of Geology* **68**, 54–74.

8 Monitoring and Evaluation of Restoration Actions

Philip Roni[1], Martin Liermann[1], Susanne Muhar[2] & Stefan Schmutz[2]

[1]Northwest Fisheries Science Center, National Oceanic and Atmospheric Administration, USA
[2]University of Natural Resources and Life Sciences, Vienna, Austria

8.1 Introduction

Monitoring and evaluation (M&E) provides critical information on restoration project effectiveness, including how physical habitat and biota respond to different restoration techniques. The need for M&E of restoration actions parallels the history of restoration itself; as restoration programs have increased in size, number, and complexity, so has the need for rigorous M&E of these efforts. However, as noted in previous chapters, limited information exists on the effectiveness of most techniques. Periodic reviews continue to show that only a small fraction of restoration projects in any country, province or watershed are adequately monitored (Tarzwell 1934; Reeves & Roelofs 1982; Roni *et al.* 2002; Ryder *et al.* 2008). This lack of rigorous M&E stems not only from inadequate funding, but also from a variety of technical and non-technical issues. These issues include a lack of clearly defined questions, improper study design, inadequate spatial and temporal replication, insensitive monitoring parameters, poor project implementation or management, and lack of periodic analysis and publication of results (Reid 2001; Roni 2005; Roni *et al.* 2008).

Monitoring is critical for evaluating whether the massive investments in restoration are meeting their objectives and providing the predicted ecological and social benefits. Restoration ecology is one of the few disciplines where numerous field experiments (i.e. restoration actions) are still undertaken without well-designed studies to evaluate them. Given the increasing number, size, cost, and complexity of restoration actions, robust M&E is needed to understand the individual and synergistic effects of projects and to guide future restoration efforts.

Funding and legal frameworks often require monitoring of at least a portion of projects. In the European Union, for example, both the Water Framework Directive (WFD) and the Habitats Directive require ecological monitoring and reporting on the status of all water bodies and evaluation of restoration measures. In contrast, some North American restoration programs limit funding of M&E in an attempt to focus funds on restoration. Unfortunately, without adequate funding, it is unlikely that robust monitoring will occur. Restoration practitioners, ecologists, managers, and others must therefore continue to demonstrate the importance of M&E to ensure that it is adequately funded as monitoring is critical to success and improvement of restoration.

Many of the obstacles discussed above can be overcome by following a series of important steps during development and implementation of an M&E program. In this chapter, we cover the key steps for designing effective monitoring and evaluation programs. As outlined in Chapter 1 (see Figure 1.1), M&E should ideally be part of the design of a restoration project and occur as early as possible in the restoration planning process and well before actions are implemented on the ground.

Stream and Watershed Restoration: A Guide to Restoring Riverine Processes and Habitats, First Edition. Philip Roni and Tim Beechie.
© 2013 John Wiley & Sons, Ltd. Published 2013 by John Wiley & Sons, Ltd.

8.2 What is monitoring and evaluation?

There are many definitions of monitoring, and it is important to distinguish among these to describe their place in the restoration process. *Monitoring* is technically defined as systematically checking or scrutinizing something for the purpose of collecting specified categories of data. In ecology, it generally refers to sampling conducted to detect changes in physical, chemical, or biological parameters (Roni 2005). Common types of monitoring

used in habitat management or restoration are baseline, status, trend, implementation, and effectiveness and validation (MacDonald *et al.* 1991; Roni 2005; Table 8.1). Baseline, status, and trend monitoring are important in the assessment, action identification, and prioritization process. They may also offer useful information for restoration design and M&E program development. For example, periodic estimates of diversity in fish or benthic macroinvertebrate are a type of status monitoring which may be available at a site prior to restoration. Such data may provide an insight into the types of habitat or conditions

Table 8.1 Definitions of monitoring types and examples of what might be monitored for a riparian fencing project to exclude livestock. Effectiveness and validation monitoring are typically the types of monitoring used to evaluate habitat restoration actions. Adapted from MacDonald *et al.* 1991 and Roni 2005.

Monitoring types	Description	Examples	Uses in restoration
Baseline	Characterizes the existing biota, chemical, or physical conditions for planning or future comparisons	Data collected on plant species and condition on site prior to implementation of management or restoration action	Help identify habitat conditions, restoration opportunities and actions, assist with prioritization of actions, and may provide data useful for design of restoration monitoring
Status	Characterizes the condition (spatial variability) of physical or biological attributes across a given area	Annual counts of fish abundance or habitat condition	Help identify habitat conditions, restoration opportunities and actions, assist with prioritization of actions
Trend	Determines changes in biota or conditions over time	Examination of the trends in annual fish abundance or habitat conditions over time	Help identify habitat conditions, restoration opportunities and actions, and provide pre-project data or assist with design of restoration monitoring
Implementation (administrative, compliance)	Determines whether project was implemented as planned	Number of miles fenced or trees planted is consistent with original proposal	Used for most any management actions, useful for understanding results of effectiveness and validation monitoring
Effectiveness	Determines whether actions had desired effects on watershed, physical processes, or habitat	Did stream temperature decline following fencing and replanting of riparian area?	Used to evaluate restoration and habitat management actions
Validation (research, sometimes considered part of effectiveness)	Evaluates whether the hypothesized cause and effect relationship between restoration action and response (physical or biological) were correct	Did reduction in stream temperatures brought on by fencing and replanting of riparian area lead to increased fish abundance?	Used to evaluate restoration and habitat management actions

that need to be restored as well as baseline data to help assess the response of biota restoration actions. Implementation (or compliance) monitoring is simply checking to determine whether a project was implemented according to design. This type of monitoring may provide insight into why an anticipated change in physical habitat and biota due to restoration has not occurred in the project area.

In this chapter, we will focus on effectiveness and validation monitoring. These two types of monitoring are intricately linked: effectiveness monitoring generally refers to assessing the primary response (i.e. did the restoration action lead to the expected change in physical habitat?), whereas validation monitoring examines the secondary or tertiary response (e.g. did the change in habitat from the restoration action lead to the expected change in biota or other conditions?). Here we used *monitoring* M&E to refer to both effectiveness and validation monitoring.

Ideally, a rigorous M&E program would be developed for all restoration projects. Unfortunately, this is not feasible or cost-effective in many cases. All projects should however include some type of simple before-and-after assessment. Oddly enough, simple things such as before-and-after drawings, photos, or site maps are often absent even from very well-designed and planned restoration projects. For example, before-and-after photos of fencing to protect stream banks clearly demonstrate changes in channel width and riparian cover following project implementation (see Chapter 5, Figure 5.5). Pre-project documentation and data can be critical in emphasizing that M&E should be included in initial restoration proposal and project design process (Chapter 7). While we do not consider these types of documentation in our definition of M&E, they are important factors to demonstrate project success that can be understood by a broad audience, including project sponsors and the general public.

8.3 Steps for developing an M&E program

Developing a rigorous and successful restoration M&E program includes several logical steps. These include: establishing project goals and objectives; defining clear

hypotheses to be tested; determining the monitoring scale; selecting an appropriate monitoring design[1]; determining the parameters to be measured; determining the number of sites and duration of monitoring; selecting a sampling methodology (scheme); implementing the M&E program; and, finally, analysis and communicating results (Roni 2005; Figure 8.1). Failure to adequately address one or more of these steps can reduce the rigorousness or likelihood of success of a monitoring program. Once these steps have been completed, the main aspects of the program should be summarized in a concise list or table (Table 8.2). We discuss each of these steps sequentially in Sections 8.3.1–8.3.6, although many may occur simultaneously. Appropriate analysis of the data collected and pitfalls to implementation are equally critical, and these are addressed in Section 8.4.

8.3.1 Defining restoration goals and monitoring objectives

The restoration goal and specific M&E objectives need to be clearly defined before discussing design or any other aspects of the program. In Chapters 3 and 7 we discussed the need for clear watershed restoration goals and project objectives. Restoration goals should be used to help define specific monitoring objectives and questions and guide the development of an M&E program. For example, the goal of a riparian fencing project may be to reduce fine sediment and improve aquatic habitat. While the monitoring *objectives* might be to detect and measure changes in bank stability, shade, and fine sediment and determine the impact these changes have on instream habitat conditions and fish abundance, the objectives may differ depending on the goals of the organization involved in funding, planning, or implementing the restoration and monitoring. A national, state, or provincial government or agency that distributes money to local restoration groups may be interested in the average effect on local habitat conditions of the hundreds of riparian, instream, or other projects across a geographic region. In contrast, a local watershed council or river trust may be interested in the watershed-scale effect of only those actions that occur within their watershed or only on a particular species. These seemingly subtle differences in objectives can lead to very different M&E programs, and it is therefore critical that restoration goals and monitor-

---

[1] We use the term 'monitoring design' to refer to the overall experimental design and 'sampling scheme' to refer to the methodology for selecting samples and collecting data within a study area.

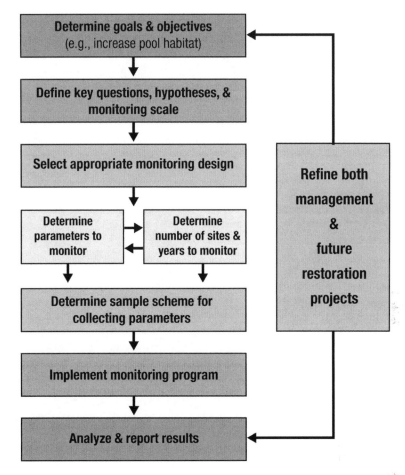

Figure 8.1 Steps for designing a monitoring program. From Roni *et al.* (2005).

ing objectives are clearly defined before moving on to other steps in the M&E design process. The SMART method (specific, measurable, attainable, realistic, and time-bound) recommended in Chapter 7 can be used to assist with defining these objectives. If M&E is for an individual project, specific and measurable objectives should have been at least partially determined in the design process.

8.3.2 Defining questions, hypotheses, and spatial scale

A common problem with both large and small monitoring programs is the lack of a clear set of questions that guide the data collection (Reid 2001). Restoration projects are in essence field experiments, and experiments by definition require a clear question or hypothesis. A critical

step in the M&E design process is therefore transforming the monitoring objectives into clear, testable questions or hypotheses (MacDonald *et al.* 1991; Roni 2005). These will obviously differ based on the monitoring objectives, type or types of restoration, and number and distribution of projects to be examined.

Despite the countless number of hypotheses that can be defined for evaluating different restoration projects, these can be categorized into a handful of key questions: (1) the spatial scale at which restoration occurs and what will be measured (region, watershed, reach); and (2) the number of projects to be monitored (Table 8.3). For example, if interested in the effect of a single riparian planting project at a reach scale, the key question would be: 'what is the effect of this riparian project on local physical and biological conditions?' In contrast, if

Table 8.2 Example of response to each monitoring step for published and ongoing evaluations for stream restoration from the western USA (modified from Roni *et al.* 2010a). EPT: extensive post-treatment; BACI = before-after control-impact; LWD: large woody debris. IMW Program is the Washington State Intensively Monitored Watershed Program, a program designed to evaluate the effects of various reach-level restoration actions (e.g. LWD placement, riparian replanting, sediment reduction) at a watershed scale.

Question or step	Monitoring program or case study		
	LWD placement	Constructed groundwater channels	IMW Program
Restoration goal	Increase pool area, instream habitat complexity, and juvenile salmon abundance	Increase spawning and rearing habitat for salmon	Restore habitat throughout watersheds
Monitoring question(s)	Is LWD placement effective at improving reach-level habitat conditions and juvenile fish abundance?	Does habitat quality and salmonid abundance differ between constructed and natural groundwater channels?	What are the effects of a suite of different restoration actions on habitat conditions and juvenile coho salmon, and steelhead survival and abundance at a catchment scale?
Scale	Reach	Reach	Watershed
Monitoring design	EPT	EPT	BACI
Parameters	LWD, physical habitat (pools and riffles, cover), summer and winter juvenile fish abundance, macroinvertebrate abundance and composition	LWD, physical habitat (pools and riffles, cover), summer and winter juvenile fish abundance, macroinvertebrate abundance and composition, temperature, water chemistry	Habitat, substrate, LWD, parr, smolt, and adult abundance and survival
Replication (no. sites and years)	30 sites; sampled once in summer and winter	11 sites sampled once in summer and winter	10 watersheds; 10+ years (2 before and several after)
Sampling scheme	Census	Systematic random sample	Census, stratified random sample
Analysis	*t*-test, regression	*t*-test, regression, ANOVA	Paired *t*-test, regression, ANOVA
Reporting	Technical report, manuscripts (Roni & Quinn 2001a)	Manuscripts (Morley *et al.* 2005)	Annual contract reports, manuscripts (Bilby *et al.* 2005)

Table 8.3 Key questions for monitoring aquatic restoration classified by scale and number of projects of interest (from Roni *et al*, 2005). These can be used as a starting point for defining more specific hypotheses for a monitoring program.

Number of projects	Spatial scale	
	Reach	Watershed or population
Single project	What is the effect of a single project on local habitat conditions or biota?	What is the effect of an individual project on watershed conditions or biota populations?
Multiple projects	What is the effect of a particular project type on local habitat conditions or biota abundance?	What are the effects of a suite of different projects on watershed conditions or biota populations?

interested in the effects of a suite of riparian, instream, and sediment reduction projects on conditions at a watershed scale, the question would be: 'what is the cumulative effect of all restoration actions on physical habitat or fish populations and other biota?' These are two fundamentally different questions in terms of scale and replication which will influence monitoring and sampling design.

8.3.2.1 Defining the spatial scale

The terminology used to describe different spatial scales varies greatly by discipline, location, region, and among countries. The terms *basin, watershed, catchment, reach, section, network, location*, and *site* all appear frequently in monitoring and restoration literature. Being consistent about these terms becomes particularly critical when we discuss monitoring. Our focus will be primarily on *watershed* and *reach* scales. A *watershed* is the entire land drainage area of a stream or river and can range in size (drainage area) from tens of square kilometers for a small stream to hundreds of thousands of square kilometers for large rivers such as the Danube, Columbia, or Mississippi. Watersheds can be further divided into sub-watersheds (or sub-basins or sub-catchments) though we avoid using that term to minimize confusion. We use the term *watershed* synonymously with *catchment, drainage* or *basin*. The term network sometimes refers to multiple reaches or an entire drainage network, and we also avoid using this term because it is synonymous with watershed. A *reach* refers to a geomorphologically similar section of stream, typically ranging from hundreds of meters to several kilometers depending upon the size of the stream. We use the term *site* or *location* to refer to a place where restoration or sampling is occurring; this could be a reach, a specific location in a reach, or an entire watershed.

Considerations of M&E scale include two major aspects: (1) the scale of inference, or the scale at which we want to draw conclusions about the restoration; and (2) the scale of effect, or the actual scale of influence of the restoration project or projects. The former helps determine the set of projects which may be drawn from. This is often determined by socio-political boundaries (nation, state, region, province, watershed, sub-watershed) or the overall goal of the restoration program, funding organization, or watershed entity. In contrast, the latter concerns the area of influence of the restoration project and the scale at which the restoration can be reasonably measured. For example, 30 instream habitat improvement projects ranging from 100 to 500 m in length may

have been implemented in many different watersheds within a region (Table 8.2). Given that habitat changes to these types of actions are not typically transmitted hundreds of meters upstream or downstream of the treatment, the reach scale is likely the most appropriate scale of measurement. It is also important to note that the restoration effects may be transmitted laterally. Defining the width of the sampling area (which may also include the riparian zone) is therefore also important (Jähnig et al. 2009).

Determining the area of influence of a project is more difficult for fishes and other mobile organisms that may move well outside the initial area impacted by the restoration. For example, restoration of longitudinal connectivity takes place at distinct sites where barriers are removed or equipped with fish passes that might provide access to upstream habitats at the watershed scale (e.g. see Chapter 6, Box 6.2 on Danube Case Study). Monitoring of fish responses to these types of restoration should focus on reaches upstream of the barrier rather than at the project site itself. Most ecological and restoration research on habitat and biota has focused on reaches or individual habitat units. This information is important, but uncertainty about movement, survival, and population dynamics of biota prevents these reach-scale studies from addressing watershed or population-level questions. Therefore, while not always feasible, monitoring of fish should ideally be undertaken at multiple scales or as broad a scale as possible in order to understand their response to even localized restoration actions.

Studies designed to assess watershed or population-level effects also provide valuable information and face multiple challenges (e.g. upstream–downstream trends or sampling logistics; Conquest 2000; Downes et al. 2002). A few well-designed monitoring programs have demonstrated that not only is it possible to monitor at a broader scale (sub-watershed or watershed), but that monitoring at broader scales can yield critical information on project effectiveness that may not be seen at a reach level (Solazzi et al. 2000; Zitek et al. 2008). For example, construction of fish passages on 11 migration barriers on the Pielach and Melk rivers in Austria demonstrated that the effectiveness of these measures depends upon the scale of measurement: the response of the fish community was only apparent at sub-watershed or watershed scale (Zitek et al. 2008). Similarly, watershed-scale evaluation of reach restoration in Colorado has demonstrated that the projects can have far-ranging effects on fish populations (Gowan & Fausch 1996). The key questions outlined in Table 8.3 are initially defined at reach and watershed scales, but can be modified to accommodate other scales (e.g. network, sub-watershed, region; see also Chapter 2 for scales). These questions can be used as a basic starting point for defining questions and hypotheses specific to an M&E program.

8.3.3 Selecting the monitoring design

Most experimental designs used in highly controlled laboratory studies (e.g. randomized block, split plot, repeated measures) are not suitable for monitoring restoration efforts. However, before-and-after or post-treatment designs are particularly useful for M&E of restoration projects. There are numerous variations of these two basic designs that vary in intensity (number of sites), extent (area or replication), and whether they include controls, references, or pre-treatment data (Hicks *et al.* 1991; Downes *et al.* 2002; Roni *et al.* 2005). Designs without spatial replication are in essence case studies where the inference (conclusion) is confined to the individual site. For designs with spatial replication (many sites), the scale of inference is about the larger population of sites rather than one individual site. The ideal design depends upon the restoration goals and monitoring objectives, although broad-scale designs with multiple sites provide some of the most robust results and useful management information (Roni *et al.* 2005). Before discussing how to select an appropriate design, we first define treatments, controls and references. We then discuss different types of designs, their strengths, weaknesses, and other factors to consider when selecting a monitoring design.

8.3.3.1 Treatments, controls, and references

All monitoring designs incorporate some combination of treatments, controls, and references. The *treatment* refers to the restoration action and, in environmental impact studies, is referred to as the *impact*. A *control* represents a location that is nearly identical to the treated location, with the exception that no treatment occurs. In contrast, a *reference* represents the desired or target condition following restoration. It is often defined in terms of the type-specific state with no, or only minor, anthropogenic alterations (see *Leitbild* or 'reference state' as described by Henry & Amoros 1995; Muhar *et al.* 1995; Jungwirth *et al.* 2002). This distinction is critical to both the design of the monitoring program and interpretation of its results.

Controls must be chosen carefully, since a poorly chosen control may not have a trajectory similar to the treatment before restoration. This dissimilarity can add noise which complicates the analysis and makes it more difficult to detect a restoration response (Murtaugh 2002; Johnson *et al.* 2005). Similarly, reference sites must accurately represent the expected condition after treatment (i.e. the restored condition). Control, reference, and treatment sites should be similar in drainage area, stream flow, geology, land use, gradient, vegetation, and potentially other factors. Selecting treatments, references, and controls for reach-scale restoration and M&E is often done by comparing reaches within the same stream or watershed. The control and reference are generally located upstream of the treatment reach to ensure that downstream translation of treatment impacts does not influence the result. If no suitable upstream sites exist and impacts or benefits of restoration are very localized, a downstream control may be appropriate.

Another factor to consider is not only whether physical changes might be transmitted among treatments and controls, but whether the monitored biota might move among study reaches and confound detection of a biological response. For example, free-swimming organisms such as fish may have large home ranges. Treatments and controls must therefore be located long distances apart to ensure they are independent samples or sites. The appropriate distance between treatments and controls depends upon the species of interest, the parameters being measured, and the size of the stream. In general, this may range from a few hundred meters in small streams (<25 m wide) to a kilometer or more on larger streams and rivers. Selecting multiple treatments and control sites within one stream must also be done with caution; selecting more than one pair of treatments and controls within the same geomorphically similar reach can result in pseudo-replication (Hurlbert 1984).

For watershed-scale restoration and M&E, the identification and selection of control or reference watersheds can be facilitated by a regional analysis that examines the degree to which the response variable covaries between sites. This of course requires existing datasets or a pilot study. For example, Bradford (1999) and Liermann & Roni (2008) showed that populations of juvenile coho salmon (*Oncorhynchus kisutch*) tended to be moderately correlated through time when separated by distances of less than 20–30 km. However, populations separated by distances of greater than 50 km were not correlated. These studies and other reviews of treatment and control watersheds (e.g. Downes *et al.* 2002) suggest that the most appropriate treatment, control, and reference watersheds or subwatersheds are typically adjacent to one another.

8.3.3.2 Before-after and before-after control-impact designs

The most common approach to evaluating restoration projects is the before-after (BA) design (Green 1979) which simply involves monitoring the treated site before and after restoration; most early evaluations of restoration used this design. To reduce the possibility of interpreting a natural trend as a treatment effect, and to reduce the effect of temporal variability, a control (or reference) site is commonly included. This is called the before-after control-impact (BACI) design, where both a control and treatment (impact) are monitored before and after restoration (Figure 8.2; Stewart-Oaten *et al.* 1986; Underwood 1991).

For both the BA and BACI designs, it is assumed that the trajectory of the difference between the control and treatment would be flat in the absence of the treatment. Murtaugh (2002) however found that, for 61 pairs of ecological time series, 20% of the pairs had measurements that differed significantly even though neither site had been treated; this is much higher than the 5% assumed by the test. Rather than statistical tests, however, he recommended relying on graphical interpretation, expert knowledge, and common sense to interpret results of before-after designs. An additional shortcoming of the BA family of designs is that they only provide inference for a single site, precluding generalization. If resources permit, both problems can be overcome by replicating in space (multiple treatments and controls) and relying on sites as opposed to years as the inferential unit of replication (i.e. test if the average effect size across sites is greater

than zero). This design and its variants are sometimes referred to as multiple BACI (MBACI) designs (Downes *et al.* 2002; see example in Table 8.2). If it is not feasible to replicate the treatment, then multiple controls can provide a larger context from which to evaluate the treatment trajectory (sometimes called an asymmetric or beyond-BACI design; Underwood 1991). Using multiple controls and nested temporal sampling has a number of advantages and can help detect and distinguish between different kinds of responses (pulse, press, and increased variability; Underwood 1991, 1994).

Another modification of the BACI design is the staircase design, which can be used when restoration actions in different watersheds or stream reaches will be spread over a long time frame (Walters *et al.* 1988). This design requires sequential or staggered treatment of several reaches or watersheds over time, with multiple reaches or watersheds serving as treatments or controls. Treatments and controls would ideally be assigned randomly. Because the staircase design presents a number of technical and logistical challenges, such as complex data analysis and staggering treatments and controls over many years, it is not commonly used to evaluate restoration actions. Other more subtle modifications of the BACI design have been proposed that differ only in the statistical models used to analyze the data and test the hypotheses (Downes *et al.* 2002).

8.3.3.3 Post-treatment designs

In many instances, data were not or cannot be collected before restoration occurs. This is probably more common

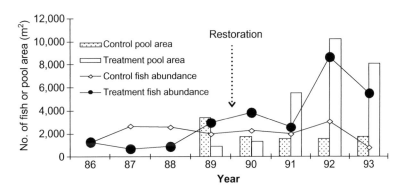

Figure 8.2 Example of before-after control-impact study design showing change in juvenile salmon coho numbers (bars) and pool area (m²) (lines) from restored and control watersheds in the Alsea basin of Oregon, before and after treatment (data from Solazzi *et al.* 2000). Measurements such as pool area, which has lower interannual variability than fish abundance, may only require one or two years of pre-project data. In the absence of a control watershed it would also be difficult to determine if changes in fish numbers are a result of restoration activities.

than most restoration practitioners would like to admit. However, planning and funding constraints often require projects to be implemented within a relatively short period of a few months with little or no funds for pre-project monitoring. In some instances a retrospective analysis, often called a post-treatment design, may be possible. This design relies on a comparison of treatment and suitable control reaches or watersheds, with the assumption that the control was similar to the treatment before restoration. The differences between the treatment and control parameters or metrics are examined after the restoration has taken place. A post-treatment design is typically done at a reach level with pairing of nearby treatment and control reaches. Applying at broader scales is possible, but logistically more difficult.

The lack of pre-project data makes it impossible to adjust for pre-treatment differences between the control and treatment. It is therefore essential that the pre-treatment values for the treatment and control are similar and that, for multiple treatment and control pairs, the differences (one for each pair) are statistically independent. In other words, if there are a large number of treatment and control pairs, the average of the differences should be nearly zero and the value of the difference for one treatment-control pair should not provide any information about the others. As with other designs, proper selection of the control sites is therefore essential.

There are two basic types of post-treatment designs: the intensive post-treatment and extensive post-treatment (IPT and EPT; Hicks *et al.* 1991). The IPT, which includes monitoring one or a few treatment and controls for multiple years post-treatment, is in essence a BACI design without the before data. The obvious shortcoming of this design is that no pre-project data are available and only one or two paired (or unpaired reaches) are monitored. It is therefore nearly impossible to tell if the difference between the treatment and the control is the result of restoration, underlying differences between treatments and controls, or other factors. For this reason, the IPT is rarely used. However, it can be an appropriate design for projects where pre-project data are not needed or where values are known to be zero prior to restoration (e.g. removal of dam blocking fish passage or creation of new floodplain habitats).

The EPT is the most commonly used post-treatment design. Rather than long-term monitoring of a few sites, multiple treatment and control pairs are examined to determine the average restoration response. It is sometimes referred to as space-for-time substitution because sampling is replicated across space (many sites) rather than time (many years). This approach has been widely used in forestry studies (riparian harvest; Grant *et al.* 1986; Hicks *et al.* 1991) and more recently in retrospective evaluation of restoration actions (e.g. Roni & Quinn 2001a; Lepori *et al.* 2005; Morley *et al.* 2005). The EPT has been shown to be a valid design for examining relationships between habitat change and salmonid fishes (Grant *et al.* 1986). Moreover, the extensive replication allows for correlation of the responses with key physical or other independent variables. For example, M&E programs using this design have correlated restoration responses to key physical differences such as habitat area created or amount of large woody debris (LWD), this providing guidance on which projects or project characteristics lead to the largest improvements in habitat and biota (e.g. Roni & Quinn 2001a; Roni 2003; Figure 8.3). In addition, because the inference is based on spatial replication (instead of one or a few projects over time), the results can be applied to a larger population of similar projects.

An EPT often requires a large number of existing projects to choose from. Many monitoring parameters or metrics require a sample size of 10 or more treatment and control pairs to detect statistically significant differences between pairs or correlations between dependent and independent variables (e.g. between biotic response and physical habitat change; Figure 8.3). Equally important is the assumption that suitable treatment and controls can be located. As with all restoration monitoring designs, controls need to be similar to the treatment prior to treatment which means locating reaches or watersheds that are similar in key physical and biological attributes (e.g. geology, land use, vegetation, flow, gradient, channel type).

Another challenge of the EPT design is the large numbers of sites or streams that need to be visited to locate those with suitable controls. For example, to identify 30 stream reaches that had appropriate treatment and controls, Roni & Quinn (2001a) examined more than 100 restoration projects (Table 8.2). Most were excluded because an adequate control did not exist. Similarly, inclusion of references that represent target or ideal conditions requires locating suitable references sites which, in many parts of the developed world, is even more difficult than finding suitable controls. For future projects, this can be partly overcome by making all restoration projects identify or set aside suitable controls or references during the design process to allow later post-treatment monitoring.

Most post-treatment designs use paired reaches within a stream because nearby stream reaches are typically

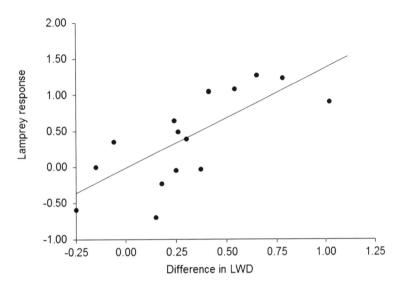

Figure 8.3 Example of correlation analysis possible with monitoring designs that utilize extensive replication (multiple projects), such as the EPT design. Nearly 60% of the variation in larval lamprey (*Lampetra* and *Entosphenus* spp.) response ($\log_{10}$(treatment density/control density)) to placement of instream structures was explained by the increase in LWD-forming pools ($\log_{10}$(treatment/control)). Sites with largest increase in LWD resulting from restoration therefore had the largest increase in larval lamprey (data from Roni 2003).

more similar than those in other streams or watersheds. It is also possible to use unpaired reaches or watersheds, but this typically introduces much higher variability since it compares the mean values of two random samples (i.e. between the control sample and the treatment sample); a larger sample size is therefore needed.

Some studies have used paired (or unpaired) reference or Leitbild reaches (Box 8.1; Muhar *et al.* 1995), with the objective being to demonstrate that the restored reaches were in fact no different than ideal habitat or are trending toward reference conditions (Lepori *et al.* 2005; Morley *et al.* 2005; Table 8.2). In other instances, a benchmark or target multimetric index value is used to represent the target or ideal value to judge the treatment success or condition. Often measurable benchmarks such as these are set during the project planning and design process discussed in Chapter 7. Ideally, a comparison of matched treatment, control, and reference reaches would be the most powerful for either post-treatment or BACI designs.

8.3.3.4 Which design is most appropriate?
Each monitoring design has its strengths and weaknesses, and a number of factors determine which design is most appropriate (Roni *et al.* 2005). The replication of the EPT and the MBACI designs allow for broadly applicable

results. In contrast, the BA and BACI designs often have little or no spatial replication and are essentially case studies that are difficult to apply outwith the area of concern. The beyond-BACI design has multiple controls, which provide additional advantages in detecting responses not found in other BACI designs but still typically have inadequate spatial replication of treatments to be broadly applicable. In contrast, the EPT design is not designed to determine the effectiveness of individual restoration projects through time, but to examine average response to a particular type of restoration. The EPT design could be replicated through time to address the lack of temporal replication and BACI and BA designs could include spatial replication (MBACI), both of which would be very useful although potentially costly.

Selecting an appropriate monitoring design depends not only upon the objective, monitoring questions, and financial constraints, but also on availability of controls, the number of treatments (projects) and pre-project data (Table 8.4). The ability to collect or availability of existing pre-project data will have a large effect on the design. If no pre-project data are available then a post-treatment design may be appropriate; the question would then focus on whether there are suitable control or reference sites and whether there are enough sites (i.e. adequate

Table 8.4 Examples of study designs typically used depending upon the number of sites sampled (one or many), whether a control exists, or whether pre-project data can be collected. BA: before-after study design; BACI: before-after control-impact; MBACI: multiple BACI; IPT: intensive post-treatment design; EPT: extensive post-treatment design; MBA: multiple before-after. Extensive design refers to a design that is spatially replicated (many study sites, reaches, or watersheds), whereas intensive design refers to monitoring at one or a few sites.

	Before and control	No before data but with control	Before data but no control
One site	BACI	IPT	BA
Many sites	MBACI	EPT	MBA

spatial replication) to use an EPT design. Determining if there are enough sites to test monitoring hypotheses will require estimates of sample size and statistical power (see Section 8.3.6 below). It also depends upon the number of projects available for restoration and monitoring, or when and where the restoration takes place can be influenced. If only one or a few treatment or control sites are available, then an IPT design comparing the treatment to a control or reference locations could be used (see Box 8.1 on River Drau case study). The scale of restoration can influence the study design. If entire watersheds rather than stream reaches are used as replicates, it may be difficult to find enough replicates for EPT or MBACI designs. If pre-project data or adequate treatments and controls are not available, then development of an effective monitoring program may not be possible.

If there is time to collect pre-project data, then the question is whether suitable control or reference sites exist. If so, then the BACI design is likely appropriate; if not, then the BA design is most appropriate. With either of these designs, multiple years of pre-project data will be needed. The volume of pre-project data required depends upon the variability of the monitoring parameters selected (see Sections 8.3.4 and 8.3.6 below). In some cases, however, it is not possible to collect data before restoration or the restoration has already occurred. If this is the case, it still may be possible to conduct a BA or BACI design if pre-project data are available or were collected as part of another monitoring effort. The success of using existing data depends upon pre-project data being compatible with post-treatment data in terms of parameters measured, collection methods, and data quality.

In practice, it may be necessary to use a combination of designs for a monitoring program (e.g. Louhi *et al.* 2011). Pre-project data on habitat conditions, for example, are often available or easily collected. In contrast, fish and other biota data are often both more labor intensive

to collect and need to be collected for multiple years before project implementation. Within a monitoring program, a BACI design might therefore be used to examine habitat changes before and after for individual projects while a EPT design might be used to examine fish response across a number of projects (see Box 8.1 on Drau River case study).

8.3.4 Parameters: Determining what to monitor

Selecting appropriate monitoring parameters is critical to a successful monitoring program, and goes hand in hand with defining hypotheses and determining the level of spatial and temporal replication. The choice of monitoring parameters will therefore often influence selection of a monitoring design. Sometimes parameters are referred to as *variables* or *metrics*. We use *parameter* here to refer to both simple measurements such as fish abundance or pool area and more complex multimetric indexes such as the index of biotic integrity (IBI; Karr & Chu 1999). Occasionally, monitoring parameters are selected because they were historically part of another monitoring program or research project, or because the investigator feels that information on a wider range of physical and biological parameters would be useful. However, collecting detailed information on numerous parameters that are not in line with the goals of the restoration project can lead to an unnecessarily cumbersome and costly monitoring program. For example, the Environmental Protection Agency has developed detailed guidelines for monitoring habitat and water quality in small streams throughout the United States (USEPA 2004). It is an ideal approaches for monitoring the water quality of short stream segments throughout the country and the methods, protocols, and parameter sensitivity are well documented (Larsen *et al.* 2001, 2004; USEPA 2004). For these reasons, some have attempted to apply the protocols

Box 8.1 Monitoring the success of restoration measures in the Drau River

During the past two decades, a comprehensive restoration project was carried out at the Drau River in Austria. The Drau is a 6th-order alpine river which had been subject to continuous straightening and channelization since the 1930s, resulting in a largely uniform and monotonous channel. Restoration activities began in 1993 and included removal of bank protection, river widening and reconnecting former side channels. Goals of the restoration were to restore river morphology, reconnect adjacent areas of the former floodplains, and achieve good ecological status as defined by the European WFD (Figure 8.4). To evaluate these restoration measures, which varied in length from 250 to 2000 m, a three-year post-treatment monitoring program was implemented. The program was designed to compare: (1) the ecological status of multiple degraded, channelized reaches (controls) with the restored reaches (treatments) and (2) the deviation of each of the treatments and controls to type-specific reference conditions (Leitbild).

Sampling included semi-quantitative and quantitative surveys to assess fish and habitat conditions (Table 8.5). A five-tiered numerical evaluation approach consistent with the WFD was used to assess hydromorphological status (Jungwirth *et al.* 2002). The value of the 'hydromorphological status' and the 'fish ecological status' was defined as the mean value of the scores of the nine habitat and seven fish criteria. These criteria were assessed by hydromorphological field mapping and field surveys of the fish fauna. Finally, the overall ecological status was calculated as the mean value of both assessments (MulFA; Schmutz *et al.* 2000). 'High ecological status' reference conditions were determined from historic maps and reports, commercial fish catch, and biological data on species distribution, as well as field data from comparable river reference sites and reference models (Kern 1992; Hughes 1995; Muhar *et al.* 1995).

The monitoring at all restored sites indicated a clear increase in the variety of type-specific habitats such as stagnant shallow zones, riffle areas, gravel and sand bars (Figure 8.5). For instance, the aquatic zone (wetted area) in one of the restored reaches increased by 12% (from 76,000 m^2 to 85,800 m^2). Four additional aquatic mesohabitat types not found in channelized reaches were created in restored reaches. The most important habitats restored were shallow-water areas with both riffles and stagnant areas, essential habitats for gravel spawners and young-of-the-year fishes. The area of these re-established habitat types is now approaching that found in reference reaches (indicated by gray arrows in Figure 8.5), although it has not fully reached the reference benchmarks.

Fish sampling demonstrated increases in European grayling (*Thymallus thymallus*), a key species that benefit from increased shallow-water habitats such as gravel bars and islands. Densities of 0+ grayling in channelized sections, where the banks had been paved with rip-rap, were extremely low (6.1 CPUE), while densities were higher in restored sites where the dominant habitat type along the bank was shallow gravel bars (39.8–117 CPUE). Although the 0+ grayling densities at Spittal (39.8 CPUE), which is impacted by hydro-peaking surges, were clearly higher than at the channelized location, the upstream values near Dellach (88.3 CPUE) and Kleblach, which are un-impacted by dams and modified flows from hydropower operations, were distinctly higher. Biomass and density of the grayling increased in the restored versus channelized sites while brown (*Salmo trutta*) and rainbow trout (*Oncorhynchus mykiss*), which prefer the rip-rapped banks, showed little difference between restored and control reaches.

Monitoring demonstrated the success of the restoration efforts; restored habitats, recovered natural processes, and re-established fish communities (biocoenoses) all improved to levels near those found in reference reaches. However, due to the rather small extent of most of the restoration projects, the benchmark of good ecological status (= 2.5 on a 1–5 scale, 5 being lowest) was only achieved in the longest and most comprehensively restored reaches (Kleblach II). The importance of areas restored and the degree of 'morphodynamic processes restored' was demonstrated by the strong correlation between the percent of habitat area restored to reference conditions and the ecological status (Figure 8.6).

Almost all restoration monitoring programs in Austria are a part of the EU Life Nature projects or monitoring in accordance to the WFD (see Section 8.1). As such, these efforts are relatively restricted in extent and replication. However, the monitoring program in the Drau River is a good example of the use of a post-treatment design to assess reach-level restoration effectiveness. Ironically, locating suitable treatment and control reaches was relatively easy due to the uniform nature of the Drau River which resulted from channel modifications.

The use of multiple criteria and ecological status based on Schmutz *et al.* (2000) demonstrated clear differences between restored and channelized reaches, which would have been difficult if only examined abundance or simple habitat metrics were examined. The monitoring program was also effective in determining which impacts were successfully addressed by the different restoration measures (e.g. riprap removal, channel widening, side-channel reconnection) and to what degree. The approach applied in this case study also demonstrates the benefits of using reference or Leitbild conditions as a benchmark to compare both treatments and controls.

(a) channelized Drau River

(b) restored site at Dellach

(c) restored site at Kleblach I

(d) restored site at Greifenburg

(e) restored site at Kleblach II

(f) restored site at Spittal

Figure 8.4 Channelized and restored reaches of the Drau River.

and parameters to monitoring of restoration projects. These efforts have met with mixed success because the parameters and protocols are time consuming and costly, and many are not sensitive to restoration actions being implemented.

Ideally, monitoring parameters should be: (1) tied to the objectives of the project; (2) relevant to the monitoring questions or hypotheses; (3) sensitive or responsive to the restoration action; (4) efficient to measure; and (5) have limited variability (Bauer & Ralph 2001; Downes

et al. 2002; Roni *et al.* 2005). The most critical of these are relevance to the monitoring questions and sensitivity to restoration action. The restoration objective should help define which parameters are the most important, and these parameters should be explicitly stated in the hypotheses. Selecting parameters that change in a measureable way to the restoration is critical to a successful monitoring program. For example, if the objectives of the project are to protect and replant riparian areas to stabilize banks, increase shade, and reduce stream temperatures, then

Table 8.5 Summary of monitoring and evaluation program for Drau River restoration measures and description of each monitoring step outlined in Figure 8.1. Details of methods can be found in Muhar *et al.* (2008).

Restoration goal	Achieve 'good ecological status' (WFD), enhance aquatic habitat conditions, and re-establish fish communities
Monitoring question(s)	Did the measures initiate/create qualitatively and quantitatively improved habitat conditions in restored reaches? Did changes in habitat due to restoration result in healthy fish community compared to target reference conditions ('high ecological status')?
Scale	Reach
Monitoring design	Post-treatment design comparing restored versus control (channelized)
Parameters	Fish: number of river-type-specific species, number of species with self-sustaining populations, fish region, number of guilds, guild composition, population size, and population age structure (MulFA; Schmutz *et al.* 2000).
	Hydromorphology: morphological river type (channel type), morphodynamics (Jungwirth *et al.* 2002), number and spatial extent of habitat types, proportion of wetted, riparian and floodplain zone, hydrological regime, flow conditions, longitudinal continuum, lateral connectivity
Replication (no. sites and years)	Five restored (three for fish monitoring) and unpaired adjacent control locations with a number of subsamples based on habitat type within each site, depending on size and diversity of investigated habitat
Sampling scheme	Habitat: area-wide habitat mapping (census) as well as detailed investigations of representative transects (Muhar *et al.* 2000)
	Fish: stratified random sampling based on habitat types. Semi-quantitative electro-fishing (CPUE) for juvenile fish in 2003, quantitative electrofishing (biomass) by strip-electrofishing method (Schmutz *et al.* 2001)
Analysis	Descriptive and graphical analyses including comparison of multimetric fish index (MuLFA)
Reporting	Project report and manuscripts (i.e. Unfer *et al.* 2004; Muhar *et al.* 2008)

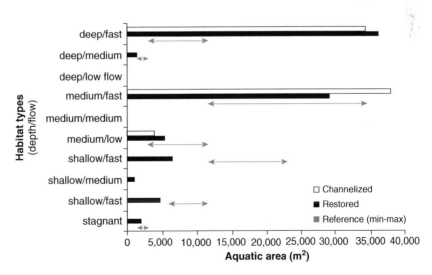

Figure 8.5 Distribution of aquatic habitat types in channelized and restored sites, as well as the reference range for each habitat type (range indicated by gray arrows) in the Drau River at Greifenburg. Total aquatic area of the reference site was 114,400 m^2; habitat area of the channelized reach was 76,000 m^2 before restoration and 85,800 m^2 after restoration.

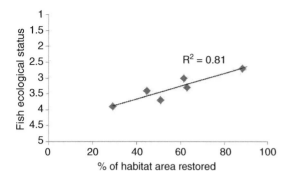

Figure 8.6 Correlation of fish ecological status (1 being highest and 5 being lowest) with the percentage of aquatic habitat area restored (100% = reference situation) at six restoration sites in the Drau River. Note that even the highest extent of aquatic habitat area restored (site Kleblach II) does not result in a fish ecological status better than 2.7. This may be partly due to disrupted longitudinal continuum caused by a chain of hydropower plants downstream and the reduced degree of lateral connectivity.

monitoring parameters such as bank stability, shade, and stream temperature would clearly test the objectives. In contrast, LWD recruitment or fish abundance are parameters that might take several decades to respond, have high interannual variability (fish abundance), or respond only indirectly to the riparian treatment.

It is difficult to provide a comprehensive list of monitoring parameters for either restoration or aquatic habitat monitoring because appropriate parameters depend on the restoration objectives, technique, the hypotheses, as well as resources available for monitoring. This is further complicated by the wide array of parameters to monitor and potential protocols for their measurement (e.g. Johnson *et al.* 2001; Roni 2005). Nevertheless, there are some obvious key parameters that should be measured for many restoration techniques. For example, M&E of restoration efforts that attempt to improve habitat for fish such as placement of instream structures, should at a minimum measure changes in habitat types such as pools and riffles and measure fish abundance and diversity. Evaluation of riparian fencing projects or replanting of riparian trees should include measurement of plant survival, composition, and potentially shade. Table 8.6 provides some examples of parameters that are typically monitored for different project types and the scale at which they are often monitored.

Many parameters and measurements are only appropriate at certain scales. For example, fine sediment or substrate size can be measured at a reach or finer scale

and are applicable for projects that focus on reducing sediment supply (e.g. road removal, fencing to exclude livestock and reduce bank erosion). In contrast, measurement of landslides or sediment supply due to road removal is best done at the watershed or sub-watershed scale. Other parameters measured at the site or reach scale (e.g. fine sediment, habitat types, fish abundance) can be scaled up or down to provide information at multiple scales.

Several biological parameters such as fish abundance have high interannual variability, which is an important factor in determining an adequate sample size. In addition, fish are highly mobile and can move among study reaches or even study watersheds. This has lead to concern that restoration may initially be concentrating fish in restored areas rather than leading to an increase in fish abundance. Few studies have specifically examined this phenomenon. Some have suggested that while movement and recruitment from nearby reaches or even nearby watersheds to restored areas may occur, it explains only a portion of the increases due to restoration (Burgess & Bider 1980; Roni & Quinn 2001b; Vehanen *et al.* 2010). It is important to note those areas vacated by fish migrating to newly restored habitats will be colonized by the next cohort (Roni *et al.* 2005, 2008). Any decreases in fish abundance in nearby reaches due to immigration to restored areas will therefore be short lived. Recent advances in fish tagging and tracking, such as passive integrated transponder (PIT) tags, continue to shed new light on the migrations and movements of stream fishes. These tagging technologies can be used to examine fish movement among study locations.

Recent studies have employed multimetric indices, such as the index biotic integrity (IBI) or multi-level concept for fish-based assessment (MuLFA), to assess restoration success (e.g. Box 8.1; Zitek *et al.* 2008). Because these indices are based on multiple biotic metrics that reflect different levels of biological organization (biodiversity, community structure, biological traits, population characteristics) they can be more robust than single metric assessment methods. The standardized sampling design and scoring procedures of these methods also allow direct comparisons between sites and give an indication of the contribution of particular restoration actions to the overall management objective (e.g. 'good ecological status' of the WFD). However, multimetric indices are often region or ecosystem specific and may need to be modified when being applied outside the area for which they were originally developed (Karr & Chu 1999).

Table 8.6 Monitoring parameters for some common types of restoration actions. See Chapter 5 for a complete description of common techniques. Modified from Roni et al. (2005).

Restoration technique	Common effectiveness monitoring parameters
Barrier removal: dams, culverts and road crossings	Physical: change in channel morphology and elevation, sediment storage and composition Biological: presence and absence of migratory fish species, seasonal species abundance and diversity, composition and age structure of riparian vegetation (dam removal)
Reconnect floodplain feature	Physical: flow connection with main channel, channel morphology and elevation, habitat, sediment storage, wood and organic retention Biological: fish abundance and diversity, fish passage and migration, macroinvertebrate and periphyton communities
Levee removal or setback	Physical: channel and floodplain morphology, topography and habitat: sediment and wood delivery and storage, duration of floodplain inundation Biological: Composition and age structure of riparian vegetation, diversity and abundance of fish, macroinvertebrates and periphyton
Aggrade incised channels	Physical: channel geometry, elevation and morphology, connectivity of floodplain habitats and their area and density, wood, water and sediment retention Biological: fish abundance and diversity, fish passage and migration, macroinvertebrate and periphyton communities
Roads (removal, resurfacing, stabilization)	In-channel: pool depth, scour, fine sediment, turbidity and water quality, stream flow, bankfull width Upslope areas: landslide rate and volume (forest roads) Biological: fish survival, macroinvertebrate diversity or abundance. Note that biological parameters are rarely measured because of difficulty relating changes in biota with road modifications
Riparian planting and grazing management	Riparian area: tree or plant survival, species composition, density and biomass; tree growth, height and diameter In-channel: shade, temperature, organic inputs (leaf litter, woody debris), bankfull width, bank stability, channel migration, channel units, pool depth
Placement of instream structures	Physical: channel morphology, habitat area and composition (pool and riffle depth, area and number), wood abundance Biological: fish abundance, diversity, growth and survival
Creation of side channels, ponds and other floodplain habitats	Physical: substrate, flow connection, habitat (channel units) Biological: fish diversity, abundance, growth and survival; macroinvertebrate and periphyton (abundance, diversity)

8.3.5 Determining how many sites or years to monitor

In conjunction with selecting parameters, we need to determine the level of spatial and temporal replication or the number of years and numbers of projects or locations (sites, reaches, watersheds) to monitor. For BA or BACI designs with no spatial replication, only the number of years needs to be considered. Conversely, for EPT designs only the number of sites must be determined. However, both the number of years and number of sites need to be considered for multiple BA and BACI designs. Initial estimates of sample size or monitoring duration can be

made from information provided in most statistics text and a hand-held calculator. However, sample size can be estimated with most statistical software packages or even by sample size estimators available on the internet. To estimate the minimum sample size, the following information is necessary:

- estimate of the spatial and temporal variability of the parameter of interest;
- desired minimum effect size (desired change in the parameter due to restoration);
- statistical power ($1 - \beta$, where β is the probability of a Type II error): the probability of detecting a difference or

change if it does exist (the probability of correctly rejecting a null hypothesis); and

• significance level (α, the probability of a Type I error): the probability of detecting a difference when it does not exist (probability of incorrectly rejecting the null hypothesis).

The first, variability, can usually be calculated from existing data or data from a pilot project, while the desired effect size, power and significance level are determined by the investigator. A typical power analysis might consist of determining the necessary sample size to have an 80% probability of detecting a 50% change given $\alpha = 0.05$ and a variance of x. Decreasing the variance or the power or increasing α or the effect size will decrease the necessary sample size. In other words, fewer samples are necessary when there is less noise (variability), a larger effect size, and you are less concerned about Type I and Type II errors.

Estimates of the variance can often be calculated using existing data from similar studies. For example, a number of studies have reviewed and provided estimates of standard error or coefficient of variation for abundance of fish, plants, and physical parameters (Gibb *et al.* 1998; Dauwalter *et al.* 2009). In addition, many agency or regional monitoring reports provide local estimates of variance in common physical and water quality parameters (e.g. Archer *et al.* 2004). Ideally, however, a pilot study should be conducted to ensure variance estimates are applicable to the specific situation and parameter of interest. The alpha-level (α) is typically assigned a value of 0.05 or 0.10 in ecological studies, while a power of 0.80 is common (see Zar 1999 or other statistical texts for more detail). Temporal and spatial variance differs considerably among monitoring parameters; minimum sample sizes calculated for these parameters will therefore also differ substantially. A power analysis based on results of Roni & Quinn (2001a) nicely illustrates the different sample sizes needed to detect different effect sizes (response) for physical and biological parameters (Figure 8.7). Notably, this analysis found that monitoring physical habitat and invertebrate communities required fewer samples than for monitoring fish densities. These differences were largely a reflection of the differences in the variability of the parameters, with fish density being much more variable than habitat or macroinvertebrate diversity.

In a power analysis of a more complex MBACI design, Liermann & Roni (2008) investigated the tradeoffs in costs between adding sites versus adding years for detecting an average response of juvenile coho salmon smolts (migrants) across multiple sites (Figure 8.8). They found

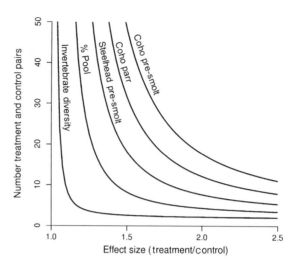

Figure 8.7 Estimated sample sizes needed to detect change in coho salmon parr and pre-smolts, steelhead pre-smolts, percent pool area, and macroinvertebrate taxa richness for an extensive post-treatment design, a power of 0.80, and an alpha of 0.05 (Roni & Quinn 2001a). The *x* axis represents the ratio of treatment to control (i.e. 1.5 equals a ratio of 3:2 or 1.5-fold increase; a 2 equals 2:1 or a two-fold increase, etc.). The *y* axis represents the number of treatment and control pairs. Modified from Roni (2005).

that unless the cost of establishing a site was much larger than the yearly monitoring cost, adding additional sites was preferable. Because of the low spatial replication in most MBACI designs, power and sample size analyses often suggest that additional replication in space is needed to increase statistical power (Downes *et al.* 2002). Estimates of power and sample size are relatively straightforward for the cases that have only temporal (BA) or only spatial replication (EPT) and use simple statistical tests such as a *t*-test or an ANOVA (analysis of variance). However, for more complex designs such as the MBACI design (Figure 8.8), consulting with a statistician is highly recommended.

As stated above, a smaller variance reduces the number of sites or years necessary to detect an effect. Variance is composed of natural variability (temporal and spatial) and measurement error. Natural variability can be reduced by pairing treatments and controls or including environmental covariates that explain the natural variability (e.g. gradient, region, geology). Measurement error can be reduced through actions such as additional staff training, instrumentation calibration, and modifications to the sampling scheme (discussed in the following section). However, natural variability is often large rela-

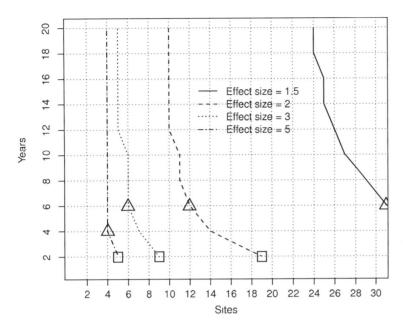

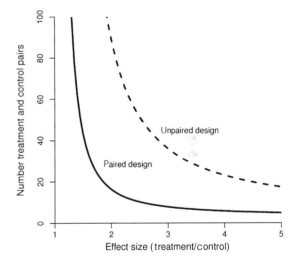

Figure 8.8 Results of the power analysis to determine optimal replication for a MBACI design to evaluate coho salmon response to restoration. Each line represents year-site combinations necessary to detect the given effect size (relative increases in the number of outmigration smolts). All values were estimated for a minimum statistical power of 0.8 and a significance level (α) of 0.1. Designs above and to the right of each line have greater statistical power. Squares represent the optimal design when there is no set-up cost to adding a site; triangles represent the scenario where the per-site set-up cost is 10 times the cost of monitoring a site for 1 year. Modified from Liermann and Roni (2008).

tive to the measurement error. If this is the case, minor reductions in measurement error are unlikely to have much effect on the analysis and efforts should probably be expended elsewhere (ideally more sites or years).

If there are large differences in the parameters of interest based on location, then pairing similar controls and treatments can substantially increase the power to detect restoration effects. For example, if part of the variation between multiple treatments and controls is explained by the location, then pairing treatments and controls within the same general location often provides more power. Figure 8.9 illustrates how pairing can dramatically reduce the sample size required. In addition, by using a paired design the population of control sites is more likely to be representative of the treatment sites which will provide more justifiable inference even if the pairing is not used in the analysis. For this and other reasons mentioned previously (see Sections 8.3.3.1–8.3.3.3), treatments and controls within the same stream are typically paired (Figure 8.9). Again, it is of course important when selecting reaches in the same stream to make sure they are independent and not interacting with each other.

Statistical considerations are important factors in the experimental design, but are not the only factors. The tradeoff between spatial and temporal replication is often constrained by available resources or logistical constraints that prevent the monitoring of additional years or sites. Furthermore, most estimates of temporal replication

Figure 8.9 Examples of the sample size needed to detect a change in juvenile trout (*O. mykiss*) abundance using a post-treatment design with paired versus unpaired treatment and control reaches. The *x* axis represents the ratio of treatment to control (effect size). Thus, 2 would represent a doubling of trout abundance, etc. The standard deviation used for the paired and unpaired tests were 0.93 and 1.64, respectively. Correlation between control and treatment sites was 0.81. Modified from Roni *et al.* (2005).

assume the restoration response will be immediate or that all restoration will occur in a short period of time. Most restoration actions in a watershed occur over many years or take time to produce change, thus having more of a 'press effect' than a rapid 'pulse effect'. Techniques such as riparian planting or reduction of sediment may require several years to result in changes to the stream channel. In these cases, additional years of monitoring beyond that predicted by power and sample size estimates may be needed to detect a response to restoration. If effort or funding is limited, long-term monitoring could be staggered across years (years 1, 3, 5, 7, etc.).

8.3.6 Sampling scheme

In conjunction with selecting parameters and determining the spatial and temporal replication, we need to determine the spatial allocation of effort used to measure particular parameters. We refer to this as the *sampling scheme*, although it is sometimes called the inference design (Stevens & Urquhart 2000). Ideally, a complete census of the study reaches or watersheds would be conducted, but for most parameters this is not feasible. For example, if monitoring one or two 500 m reaches before and after restoration in a small- to medium-sized river, a complete census (measurement) of all habitat units and fish populations might be feasible. In contrast, a complete census quantifying substrate size, riparian vegetation composition, or aquatic macroinvertebrates throughout the same reaches would not be feasible and would require some type of sampling or subsampling.

As the temporal and spatial scale of the monitoring and number of restoration projects increases, the need for sampling becomes more pronounced. Common schemes for spreading samples in space and time include simple random, stratified random, systematic, cluster, multi-stage, double, and generalized random–tessellation stratified (GRTS) sampling (Table 8.7; Thompson 1992; Buckland *et al.* 2001; Stevens & Olson 2004). Stratified random sampling can be used to ensure a more even distribution of samples across space or time, or focus sampling efforts in specific areas such as those where fish density is thought to be higher or along a gradient such as elevation. Systematic and GRTS sampling ensure a more even distribution of samples across space or time. Cluster and multi-stage sampling are useful in situations where the cost of visiting sites is high. For example, it may be less expensive to randomly select a subset of watersheds and then select a sample of restoration projects within those watersheds as opposed to just randomly sampling from all restoration projects in all watersheds. Double sampling is useful when there is an expensive

Table 8.7 Common sampling approaches for measuring parameters within and across study areas (from Roni *et al.* 2005).

Type of sampling	Description
Census	All units are sampled
Simple random	A subset of randomly selected units are sampled within the study location in such a way that each subunit has an equal chance of being selected
Systematic	An initial unit (or units) is chosen randomly and then every *k*th unit (starting from that unit) is sampled
Stratified random	The units are broken into strata (categories) based on other available covariates (variables) such as geology, gradient, channel type, etc. Random sampling then occurs in each stratum
Multistage	Sampling units are selected and then the selected units are further divided into units and a subset of those units sampled
Double sampling	Selected sites are first sampled using a low-cost method and then a subset of these sites are resampled with a more costly and accurate method. The relationship between the low- and high-cost methods is used to correct estimates from sites that were only sampled with the low-cost method

accurate method and an inexpensive but likely biased alternative. Values from a larger sample using the less expensive method can be corrected based on a smaller subsample in which the more accurate and costly method is also used.

These different sampling schemes are not mutually exclusive. For example, GRTS samples could be taken within the different strata from a stratified random sample. Long-term monitoring programs will also require some combination of spatial and temporal stratification such as a pulsed monitoring or sampling, where some factors and sites are monitored in some seasons and years and not in others (Bryant 1995). It is important to realize that no single sampling scheme is appropriate for all situations. The optimal sampling strategy will depend on factors such as the spatial distribution of the organism and the logistics of moving the field staff among locations, collecting samples or observing the organism.

8.4 Guidelines for analyzing and summarizing data

The appropriate type of data analysis will be dictated largely by the monitoring design, sampling scheme, parameters selected, level of spatial and temporal replication, and assumed response type (i.e. press or pulse). There are a multitude of statistical approaches, models, and software that can be used for examining monitoring data from restoration projects. For designs with replication in only time or space (e.g. BACI or EPT), analysis is typically straightforward. Simple graphical analysis and common statistical approaches such as t-tests, ANOVA, regression, and correlation analysis can be used and are part of most basic spreadsheet or statistical packages. If the variance in the data or transformed data is constant and they have a roughly normal bell-shaped distribution, a simple t-test is generally appropriate to determine differences in means between controls and treatments. For designs with spatial and temporal replication, fixed or mixed effects linear regression, ANOVA and ANCOVA (analysis of covariance) models are typically sufficient. For multivariate responses (common in invertebrate data, for example), similar methodologies are available based on permutation tests (Anderson 2001). Model specification and testing is often somewhat subtle however (e.g. nested versus crossed designs, random versus fixed effects, non-linearity, non-constant variance). We therefore recommend consulting with a statistician when designing the monitoring program and analyzing and interpreting the results.

The results of any statistical analysis must always be considered in the context of the design. Tests applied to data collected using the BA and BACI designs could easily be overly optimistic; some authors have suggested that, for BACI designs, detecting statistical differences in population data is often difficult due to low sample sizes and inherent variability (Osenberg *et al.* 1994; Murtaugh 2002). For BA or BACI designs, particularly those with little spatial replication, emphasis should be initially placed on the graphical interpretation of the data rather than statistical analysis (Conquest 2000).

For the EPT designs, significant results can be introduced through flawed selection of controls (i.e. selecting controls that are clearly different from treatments). Examination of important environmental covariates collected at all sites can help to support the hypothesis that restoration is indeed causing the observed differences. This allows an independent assessment of the controls (are they similar to the treatments in terms of the available covariates?) and possibly further elucidation of treatment effects. For example, does the magnitude of the post-treatment difference in wood loading correlate with the effects size (Figure 8.3)? The following steps can help with initial analysis of monitoring data.

1. Confirm the hypotheses that are to be tested.
2. Proof data to ensure that it has been entered correctly. Considerable analysis time can be squandered if this step is not done carefully. Often conducting queries of a database or calculating basic statistics will demonstrate errors in data entry.
3. Plot the data to highlight any unusual observations and to provide guidance for potential transformations.
4. Transform the data as necessary. In particular, many parameters tend to be right-skewed with variance that increases with the mean. This can generally be remedied with a log or square-root transformation. If the parameter is a response variable, consider using the different distributions available in many modeling frameworks as an alternative to transformation (e.g. Generalized Linear Models).
5. Generate figures that provide a visual or graphical test of the specified hypotheses.
6. Test the hypotheses. Avoid picking through many statistical test results to highlight a few significant results (e.g. if you do 20 analyses with a $p < 0.05$ you are likely to find a significant result by chance alone).
7. Use the figures and graphs in step 5 above and the residuals from step 6 to confirm that the data meet the assumptions of the tests (many tests assume normal distribution).
8. Report the results using estimates of effect size and confidence intervals (in addition to test results). This will allow the reader to assess biological as well as statistical significance (e.g. Bradford *et al.* 2005).

8.5 Monitoring of multiple restoration actions at a watershed scale

Much of the discussion above and most published studies focus on determining the effectiveness of one or a few types of restoration actions. Designing M&E for individual restoration actions, or multiple restoration actions of the same type, is relatively straightforward compared to evaluating a watershed-scale response to multiple restoration actions. Many monitoring programs that evaluate individual or multiple projects in a watershed also assume that restoration actions are independent rather than synergistic, which may not be the case. Restoration ecologists are being asked with increasing frequency to design monitoring programs to evaluate multiple actions throughout a watershed. The goal of these efforts is often to determine which restoration actions are producing the largest changes in habitat and biota at a watershed scale.

Initial efforts to do watershed-scale evaluation have often tried to apply a single monitoring design and sampling program to evaluate responses of biota and habitat at both a watershed and reach scale (Bilby *et al.* 2005; Roni *et al.* 2010b). This can be less than optimal because the design and sampling scheme for measuring changes in habitat or fish at a project (reach level) may be very different from that needed at a watershed scale. In essence, this type of approach seeks to answer multiple monitoring questions outlined in Table 8.3. It therefore requires that a monitoring program for each major question be developed and then integrated to make M&E as efficient as possible. It will likely require sampling some locations solely to answer watershed-scale questions and others only to evaluate individual restoration actions. Some locations will hopefully overlap and some samples will be suitable for evaluating responses at both scales (Figure 8.10). Designing an efficient and effective program will require extensive coordination during both development of the monitoring and collection of field data.

Even for the most rigorous of monitoring programs, the intensity of the restoration both at a reach or watershed scale can also influence the ability to detect a response due to restoration (Roni & Quinn 2001a; Roni *et al.* 2010b). If the intensity of restoration in a reach is low or percentage of area restored in a watershed is small, it may be difficult to detect a response due to restoration (Roni *et al.* 2010a). This is particularly critical for monitoring watershed or population level responses to restoration: the larger the watershed and more variability in the parameters measured, the more restoration will be needed to be able to detect a response.

8.6 Implementation: Design is not enough

The previous sections laid out clear steps for designing effective monitoring of restoration, but proper imple-

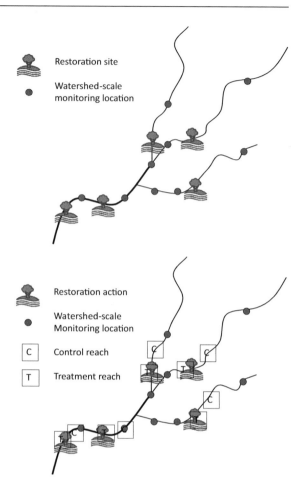

Figure 8.10 Stylized watershed diagram illustrating (upper) sampling locations needed to evaluate whole watershed response to multiple restoration actions (tree symbols) and (lower) treatment (T) and control (C) locations needed to evaluate effectiveness of individual restoration actions. Circles represent randomly selected locations for monitoring watershed response to restoration. In contrast, restoration, treatment, and control locations typically cannot be randomly selected. If monitoring parameters and protocols are similar for both project (action) and watershed-scale monitoring, data from evaluation of individual restoration projects could be included in analysis of watershed level response. Note that controls are typically located upstream of treatments for control-treatment pairs.

mentation can 'make or break' a monitoring program. Many monitoring efforts fall short of objectives or are never completed due to non-technical issues. These can include inability to control other management actions, improper data collection, poor data management, and failure to publish results. For instance, in many cases it is difficult to control other ongoing management actions in an area. Management actions such as the stocking of native or non-native fishes, or changes in fishing regulations and therefore fishing effort, can make it nearly impossible to detect changes in fish abundance due to restoration (Rinne 1999; Medina *et al.* 2005). In other cases, overzealous restoration practitioners implement restoration actions in control reaches or before completion of pre-project data collection, which can compromise even the most well-designed M&E program. Similarly, other large changes in land use or environmental regulations can hamper efforts to monitor projects and detect changes due to restoration.

These problems emphasize the need to communicate and coordinate with natural resource agencies and other groups within the study area to ensure that these types of actions do not occur or are implemented in a way that will not compromise the effectiveness of the monitoring program. The larger the geographic extent covered by the monitoring program, the more difficult this becomes and the greater is the need for periodic meetings with natural resource management partners and groups within the project area.

The importance of training and overseeing those collecting data cannot be overemphasized. The project leader should periodically participate in field data collection to ensure that crews are collecting data properly. A review of monitoring programs in the USA demonstrated that many failed because the principal investigator did not train the field staff properly and did not participate in the data collection (Reid 2001). Entering and summarizing field data is also critical. Delays in data entry can allow errors in field protocols or data collection to go undetected for years. Annual or more frequent data summaries allow detection of these problems. Frequent summaries also facilitate adaptive management, such as modifying protocols and procedures or examining parameters collected to see if some need to be added or eliminated.

Periodically summarizing data not only helps maintain a rigorous monitoring program, but also provides an instrument for communicating results to the funders and the public. Most funding entities require regular reporting of project activities on an annual or semi-annual basis. If not required, project leaders should still ensure that annual contract reports are completed. Maintaining and expanding funding for monitoring and evaluation is one of the larger challenges faced by the restoration community. When funding entities can see the results of restoration and M&E documented in periodic reports, they are more likely to support continuation or expansion of the monitoring program and less likely to reduce funding. Despite the fact that many scientists, biologists, and technicians may like to ignore it, 'marketing' of projects through frequent presentation of goals, objectives, and results is a critical part of maintaining a monitoring program.

These non-technical issues are particularly acute for monitoring programs evaluating watershed-scale responses. Evaluating restoration actions across a whole watershed or multiple watersheds can be expensive, involve many groups or agencies and require extensive coordination. Well-defined data collection, entry, summarization, and rigorous reporting procedures are therefore necessary. This will require constant and consistent attention throughout the duration of the project, which can be difficult for long-term (>5 years) monitoring programs that may see changes in project personnel and funding.

It is important to be able to clearly and concisely explain each important step in the monitoring program to a variety of audiences. Following completion of the steps outlined in this chapter, we should be able to provide a concise summary of key aspects of an M&E program similar to those outlined in Tables 8.2 and 8.5. Finally, publication of completed M&E programs is a critical step. With the exception of instream habitat improvement structures (see Roni *et al.* 2008; Whiteway *et al.* 2010), there is relatively little published information on the effectiveness of most restoration techniques and almost no long-term evaluation (>10 years). There is also increasing concern over a potential bias in the literature toward studies that find positive results, because studies that show no response to restoration are either not submitted or not accepted for publication. Regardless of whether this is true it is clear that, without publication of monitoring findings (both good and bad), we cannot learn from others to improve both our restoration techniques or our M&E programs.

8.7 Summary

Monitoring and evaluation is critical for improving our understanding of restoration, and is a key part of the restoration process. This chapter outlined the key considerations for developing a rigorous and effective monitoring program. These include determining goals and objectives, hypotheses, monitoring design, parameters, replication, sampling scheme, as well as implementing the monitoring program and reporting the results. Defining clear, testable hypotheses that succinctly outline the scale, study design, and parameters to be measured is absolutely essential. Monitoring parameters should be chosen judiciously to focus on those that respond to the restoration action and address the questions outlined in the hypotheses. Spatial and temporal replication (sample size) should be estimated to optimize study design and refine the list of monitoring parameters. In addition to these technical aspects, successful implementation and completion of monitoring depends on many non-technical issues: the importance of good project management should not be overlooked. Furthermore, design of M&E should occur during initial restoration design or as early in the design process as possible. These and other factors are critical for a successful M&E program and improving future restoration design and effectiveness. Finally, to address the continual challenge of M&E funding, restoration practitioners, managers, and scientists must continue to demonstrate its importance by completing and reporting results and observations in a variety of media.

8.8 References

Anderson, M.J. (2001) A new method for non-parametric multivariate analysis of variance. *Austral Ecology* **26**, 32–46.

Archer, E.K., Roper, B.B., Henderson, R.C., Bouwes, N., Mellison, S.C. & Kershner, J.L. (2004) *Testing common stream sampling methods for broad-scale, long-term monitoring.* General Technical Report RMRS-GTR-122. United States Department of Agriculture, Fort Collins, Colorado.

Bauer, S.B. & Ralph, S.C. (2001) Strengthening the use of aquatic habitat indicators in Clean Water Act programs. *Fisheries* **26**(6), 14–24.

Bilby, R., Ehinger, B., Quinn, T. *et al.* (2005) Study evaluates fish response to management actions. *Western Forester* **50**, 14–15.

Bradford, M.J. (1999) Temporal and spatial trends in the abundance of coho salmon smolts from western North America. *Transactions of the American Fisheries Society* **128**, 840–846.

Bradford, M.J., Korman, J. & Higgins, P.S. (2005) Using confidence intervals to estimate the response of salmon populations (*Oncorhynchus spp.*) to experimental habitat alterations. *Canadian Journal of Fisheries and Aquatic Sciences* **62**, 2716–2726.

Bryant, M.D. (1995) Pulsed monitoring for watershed and stream restoration. *Fisheries* **20**(11), 6–13.

Buckland, S.T., Anderson, D.R., Burnham, K.P., Laake, J.L., Borchers, D.L. & Thomas, L. (2001) *Introduction to Distance Sampling: Estimating Abundance of Biological Populations.* Oxford University Press, Oxford, United Kingdom.

Burgess, S.A. & Bider, J.R. (1980) Effects of stream habitat improvements on invertebrates, trout populations, and mink activity. *Journal of Wildlife Management* **44**, 871–880.

Conquest, L.L. (2000) Analysis and interpretation of ecological field data using BACI designs: discussion. *Journal of Agricultural, Biological, and Environmental Statistics* **5**, 293–296.

Dauwalter, D.C., Rahel, F.J. & Gerow, K.G. (2009) Temporal variation in trout populations: implications for monitoring and trend detection. *Transactions of the American Fisheries Society* **138**, 38–51.

Downes, B.J., Barmuta, L.A., Fairweather, P.G., Faith, D.P., Keough, M.J., Lake, P.S., Mapstone, B.D. & Quinn, G.P. (2002) *Monitoring Ecological Impacts: Concepts and Practice in Flowing Waters.* Cambridge University Press, Cambridge, U.K.

Gibbs, J.P., Sam, D. & Eagle, P. (1998) Monitoring populations of plants and animals. *BioScience* **48**, 935–940.

Gowan, C. & Fausch, K.D. (1996). Long-term demographic responses of trout populations to habitat manipulations in six Colorado streams. *Ecological Applications* **6**, 931–946.

Grant, J.W.A., Englert, J. & Bietz, B.F. (1986) Application of a method for assessing the impact of watershed practices: effects of logging on salmonid standing crops. *North American Journal of Fisheries Management* **6**, 24–31.

Green, R.H. (1979) *Sampling Design and Statistical Methods for Environmental Biologists.* John Wiley & Sons, New York.

Henry, C.P. & Amoros, C. (1995) Restoration ecology of riverine wetlands: I. A scientific base. *Environmental Management* **19**, 891–902.

Hicks, B.J., Hall, J.D., Bisson, P.A. & Sedell, J.R. (1991) Responses of salmonids to habitat changes. In: Meehan, D.H. (ed.) *Influences of Forest and Rangeland Management on Salmonid Fishes and Their Habitats.* American Fisheries Society Special Publication 19, American Fisheries Society, Bethesda, Maryland, pp. 483–518.

Hughes, R.M. (1995) Defining acceptable biological status by comparing with reference conditions. In:

Davis, W.S. & Simon, R.F. (eds) *Biological Assessment and Criteria; Tools for Water Resource Planning and Decision Making*. Lewis Publishers, Boca Raton, Florida, pp. 31–47.

Hurlbert, S.H. (1984) Pseudoreplication and the design of ecological field experiments. *Ecological Management* **54**, 187–211.

Jähnig, S.C., Brunzel, S., Gacek, S. *et al.* (2009) Effects of re-braiding measures on hydromorphology, floodplain vegetation, ground beetles and benthic invertebrates in mountain rivers. *Journal of Applied Ecology* **46**, 406–416.

Johnson, D.H., Pittman, N., Wilder, E., Silver, J.A., Plotnikoff, R.W., Mason, B.C., Jones, K.K., Roger, P., O'Neil, T.A. & Barrett, C. (2001) Inventory and monitoring of salmon habitat in the Pacific Northwest: directory and synthesis of protocols for management/research and volunteers in Washington, Oregon, Idaho, Montana, and British Columbia. Washington Department of Fish and Wildlife, Olympia, Washington.

Johnson, S.L., Rodgers, J.D., Solazzi, M.F., & Nickelson, T.E. (2005) Effects of an increase in large wood on abundance and survival of juvenile salmonids (*Oncorhynchus spp.*) in an Oregon coastal stream. *Canadian Journal of Fisheries and Aquatic Sciences* **62**, 412–424.

Jungwirth, M., Muhar, S. & Schmutz, S. (2002) Reestablishing and assessing ecological integrity in riverine landscapes. *Freshwater Biology* **47**(4), 867–888.

Karr, J.R. & Chu, E.W. (1999) *Restoring Life in Running Waters: Better Biological Monitoring*. Island Press, Washington, DC.

Kern, K. (1992) Rehabilitation of streams in South-West Germany. In: Boon, P.J., Calow, P. & Petts, G.E. (eds) *River Conservation and Management*. John Wiley & Sons Ltd., Chichester, United Kingdom, pp. 321–336.

Larsen, D.P., Kincaid, T.M., Jacobs, S.E. & Urquhart, N.S. (2001) Designs for evaluating local and regional scale trends. *Bioscience* **51**(12), 1069–1078.

Larsen, D.P., Kaufmann, P.R., Kincaid, T.M. & Urquhart, N.S. (2004) Detecting persistent change in the habitat of salmon-bearing streams in the Pacific Northwest. *Canadian Journal of Fisheries and Aquatic Sciences* **61**(2), 283–291.

Lepori, F., Palm, D., Braennaes, E. & Malmqvist, B. (2005) Does restoration of structural heterogeneity in streams enhance fish and macroinvertebrate diversity? *Ecological Applications* **15**(6), 2060–2071.

Liermann, M. & Roni, P. (2008) More sites or more years? – optimal study design for monitoring fish response to watershed restoration. *North American Journal of Fisheries Management* **28**, 935–943.

Louhi, P., Mykrä, H., Paavola, R. *et al.* (2011). Twenty years of stream restoration in Finland: little response by benthic macroinvertebrate communities. *Ecological Applications* **21**(6), 1950–1961.

MacDonald, L.H., Smart, A.W. & Wissmar, R.C. (1991) *Monitoring guidelines to evaluate effects of forestry activities on streams in the Pacific Northwest and Alaska*. U.S. Environmental Protection Agency, Region 10. Seattle, WA.

Medina, A.L., Rinne, J.N. & Roni, P. (2005) Riparian restoration through grazing management: considerations for monitoring project effectiveness. In: Roni, P. (ed.) *Monitoring Stream and Watershed Restoration*. American Fisheries Society Bethesda, Maryland, pp. 97–126.

Morley, S.A., Garcia, P.S., Bennett, T.R. & Roni, P. (2005) Juvenile salmonid (*Oncorhynchus spp.*) use of constructed and natural side channels in Pacific Northwest rivers. *Canadian Journal of Fisheries and Aquatic Sciences* **62**(12), 2811–2821.

Muhar, S., Schmutz, S. & Jungwirth, M. (1995) River restoration concepts – goals and perspectives. *Hydrobiologia* **303**, 183–194.

Muhar, S., Schwarz, M., Schmutz, S. & Jungwirth, M. (2000) Identification of rivers with high and good habitat quality: methodological approach and applications in Austria. *Hydrobiologia* **422–423**, 343–358.

Muhar, S., Jungwirth, M., Unfer, G. *et al.* (2008) Restoring riverine landscapes at the Drau River: successes and deficits in the context of ecological integrity. In: Habersack, H. Piègay, H. & Rinaldi, M. (eds) *Gravel-Bed Rivers VI: From Process Understanding to River Restoration*. Elsivier, Amsterdam, The Netherlands, pp. 779–803.

Murtaugh, P.A. (2002) On rejection rates of paired intervention analysis. *Ecology* **83**, 1752–1761.

Osenberg, C.W., Schmitt, R.J., Holbrook, S.J., Abusaba, K.E. & Flegal, A.R. (1994) Detection of environmental impacts – natural variability, effect size, and power analysis. *Ecological Applications* **4**(1), 16–30.

Reeves, G.H. & Roelofs, T.D. (1982) *Rehabilitating and enhancing stream habitat: 2. Field applications*. General Technical Report PNW-GTR-140, United States Department of Agriculture Forest Service Pacific Northwest Research Station, Portland, Oregon.

Reid, L.M. (2001) The epidemiology of monitoring. *Journal of American Water Resources Association* **34**(4), 815–820.

Rinne, J.N. (1999) Fish and grazing relationships: the facts and some pleas. *Fisheries* **24**(8), 12–21.

Roni, P. (2003) Responses of benthic fishes and giant salamanders to placement of large woody debris in small Pacific Northwest streams. *North American Journal of Fisheries Management* **23**, 1087–1097.

Roni, P. (2005) *Monitoring Stream and Watershed Restoration*. American Fisheries Society, Bethesda, Maryland.

Roni, P. & Quinn, T.P. (2001a) Density and size of juvenile salmonids in response to placement of large woody debris in western Oregon and Washington streams. *Canadian Journal of Fisheries and Aquatic Sciences* **58**(2), 282–292.

Roni, P. & Quinn, T.P. (2001b) Effects of wood placement on movements of trout and juvenile coho salmon in natural and artificial stream channels. *Transactions of the American Fisheries Society* **130**, 675–685.

Roni, P., Beechie, T.J., Bilby, R.E., Leonetti, F.E., Pollock, M.M. & Pess, G.R. (2002) A review of stream restoration techniques and a hierarchical strategy for prioritizing restoration in Pacific Northwest watersheds. *North American Journal of Fisheries Management* **22**, 1–20.

Roni, P., Liermann, M.C., Jordan, C. & Steel, E.A. (2005) Steps for designing a monitoring and evaluation program for aquatic restoration. In: P. Roni (ed.) *Monitoring Stream and Watershed Restoration*. American Fisheries Society, Bethesda, Maryland, pp. 13–34.

Roni, P., Hanson, K. & Beechie, T. (2008) Global review of the physical and biological effectiveness of stream habitat rehabilitation techniques. *North American Journal of Fisheries Management* **28**, 856–890.

Roni, P., Pess, G., Beechie, T. & Morley, S. (2010a) Estimating changes in coho salmon and steelhead abundance from watershed restoration: how much restoration is needed to measurably increase smolt production? *North American Journal of Fisheries Management* **30**, 1469–1484.

Roni, P., Pess, G.R. & Morley, S.A. (2010b) Monitoring salmon stream restoration: guidelines based on experience in the American Pacific Northwest. In: Kemp, P. (ed.) *Salmonid Fisheries: Freshwater Habitat Management*. Wiley-Blackwell, Oxford, United Kingdom, pp. 119–147.

Ryder, D., Brierley, G.J., Hobbs, R., Kyule, G. & Leishman, M. (2008) Vision generation: why do we seek to achieve in river rehabilitation. In: Brierley, G.J. & Fryirs, K.A. (eds) *River Futures: An Integrative Scientific Approach to River Repair*. Island Press, Washington, DC, pp. 16–27.

Schmutz, S., Kaufmann, M., Vogel, B., Jungwirth, M. & Muhar, S. (2000) A multi-level concept for fish-based, river-type-specific assessment of ecological integrity. *Hydrobiologia* **422–423**, 279–289.

Schmutz, S., Zauner, G., Eberstaller, J. & Jungwirth, M. (2001) Die Streifenbefischungs-methode: eine methode zur quantifizierung von fischbeständen mittelgroßer fließgewässer. *Österreichs Fischerei* **54**, 14–27.

Solazzi, M.F., Nickelson, T.E., Johnson, S.L. & Rodgers, J.D. (2000) Effects of increasing winter rearing habitat on abundance of salmonids in two coastal Oregon streams. *Canadian Journal of Fisheries and Aquatic Sciences* **57**, 906–914.

Stewart-Oaten, A., Murdoch, W.W. & Parker, K.R. (1986) Environmental impact assessment: "pseudoreplication" in time? *Ecology* **67**, 929–940.

Stevens, D.L. & Olsen, A.R. (2004) Spatially balanced sampling of natural resources. *Journal of the American Statistical Association* **99**, 262–278.

Stevens, D.L. Jr. & Urquhart, N.S. (2000) Response designs and support regions in sampling continuous domains. *Environmetrics* **11**(1), 13–41.

Tarzwell, C.M. (1934) *Stream Improvement Methods*. United States Bureau of Fisheries, Division of Scientific Inquiry, Ogden, Utah.

Thompson, S.K. (1992) *Sampling*. John Wiley & Sons, New York.

Underwood, A.J. (1991) Beyond-BACI: experimental designs for detecting human environmental impacts on temporal variations in natural populations. *Australian Journal of Marine and Freshwater Research* **42**, 569–587.

Underwood, A.J. (1994) On beyond BACI: sampling designs that might reliably detect environmental disturbances. *Ecological Applications* **4**, 3–15.

Unfer G., Schmutz, S., Wiesner, C., Habersack, H. *et al.* (2004) The effects of hydro-peaking on the success of river-restoration measures within the LIFE-project "Auenverbund Obere Drau". In: Garcia de Jalon, D. & Martinez, P.V. (eds) *Proceedings of the Fifth International Conference on Ecohydraulics – Aquatic Habitats: Analysis and Restoration*. IAHR-AIRH Madrid, Spain, pp. 741–746.

USEPA (2004) *Wadeable Stream Assessment: Field Operations Manual*. EPA841-B-04-004. United States Environmental Protection Agency, Office of Water and Office of Research and Development, Washington, DC.

Vehanen, T., Huusko, A., Mäki-Petäys, A., Louhi, P., Myrka, H. & Muotka, T. (2010) Effects of habitat rehabilitation on brown trout (*Salmo trutta*) in boreal forest streams. *Freshwater Biology* **55**, 2200–2214.

Walters, C.J., Collie, J.S. & Webb, T. (1988) Experimental designs for estimating transient responses to management disturbances. *Canadian Journal of Fisheries and Aquatic Sciences* **45**, 530–538.

Whiteway, S.L., Biron, P.M., Zimmermann, A. *et al.* (2010) Do in-stream restoration structures enhance salmonid abundance? A meta-analysis. *Canadian Journal of Fisheries and Aquatic Sciences* **67**, 831–841.

Zar, J.H. (1999) *Biostatistical Analysis*. Prentice Hall, Upper Saddle River, New Jersey.

Zitek, A., Schmutz, S. & Jungwirth, J. (2008) Assessing the efficiency of connectivity measures with regard to the EU-Water Framework Directive in a Danube-tributary system. *Hydrobiologia* **609**, 139–161.

9 Synthesis: Developing Comprehensive Restoration Programs

Tim Beechie, Philip Roni & George Pess

Northwest Fisheries Science Center, National Oceanic and Atmospheric Administration, USA

9.1 Introduction

In this book we have laid out a systematic approach to planning, implementing, and monitoring restoration programs for watersheds, encompassing the key restoration steps presented in Chapters 2–8. Each of these steps contributes to the development of a comprehensive restoration program, and supports the development of restoration proposals and project designs. In addition, the steps outlined in this book can be used to assist funders or regulatory agencies with a review of restoration plans, proposals, and project designs. In this final chapter we synthesize the key purposes and products of each of the restoration steps, and discuss how they support the development of a comprehensive restoration program and individual restoration proposals. Finally, we discuss how the development of a comprehensive plan using the steps outlined in this book (1) informs the design and review of individual restoration projects or suites of projects, and (2) facilitates a transition from opportunistic to strategic restoration.

9.2 Components of a comprehensive restoration program

A restoration program developed by following the steps outlined in Chapter 1 follows a logical path that links resto-ration goals, watershed assessments, identification of restoration needs, selection and prioritization of actions, design of projects, and development of a monitoring program (Figure 1.1). Each step has a specific set of purposes and products that inform the next step in the process (Table 9.1). The products of steps 2 and 3, for example, are the desired outputs of a watershed assessment, which identifies the key restoration needs for a watershed and informs the selection and prioritization of actions (Steps 4 and 5). Together Steps 1–5 provide the information needed to develop a comprehensive restoration plan, which then serves as the basis for designing restoration projects and monitoring programs (Steps 6 and 7). In most cases the comprehensive restoration plan (Steps 1–5) is summarized in a single report, while restoration proposals and monitoring programs (Steps 6 and 7) are produced as separate proposals or reports. Nevertheless, the complete restoration program should include the full sequence of components from watershed assessment to monitoring program, and the components should be clearly linked even if they are in separate reports.

The development of a restoration plan is not a static but an iterative process. Ideally, a synthesis of the restoration program components will strengthen feedback loops among them, so that learning from each step can be used to improve the restoration effort through time (i.e. it will facilitate adaptive management). For example, monitoring

Stream and Watershed Restoration: A Guide to Restoring Riverine Processes and Habitats, First Edition. Philip Roni and Tim Beechie.
© 2013 John Wiley & Sons, Ltd. Published 2013 by John Wiley & Sons, Ltd.

Table 9.1 Products and purposes of the restoration steps (shown in Figure 1.2). Steps 1–5 are the components of a restoration plan and provide the supporting documentation for development of project proposals and designs (Step 6) and monitoring plans (Step 7). Together these components help guide a successful restoration program.

Program component	Step	Product(s)	Purpose
COMPREHENSIVE RESTORATION PLAN	1. Define the restoration goal	Clearly defined goal that specifies biological aims of restoration	The restoration goal guides the design of watershed assessments and choice of prioritization criteria
	2. Assess watershed processes, habitats, and biota	Maps and tables indicating the causes of habitat change, which habitats have been most altered, which habitat losses most influence biological declines	Watershed assessment results are the basis for identifying restoration needs, and inform the design of monitoring programs
	3. Identify restoration needs	Maps and tables indicating which kinds of restoration actions are needed and where	Restoration needs focus the selection of actions on those that are most important for achieving the biological aims of restoration
	4. Select restoration actions	Specific techniques that address identified needs and follow the process-based restoration principles	Restoration actions are selected to effectively address identified problems
	5. Prioritize restoration actions	Prioritization approach, and ranked restoration actions or types of actions	Prioritization should improve success and cost-effectiveness of restoration efforts (i.e. actions are more focused on key problems)
PROJECT PROPOSAL (IMPLEMENTATION AND MONITORING)	6. Design restoration project and monitoring program	Project designs that accommodate driving processes and address high-priority restoration needs	Produce restoration actions that focus on effectiveness and sustainability
	7. Implement restoration and monitoring	Monitoring program that focuses on determining effectiveness of restoration action types or suites of restoration actions	Provide information needed to adjust restoration plans or designs based on biological effectiveness of restoration actions

specific project types might show that some types of projects are more effective or robust to environmental perturbations than others, which might lead to changes in the restoration plan or priorities. Similarly, updated or improved assessment of watershed processes, habitats, or biota might lead to the discovery of previously unidentified impacts or mechanisms, again leading to changes in plans or priorities or even changes to the design of monitoring efforts. Ensuring that mechanisms are in place for incorporating new information and making adjustments to the restoration program will help to increase cost-effectiveness of future restoration efforts. In this section we briefly summarize the important outcomes of each step, its relationship with other steps, and how it might inform an adaptive restoration program.

9.2.1 Goals, assessments, and identifying restoration actions

The restoration goal is the foundation of any restoration plan or program as it guides the development of assessments, selection of appropriate actions, prioritization of actions, and monitoring of program effectiveness. In practice, goals for large restoration programs are often driven by legislation or the goals of a particular funding organization, which may be focused on recovery of individual species or more broadly defined indicators of biological or ecosystem integrity (Chapter 3). Species-focused goals require watershed assessments that focus on habitat requirements of focal species and identification of which habitats and driving processes need to be restored in order to achieve recovery of that species. By contrast, ecosystem-focused goals require a broader range of biological assessments and identify habitats and processes that will most contribute to restoration of biological integrity if restored. Regardless of the legislation driving restoration, however, the assessment of causes of habitat loss or degradation examines a common set of watershed-scale and reach-scale processes.

The purpose of watershed assessments is to identify where and what kinds of restoration are needed to achieve the restoration goal (Table 9.1). Watershed assessments therefore focus on answering two key questions: 'how have habitats changed and affected biota?' and 'what are the causes of those habitat changes?' Answers to these two questions identify which kinds of habitat most need to be restored, as well as the key causes of degradation that can be addressed through restoration. A restoration strategy can then be developed, focusing on identifying the restoration action types that are most important for achieving a restoration goal (e.g. removing migration barriers, restoring floodplain habitats, restoring streamflow regimes, etc.). The proposed restoration actions should focus on restoration of processes to the extent possible, although habitat improvement or creation efforts can be used where constraints prohibit restoration or where restoration might require decades to achieve benefits (Chapter 2). The watershed assessment results should also include inventories of specific restoration actions that are needed to address the key causes of degradation. For example, inventories of barriers to fish migration are needed to identify which barriers block the most habitat and would provide the greatest restoration benefit.

9.2.2 Prioritizing restoration actions or watersheds

A variety of approaches can be used to prioritize restoration actions including project type, refugia, benefit to species or habitat types, cost-effectiveness, scoring methods, or computer models and conservation software (Chapter 5). Each approach has advantages and disadvantages, but perhaps the most common and transparent approach is to score projects using multiple criteria that reflect the restoration goal and the values of local stakeholders. This approach usually involves scoring a variety of individual criteria (e.g. number of species that benefit, change in multimetric biological indicator, number of fish produced, cost, education value, etc.), then summing the criteria to arrive at a project score. Because implementation of restoration actions is often partly driven by cooperation of landowners or by availability of funding, it is often useful to group scored projects into categories of high, medium, and low priority. This approach recognizes that lack of funding or landowner cooperation might inhibit the implementation of some of the high-priority projects, and that only a subset of the high-priority projects will be implemented in the near term regardless of priority score. Nevertheless, having a clear understanding of high-priority projects is a key part of the restoration plan that will help move the restoration effort toward greater effectiveness over time.

Where a restoration effort may encompass a large number of watersheds, it may also be important to prioritize among watersheds to focus restoration efforts. In general, the same approaches listed above for ranking actions can be used to rank watersheds, but the data that support prioritization are more general and at a coarser resolution. For example, limited funding often leads to distribution of restoration effort across a large number of watersheds, which results in too little action in each watershed to detect progress toward the restoration goals. Prioritizing watersheds therefore helps to focus more restoration on a limited number of key watersheds, which increases the likelihood that restoration efforts can result in measureable biological gains. Ultimately, prioritizing both watersheds and projects is an important part of a restoration plan that includes goals, assessment and identification of restoration opportunities, and prioritization of actions needed to achieve the biological goals of restoration.

9.2.3 Selecting restoration techniques and designing restoration actions

Once the restoration needs have been identified and prioritized based on the watershed assessment results, specific restoration techniques can be chosen to address those needs. First and foremost, any chosen technique must be appropriate for the problem identified (Chapter 5). For example, where an erosion problem from agricultural fields has been identified, a number of techniques might be selected to address that problem including altering tilling practices to reduce erosion or developing riparian buffers that interrupt sediment delivery to the stream. In practice, the selection of such restoration techniques may involve considerable interaction with local stakeholders to identify actions that both address the problem and minimize economic impact (Chapter 4). Alternatively, the selection of a restoration technique might take place during the project design phase, in which several possible solutions are evaluated for addressing a specific restoration need (Chapter 7).

In selecting restoration actions, it is important to maintain a clear distinction between infrastructure protection (river engineering) and river restoration. River engineering actions, such as bank armoring to protect property or infrastructure, do not have restoration as their purpose. By contrast, restoration or habitat improvement actions are intended to restore processes or habitats that drive ecosystem functions. We have not discussed river engineering actions in this book. Rather, we have focused on actions that are intended to restore ecosystem processes and functions and habitat, and provided a set of guiding principles for designing restoration actions: (1) treat the root cause of degradation; (2) design the project to be consistent with local driving processes; (3) address problems at the appropriate scale; and (4) be specific about expected outcomes (Chapter 2). These principles are intended to focus restoration actions primarily on addressing causes of degradation (full or partial process restoration), and secondarily on actions that improve or create habitat where human constraints such as land uses or infrastructure preclude treating the root cause. The principles also guide the selection and design of habitat creation actions to be consistent with local driving processes and to address problems at an appropriate scale. Finally, the principles ask for realistic predicted outcomes of restoration actions. Each of these principles is intended to guide restoration plans toward actions that are effective and do not require long-term maintenance.

The design process for individual restoration actions has several key steps: (1) identify the cause of the problems to be addressed; (2) understand the watershed context and constraints; (3) clearly define project goals and objectives; (4) evaluate alternative strategies and select a preferred alternative; (5) design the project; (6) implement the project; and (7) monitor outcomes (Chapter 7). These steps ensure that the final project design successfully treats the cause of degradation and is compatible with the local setting and processes (particularly for in-channel restoration or habitat creation actions). At each step, the process-based principles help to guide the design process toward solutions that restore key processes and will therefore sustain themselves over the long term without continual intervention. The same steps and principles apply in designing a habitat improvement or creation project, but the project does not address the root cause of degradation. However, the project should still be consistent with local driving processes, address the problem at an appropriate scale, and have clearly stated expected outcomes.

9.2.4 Monitoring

Monitoring programs should evaluate both the effectiveness of restoration projects and the progress of a restoration program toward achieving the restoration goal. For monitoring restoration project effectiveness, it is often most cost-effective to evaluate each project type with a focused monitoring program (Chapter 8) rather than have limited and separate monitoring projects for each individual restoration action. By contrast, monitoring to evaluate whether a suite of diverse restoration actions are achieving restoration goals (i.e. the desired biological response) requires a watershed-scale monitoring approach that measures key biological indicators as well as habitat changes.

Each type of monitoring follows a common set of planning and implementation steps including identification of: restoration goals and monitoring objectives; key questions or hypotheses; monitoring design parameters; spatial and temporal replication; and the sampling scheme. Understanding the goals of restoration actions to be monitored is the crucial first step in setting up a monitoring plan. These goals and objectives guide the identification of key questions or hypotheses to be addressed through monitoring. Defining clear, testable hypotheses that outline the scale, study design, and parameters to be measured is absolutely essential. Monitoring parameters should be selected judiciously and sample sizes estimated to help refine spatial and temporal replication needed. In addition to technical issues, proper project implementation and management are critical; even the most well-designed monitoring programs can fail if poorly implemented. Lastly, when the monitoring is underway or has been completed, analysis of the data and reporting results is a critical step. Without this final step, the monitoring effort cannot contribute to improving the identification of necessary restoration actions or the appropriate design of restoration projects.

The total amount of restoration also needs to be considered when developing the restoration and monitoring programs. Many restoration actions are relatively small scale, but achieving restoration goals often requires restoring a large percentage of a watershed (Roni et al. 2010). The intensity and amount of restoration can influence whether even the most rigorous monitoring program can detect a response. This is particularly critical for watershed-scale monitoring: the larger the watershed and more variability in parameters measured, the more restoration that will be needed to detect a population- or watershed-level response.

9.2.5 Examples of bringing the components together

At various points in this book we have provided examples of assessments, restoration actions, and project designs from the Skagit River in Washington State, USA and the River Eden, UK. Ultimately each of the restoration program components described above leads to design and implementation of individual projects, beginning with assessments to identify key problems, proceeding to selecting and prioritizing actions, then to designing and implementing projects, and finally to monitoring. Here we return to these two examples to illustrate how the components are linked to each other.

In Chapter 3 we illustrated how watershed analyses identify key processes and habitats in need of restoration for each of the two watersheds. For example, the assess-

Table 9.2 Example of linking restoration planning and design steps, Skagit River, USA.

Restoration step	Process	
	Watershed-scale: erosion and sediment supply	*Reach-scale*: floodplain dynamics
Watershed planning		
Restoration goal	Restore salmon populations to viable and harvestable levels	Restore salmon populations to viable and harvestable levels
Assessment and problem identification	Logging and roads have increased the frequency and magnitude of landslides; most sediment comes from stream-crossing fill failures	Loss of floodplain habitats due to levees and land conversion has significantly reduced juvenile salmon rearing habitat
Priority	Moderate priority	High priority
Project development		
Goal and objectives	Reduce road-related landslide rates to background levels (c. 150 m²/km²/yr) in Finney Creek over the next 10 years	Reconnect two of the three major floodplain channels to increase floodplain rearing habitats by over 50%
Alternatives	Alternative solutions include road removal or installing bridges, larger culverts, or armored fords	Alternatives include levee removal, levee setback, riprap removal, or habitat construction
Selected design	Remove undersized culverts and install bridge over large streams and larger culverts on small streams	Full levee removal on first channel, partial levee setback and off-channel habitat construction on second channel
Permitting	Forest practice permit from Department of Natural Resources and Hydraulics permit from Department of Fish and Wildlife	Army Corps of Engineers and Skagit County fill and grade permits, Hydraulics permit from Department of Fish and Wildlife, Endangered Species Act permit
Implementation	Construct during the 'dry' season (15 July–15 September)	Construct during the 'dry' season (15 July–15 September)
Monitoring	Implementation and effectiveness monitoring assess whether road landslide rates are similar to 'natural' levels	Implementation and effectiveness monitoring will examine the increase in habitat capacity

Table 9.3 Example of linking restoration planning and design steps, River Eden, UK.

Restoration step	Process	
	Watershed-scale: erosion and sediment supply	*Reach-scale*: riparian condition
Watershed planning		
Restoration goal	Restore Atlantic salmon and brown trout populations	Restore Atlantic salmon and brown trout populations
Assessment and problem identification	Heavily used pasture lands are the most likely sediment sources and cause of age 0+ Atlantic salmon declines	Degraded riparian areas are the most likely cause of age 0+ brown trout declines
Priority	High priority	High priority
Project development		
Goal and objectives	Reduce sediment to streams to near background levels	Increase shade and intercept sediment delivery to streams from high-risk pastures locations
Alternatives	Alter pasture management practices or restore riparian buffers to intercept sediment delivery	Establish riparian buffer and/or fence livestock away from stream to filter sediment and reduce bank erosion
Selected design	Reduce grazing pressure on specified high-risk pastures	Establish riparian buffer to reduce bank erosion
Permitting	None	None
Implementation	Local landowners select altered management practices to reduce sediment supply	Local landowners select buffers to reduce bank erosion
Monitoring	Monitor status of Atlantic salmon and brown trout in impacted reaches to track restoration effectiveness	Monitor status of Atlantic salmon and brown trout in impacted reaches to track restoration effectiveness

ments of watershed-scale processes showed that sediment supply was elevated in both the Skagit River and the River Eden, but that sediment supply was largely driven by logging-related landslides in the Skagit and surface erosion in the Eden (Tables 9.2 and 9.3). The reach-scale assessments of riparian conditions showed that riparian vegetation was also in need of restoration in both basins. In the Skagit River basin, degraded riparian areas consist of either small trees or no trees, and both wood recruitment and shade were reduced compared to natural riparian forests. In the River Eden, degraded riparian areas generally had no vegetation and demonstrated significant bank erosion caused by livestock. Based on these assessments, the watershed analyses identified important restoration actions focusing on reducing sediment supply and improving riparian conditions in both river basins (see also Figures 3.25 and 3.28).

Once the assessments have identified restoration actions and set priorities, the design process for individual projects begins with setting project goals and objectives. For example, in the Skagit River basin,

one project goal is to reduce landslide-related sediment delivery to near-background levels in the Finney Creek subwatershed (c. 150 m³/km²/yr). Alternatives for rehabilitating problem road segments include removing roads and recontouring slopes; replacing failure-prone stream-crossing fills with bridges, larger culverts, or fords; and removing sidecast from steep slopes. For a given road segment, the selected alternative will address key landslide hazards in a cost-effective manner (e.g. usually inexpensive bridges over larger streams and large culverts on small streams). Similarly, alternatives analyses can be conducted for each of the other restoration projects, focusing on evaluating the relative merits of alternative techniques for addressing project goals and objectives (see Chapter 5 for more details on techniques and Chapter 7 for alternatives analysis). In the Skagit River basin, for example, alternatives for restoring floodplain habitats in a selected reach include levee removal, levee setback, removal of bank armoring, and construction of off-channel habitats. By contrast, objectives and alternatives for the River Eden include reducing sediment, nutrient, and pollutant delivery to streams from pastures that are hydrologically connected to streams, or restoring riparian areas to filter sediment and pollutants from runoff (Tables 9.2 and 9.3).

A final design for watershed- and reach-scale processes can be developed after the alternatives analysis. Each design considers the project objectives, effectiveness of selected techniques, and costs. Permitting, implementation and monitoring are the last phases of the project development process. Permitting should occur in a timely fashion and include each of the multiple permits that are typically needed for water, soil movement, and impacts on protected species. Implementation includes logistics and plans for dealing with constraints and focuses on when the work can occur with least impact to biota, as well as what kinds of mitigation steps might be necessary to prevent or reduce impacts on streams during construction. Finally, monitoring should confirm that the project was completed as designed (implementation monitoring) and also evaluate its overall effectiveness at restoring watershed processes, habitats, and biota (effectiveness monitoring).

9.3 Developing proposals and evaluating projects for funding or permitting

Beyond guiding the overall restoration effort, a comprehensive restoration plan provides a solid foundation for the development of project proposals and facilitates the evaluation of those proposals in funding and permitting processes. Most review processes require evidence to support claims that restoration efforts are addressing key needs, and also that the final design is an appropriate and cost-effective alternative for solving the problem. Restoration proposals should therefore include evidence from the watershed assessment and design report that: (1) the project directly addresses one of the identified problems; (2) the design is consistent with local driving processes; and (3) the chosen alternative maximizes benefits and minimizes costs given local constraints. The first of these – that the objectives of each project should be clearly focused on one of the key restoration needs – is the most important piece of supporting evidence for any restoration proposal. When the proposed project directly restores key driving processes identified in the watershed assessment, this evaluation step is straightforward. However, evaluating whether habitat improvement or creation projects address key restoration needs can be more ambiguous. In those cases, the watershed assessment should provide evidence that the proposed project addresses a biologically important habitat need, even when it does not address the root cause of that habitat loss (Chapter 7).

The second piece of evidence needed in restoration proposals is clear explanation of the project context so that reviewers can be confident that the project will not fail because it lacks an understanding of local driving processes. For example, channel design projects have often attempted to create a uniform meandering channel pattern, regardless of what the local hydrology, sediment supply, and riparian conditions can support. One such design in California attempted to create a meandering channel where a braided channel existed; within a few years however a moderate flood event destroyed the constructed meanders and recreated the braided channel form (Kondolf 2006). To avoid such failures, restoration proposals should include sufficient analysis of watershed- and reach-scale processes to ensure that the project design will be sustained under current process regimes, even when they have been altered by land- or water-use impacts that alter flow or sediment conditions at the restoration site.

Finally, project funders are often interested in comparing the cost-effectiveness of various projects, which requires that proposals have some estimate of the project's benefit relative to the goal. When proposals include evaluation of several alternative restoration actions or designs (following the planning process in Chapter 7), comparisons among them are straightforward for reviewers. However, reviewers will likely also want to compare projects put forward by multiple applicants, so each project will need some estimate of both benefit and cost to facilitate review and increase the chance of receiving funding. Reviewers for project permitting are less likely to be interested in cost-effectiveness of projects than reviewers interested in funding or prioritizing projects.

Putting all of these review components into a simple framework that lays out the reviewer questions can help to increase consistency across reviewers and agencies, and also help proposal developers streamline their proposals to more efficiently get through a review process (Figure 9.1). For example, the River Restoration Analysis Tool (RiverRAT, Skidmore et al. 2011) is a systematic review process used by agencies reviewing restoration projects in the Pacific Northwest, USA which lays out the key questions that should be answered in a project proposal. The questions are organized by project phase (planning, implementation, and monitoring) and by seven plan elements within those phases (Figure 9.1). For each

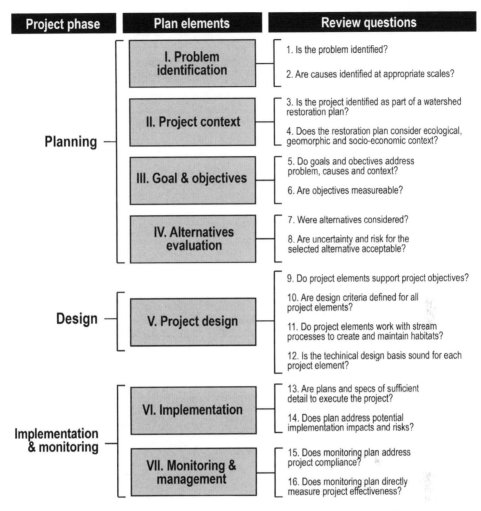

Project phase	Plan elements	Review questions

Planning

I. Problem identification
1. Is the problem identified?
2. Are causes identified at appropriate scales?

II. Project context
3. Is the project identified as part of a watershed restoration plan?
4. Does the restoration plan consider ecological, geomorphic and socio-economic context?

III. Goal & objectives
5. Do goals and obectives address problem, causes and context?
6. Are objectives measureable?

IV. Alternatives evaluation
7. Were alternatives considered?
8. Are uncertainty and risk for the selected alternative acceptable?

Design

V. Project design
9. Do project elements support project objectives?
10. Are design criteria defined for all project elements?
11. Do project elements work with stream processes to create and maintain habitats?
12. Is the techinical design basis sound for each project element?

Implementation & monitoring

VI. Implementation
13. Are plans and specs of sufficient detail to execute the project?
14. Does plan address potential implementation impacts and risks?

VII. Monitoring & management
15. Does monitoring plan address project compliance?
16. Does monitoring plan directly measure project effectiveness?

Figure 9.1 Example set of questions for reviewing a restoration project proposal and plan, arranged by project phase and plan element (based on Skidmore *et al.* 2011). Using a common set of review questions at planning, funding, and permitting phases can help streamline the restoration process.

element of the plan, the review questions focus the reviewer's attention on whether the plan adequately addresses a key restoration need, has a design that is sufficiently rigorous and likely to succeed, and has adequate implementation and monitoring plans. Not only does this structure provide a consistent review process for multiple permitting agencies and funding organizations, it also clarifies expectations for planners and designers of restoration projects so that they can prepare proposals to address each of the key questions that reviewers will ask. Moreover, adoption of a similar structure by funding agencies would then give project proponents, funders, and permit reviewers a common set of guidelines for the structure and content of proposals, which would streamline the advancement of projects from concept to implementation.

9.4 Moving from opportunistic to strategic restoration

One of the most important outcomes of developing a comprehensive plan for watershed restoration is the transformation of restoration implementation from reliance on opportunistic restoration projects to strategic selection and sequencing of restoration actions (Beamer *et al.* 2000). This shift in thinking will guide the bulk of restoration actions toward those that have been identified as the most important actions needed to achieve biological objectives, and other less important restoration actions will become less common over time. In practice, this shift might take several years (if not a decade or two) because high-priority projects may be large or expensive, and opportunistic restoration may continue for many years if practitioners are reluctant to abandon lower-priority restoration projects or, as is often the case, landowners are unwilling to participate.

In the early stage of this transition – when most restoration is still opportunistic – the watershed assessment results and restoration priorities can be used as filters to determine whether proposed actions are consistent with restoration needs (Figure 9.2). For example, where the assessment results indicate that urbanization has severely altered stream hydrology and water quality, continued restoration of habitat structure through wood or boulder placement may produce limited benefits because the root causes of biological decline are not addressed. Restoration proponents or funders should therefore be willing to shift resources away from such projects toward those that are consistent with long-term restoration needs. Projects that are more consistent with restoration needs might include stormwater detention efforts that store and filter water to address key environmental problems, or even shifting actions to less degraded parts of the watershed that will provide a greater benefit.

A comprehensive restoration plan can help restoration practitioners and funders become more strategic by providing a clear basis for selecting the most important actions. The plan should contain a substantial list of high- and medium-priority projects that address the key

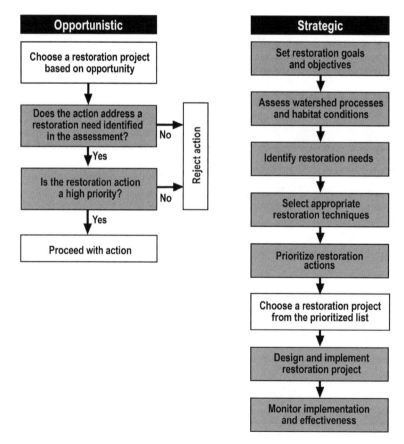

Figure 9.2 Left: illustration of pathways for evaluating opportunistic restoration proposals using the watershed assessment results and priorities and right: the alternative strategic approach, which uses the watershed assessment results and priorities to drive the selection of restoration projects. In the opportunistic approach, actions are often implemented with no evaluation relative to watershed assessments or priorities. Partly based on Beamer *et al.* (2000).

problems and have high biological benefits. Restoration groups can then choose projects from that list, focusing on those that match their skills and techniques. For example, some groups focus on riparian fencing and planting projects, and the list of priority actions should then show where those types of actions will achieve the greatest benefit. This will help guide the restoration group to beneficial locations, rather than choosing potentially low-value locations based on perceived conditions or landowner opportunities. In some cases, implementing lower-priority projects to build trust and goodwill with landowners may of course be necessary to set the stage for future implementation of higher-priority or larger projects (Chapter 4). Large projects that are usually implemented by local, provincial, or national agencies can also use the restoration plan to guide the selection of restoration actions. These larger projects often focus on actions such as levee removal or setback, dam removal, or restoration of environmental flows at large dams. Such actions are often the most expensive and require the most technical expertise to design and implement, but also often achieve the largest biological benefits. Importantly, both small and large restoration actions should focus on addressing causal mechanisms to the extent possible, and resort to habitat improvement or creation only when infrastructure or other constraints preclude restoration of processes.

9.5 Conclusions

Our purpose in writing this book was to provide a comprehensive overview of the restoration planning, implementation, and monitoring process, and to provide sufficient detail on each of the restoration steps so that restoration practitioners could develop defensible and effective watershed restoration programs. We chose this flexible approach because there can be no single cookbook for watershed assessment, prioritization of actions, project design, or monitoring programs. Rather, each watershed has unique restoration planning needs, based partly on local physiography and ecology and partly on legal and socioeconomic factors that either drive or constrain river restoration. While all restoration programs can follow a common set of development steps, the implementation of each step requires locally adapted methods and techniques. We have therefore attempted to provide not only an overview of available methods for each step, but also important guiding principles to help practitioners choose appropriate methods for their needs. We hope that this book proves useful in improving watershed and river restoration efforts.

9.6 References

Beamer, E, Beechie, T., Perkowski, B. & Klochak, J. (2000) *River Basin Analysis of the Skagit and Samish River Basins: Tools for Salmon Habitat Restoration and Protection.* Skagit Watershed Council, Mount Vernon, Washington.

Kondolf, G.M. (2006) River restoration and meanders. *Ecology and Society* **11**, 42. Available at: http://www.ecologyandsociety.org/vol11/iss2/art42/.

Roni, P., Pess, G., Beechie, T. & Morley, S. (2010) Estimating changes in coho salmon and steelhead abundance from watershed restoration: how much restoration is needed to measurably increase smolt production? *North American Journal of Fisheries Management* **30**, 1469–1484.

Skidmore, P.B., Thorne, C.R., Cluer, B., Pess, G.R., Castro, J., Beechie, T.J. & Shea, C.C. (2011) *Science base and tools for evaluating stream engineering, management and restoration proposals.* US Department of Commerce, NOAA Technical Memorandum, NMFS-NWFSC-112. http://www.restorationreview.com.

Index

---

water
  budget, *78*
  drawdown of impounded, 149
  quality of, 157–8, 176
    assessment, 84–6
    impairments, 217–18
    metrics, 86
    parameters, 84
    restoration, 146
  storage, 66
  temperature, 74, 84–5, 164–5, 167, 173, 177, 219, 224
  threats to supply, 4
Water Erosion Prediction Project (WEPP), 63
Water Flow and Balance Simulation Model (WaSimETH), 65
Water Framework Directive (WFD), 5, 86, 133, 192–3, 208, 254
Water Pollution Control Act, 5
waterpower, harnessing of, 3
watershed councils, 133
watersheds, 259
  alterations to functions, 31–5
  assessments of *see* assessment
  hierarchical structure of, 13–17
  holistic approaches to, 6
  monitoring of *see* monitoring, at watershed-scale
  prioritization of *see* prioritization, of watersheds

processes of, 11–49, *12*, *14*
  alterations to, 31–5, **32**
  see also processes, watershed-scale
  restoration history of, 5–7
  templates of, 57–61
weathering, 22
Weber, Max, 126
weevils, 165
weirs, 5, 87, 147, 151, 169, 208
  abstraction, 152
  boulder, 134, *168*, 227
  construction of, 3, 149
  jump, 134
  log, 167, 169
  removal of, 160
  submerged, 153
West Fork Lake Creek, Washington State, *222–3*, *225*
West Fork Smith River tributary, Oregon, *148*
wetlands, 22, 172–3
  communities on, 23
  draining of, 3–4
  filling of, 35
  floodplain, 153
  inundation of, *228*
Weyerhaeuser Timber Company (WeyCo), 134–5, *135*
Wilderness Act, 5
Wilderness Society, 4
wildfire, 20, 22

wildlife, protection of, 4
willows, 163, 174, 245
  pioneer, 23
  planting of, *163*
  riparian, 22, *246*
Wisconsin, 68, *69*, *166*
withdrawal *see* abstraction
wolves, 166
Wondolleck, Julia, 139
wood, 25, 30, 35, 157, 174
  abundance, 26, 39, 80–1, 225
  clearance of, 35
  debris, *27*, 79
  delivery, 23–5
  dynamics, 25
  importance of, 6
  jams, 17, 30
    see also logjams
  loading, 35, 273
  placement, 168, 170, 195, 199, 288
  recruitment, 29, 37, 39, 285
  supply, 13, *16*, 23, 40, 57, 70, 73–4
World Water Council, 4
Wupperverband, 133

Yaffee, Steve, 139

zealots, 126–31

Printed and bound by CPI Group (UK) Ltd, Croydon, CR0 4YY